Springer Series in Astrostatistics

Editor-in-chief: Joseph M. Hilbe, Jet Propulsion Laboratory, and
Arizona State University, USA

Jogesh Babu, The Pennsylvania State University, USA
Bruce Bassett, University of Cape Town, South Africa
Steffen Lautitzen, Oxford University, UK
Thomas Loredo, Cornell University, USA
Oleg Malkov, Moscow State University, Russia
Jean-luc Starck, CEA/Saclay, France
David van Dyk, Imperial College, London, UK

Springer Series in Astrostatistics, #1
For further volumes:
http://www.springer.com/series/1432

Joseph M. Hilbe
Editor

Astrostatistical Challenges for the New Astronomy

Editor
Joseph M. Hilbe
Arizona State University
Tempe, AZ, USA

ISBN 978-1-4614-3507-5   ISBN 978-1-4614-3508-2 (eBook)
DOI 10.1007/978-1-4614-3508-2
Springer New York Heidelberg Dordrecht London

Library of Congress Control Number: 2012948130

© Springer Science+Business Media New York 2013
This work is subject to copyright. All rights are reserved by the Publisher, whether the whole or part of the material is concerned, specifically the rights of translation, reprinting, reuse of illustrations, recitation, broadcasting, reproduction on microfilms or in any other physical way, and transmission or information storage and retrieval, electronic adaptation, computer software, or by similar or dissimilar methodology now known or hereafter developed. Exempted from this legal reservation are brief excerpts in connection with reviews or scholarly analysis or material supplied specifically for the purpose of being entered and executed on a computer system, for exclusive use by the purchaser of the work. Duplication of this publication or parts thereof is permitted only under the provisions of the Copyright Law of the Publisher's location, in its current version, and permission for use must always be obtained from Springer. Permissions for use may be obtained through RightsLink at the Copyright Clearance Center. Violations are liable to prosecution under the respective Copyright Law.
The use of general descriptive names, registered names, trademarks, service marks, etc. in this publication does not imply, even in the absence of a specific statement, that such names are exempt from the relevant protective laws and regulations and therefore free for general use.
While the advice and information in this book are believed to be true and accurate at the date of publication, neither the authors nor the editors nor the publisher can accept any legal responsibility for any errors or omissions that may be made. The publisher makes no warranty, express or implied, with respect to the material contained herein.

Printed on acid-free paper

Springer is part of Springer Science+Business Media (www.springer.com)

Preface

Although statistical analysis was first applied to agricultural, business, and architectural data as well as to astronomical problems at least thirty to forty centuries ago, astronomers have largely been estranged from statistics throughout the past two centuries. Except for a few astronomers, the majority have paid little attention to the advances made in statistical theory and application since the work of Laplace and Gauss in the late 18th and early 19th centuries. Descriptive statistical analysis, of course, has continued to be used throughout this period, but inferential methods were largely ignored. It was not until the last decades of the 20th century that computing speed and memory allowed the development of statistical software that was rigorous enough to entice astronomers to again become interested in inferential statistics. During this time small groups of astronomers and statisticians joined together developing both collaborations and conferences to discuss statistical methodology as it applies to astronomical data. There are a variety of reasons for this regained interest, as well as for the two-century breach. The initial monograph in the text addresses the reasons for the separation, and for the renewed interest.

Newly developed instruments are - and will be - gathering truly huge amounts astronomical data which will need to be statistically evaluated and modeled. Much of the data being generated appears to be complex, and not amenable to straightforward traditional methods of analysis. Many astrostatisticians -- astronomers, statisticians and computation theorists having an interest in the statistical analysis of astronomical/cosmological data -- have recognized the inherent multifarious nature of astronomical events, and have recently turned to Bayesian methods for analyzing astronomical data, abandoning at least in part the traditional frequentist methods that characterized the discipline during the 19th and 20th centuries. However, frequentist-based analyses still have an important role in astrostatistics, as will be learned from this volume.

Except for the first two introductory chapters, the collection of monographs in this volume present an overview of how leading astrostatisticians are currently dealing with astronomical data. The initial article is by the editor, providing a brief history of astrostatistics, and an overview of its current state of development. It is an adaptation of a presentation delivered at Los Alamos National Laboratory in October, 2011 as a seminar in the Information Science and Technology Center seminar series. The second contribution is by Thomas Loredo of Cornell University who presents an overview of both frequentist and Bayesian methodologies and details how each has been misunderstood. He then outlines considerations for future astrostatistical analyses. Prof Loredo's contribution is an adaptation of a presentation he made on this topic at the Statistical Challenges in Modern Astronomy V conference held at Pennsylvania State University in June, 2011. The other monographs describe ongoing state-of-the-art astrostatistical research. Contributions to this volume are largely adaptations of selected invited and special topics session presentations given by the authors at the 2011 ISI World Statistics Congress in Dublin, Ireland. Several monographs have been completely re-written from their original form for this volume, and some are entirely new. Nearly all of the contributors as well as co-authors are recognized as leading -- if not the leading -- astrostatisticians in the area of their research and contribution to this book.

The value of a volume such as this rests in the fact that readers get the opportunity to review the work of top astrostatisticians. Each selection employs a different manner of applying statistical methods to the data being evaluated. Articles using traditional methods utilize techniques that are common among statisticians who model astronomical data. Together with a history and overview of astrostatistics as a discipline, the research shared in this volume will provide readers with a good sense of the current state of astrostatistics. The articles may also encourage readers themselves to contribute to this new area of statistics. The challenges are great, and may involve developing new methods of statistical analysis. The answers that may be gained, though, address questions that are central to astronomy and cosmology. In fact, substantial statistical work has already been done in such astrophysical areas as high-energy astronomy (e.g., X-ray, gamma, and cosmic ray astronomy), neutrino astrophysics, image analysis, and both extra-solar and early galaxy formation. Even elusive areas such as understanding dark matter and dark energy, and if a multiverse exists, are queries that may ultimately be answered using novel applications of current statistical functions, or they may require the application of new and innovative astrostatistical methods yet to be developed. It will be clear on reviewing the monographs in this text that astrostatistics is a discipline coming to its own. It is involved with evaluating problems of cosmic concern, and as it advances it will demand the most of both computational and statistical resources.

I wish to thank Marc Strauss, Springer statistics editor, for suggesting the initiation of a Springer Series in Astrostatistics, and for his support for this text. Without his assistance this book would never have been constructed. I also must acknowledge Rajiv Monsurate for his fine work in setting up the book for publication, and Hannah Bracken, for her editorial support. Their efforts were also essential for this book's completion. Finally I dedicate this book to my late father, Rader J Hilbe, who fostered my interest in both mathematics and astronomy.

Update:

On August 30, 2012 the Executive Board of the *ISI Astrostatistics Network* approved its being re-organized as the *International Astrostatistics Association* (IAA), the first professional association for the discipline of astrostatistics. The initial officers of the IAA were also approved. Membership includes researchers from astronomy/astrophysics, physics, statistics, the computer-information sciences and others who are interested in the statistical analysis of astronomical data. The IAA has three classes of membership, with no difference in benefits: regular, Post Doc, and student. As the professional association for the discipline, the IAA fosters the creation of astrostatistics committees and working groups within both international and national astronomical and statistical organizations. During the same week that the IAA was created, the International Astronomical Union approved an *IAU Astrostatistics Working Group*. The IAA, IAU Astrostatistics WG, ISI astrostatistics committee, and LSST share a common *Astrostatistics and AstroInformatics Portal* web site: *https://asaip.psu.edu*

The IAA and ISI are co-sponsors of the Springer Series in Astrostatistics.

Contents

Preface v

1 Astrostatistics: A brief history and view to the future 1
   Joseph M. Hilbe, Jet Propulsion Laboratory and Arizona State University

2 Bayesian astrostatistics: A backward look to the future 15
   Thomas J. Loredo, Dept of Astronomy, Cornell University

3 Understanding better (some) astronomical data using Bayesian
   methods 41
   S. Andreon, INAF-Osservatorio Astronomico di Brera

4 BEAMS: separating the wheat from the chaff in supernova
   analysis. 63
   Martin Kunz, Institute for Theoretical Physics, Univ of Geneva, Switz;
   Renée Hlozek, Dept of Astrophysical Sciences, Princeton Univ, NJ;
   Bruce A. Bassett, Dept of Mathematics and Applied Mathematics,
   Univ of Cape Town, SA; Mathew Smith, Dept of Physics, Univ of Western
   Cape, SA; James Newling, Dept of Mathematics and Applied
   Mathematics, Univ of Cape Town, SA; Melvin Varughese, Dept of
   Statistical Sciences, Univ of Cape Town, SA

5 Gaussian Random Fields in Cosmostatistics. 87
   Benjamin D. Wandelt, Institut d'Astrophysique de Paris, Université
   Pierre et Marie Curie, France

6 Recent advances in Bayesian inference in cosmology
   and astroparticle physics thanks to the Multinest Algorithm 107
   Roberto Trotta, Astrophysics Group, Dept of Physics, Imperial College
   London; Farhan Feroz (Cambridge), Mike Hobson (Cambridge),
   and Roberto Ruiz de Austri (Univ of Valencia, Spain)

7 Extra-solar Planets via Bayesian Fusion MCMC. 121
   Philip C. Gregory, Dept of Astronomy, University of British
   Columbia, Canada

8 **Classification and Anomaly Detection for Astronomical Survey Data** 149
   Marc Henrion, Dept of Mathematics, Imperial College, London, UK; Daniel J. Mortlock (Imperial), David J. Hand (Imperial), and Axel Gandy (Imperial))

9 **Independent Component Analysis for dimension reduction classification: Hough transform and CASH Algorithm.** 185
   Asis Kumar Chattopadhyay, Dept of Statistics, University of Calcutta, India; Tanuka Chattyopadhyay (Univ Calcutta), Tuli De (Univ Calcutta), and Saptarshi Mondal (Univ Calcutta)

10 **Improved cosmological constraints from a Bayesian hierarchical model of supernova type Ia data** 203
   Marisa Cristina March, (University of Sussex); Roberto Trotta (Imperial), Pietro Berkes (Brandeis Univ), Glenn Starkman (Case Western Reserve Univ), Pascal Vaudrevange (DESY, Hamburg, Germany)

Contributors

Stefano Andreon Astronomer, INAF-Osservatorio Astronomico di Brera, Brera, Italy; Prof of Bayesian Methods, Universita' degli Studi di Padova, Italy
andreon@brera.mi.astro.it; http://www.brera.mi.astro.it/~andreon

Bruce A. Bassett Prof of Applied Mathematics, Dept. of Mathematics and Applied Mathematics, Univ of Cape Town, SA; African Institute for Mathematical Sciences, 68 Melrose Road, Muizenberg 7945, SA; South African Astronomical Observatory, Observatory, Cape Town, SA;
bruce.a.bassett@gmail.com

Pietro Berkes Researcher, Volen Center for Complex Systems, Brandeis Univ.
berkes@brandeis.edu

Asis Kumar Chattopadhyay Prof of Statistics, Dept. of Statistics, Univ. of Calcutta, India; Coordinator, Inter University Centre for Astronomy and Astrophysics,
akcstat@caluniv.ac.in

Tanuka Chattyopadhyay Assoc Prof, Dept. of Astrophysics/Astrostatistics, Calcutta Univ, Kolkata, India
tanuka@iucaa.ernet.in

Tuli De Dept of Statistics, Calcutta Univ, Kolkata
tuli_stat5@yahoo.co.in

Farhan Feroz Jr Research Fellow, Cavendish Astrophysics, Univ. of Cambridge, Cambridge, CB3OHE, UK
f.feroz@mrao.cam.ac.uk; http://www.mrao.cam.ac.uk/~ff235

Axel Gandy Lecturer, Statistics Group, Dept. of Mathematics, Imperial College, London, S Kensington Campus, London, SW7 2AZ
a.gandy@imperial.ac.uk; www2.imperial.ac.uk/~agandy

Philip C. Gregory Emeritus Prof, Dept. of Physics and Astronomy Dept. of Astronomy, Univ. of British Columbia, 6224 Agricultural Rd., Vancouver, BC V6T 1Z1 Canada
gregory@phas.ubc.ca

David J. Hand Prof of Statistics, Dept. of Mathematics. Imperial College, London, S Kensington Campus, London, SW7 2AZ
d.j.hand@imperial.ac.uk

Marc Henrion Statistics Section, Dept. of Mathematics, Imperial College London, S Kensington Campus, London, SW7 2AZ, London, UK
marc.henrion03@imperial.ac.uk; http://www.ma.imperial.ac.uk/~myh03/

Joseph M. Hilbe Solar System Ambassador, Jet Propulsion Laboratory, California Institute of Technology; Adj Prof of Statistics, SSFD, Arizona State Univ; Emeritus Prof, Univ of Hawaii
hilbe@asu.edu; http://works.bepress.com/joseph_hilbe/; http://ssfd.clas.asu.edu/hilbe

Renée Hlozek Lyman Spitzer Jr. Postdoctoral Fellow, Dept. of Astrophysical Sciences, Princeton Univ, Peyton Hall, Ivy Lane, NJ 08544
rhlozek@astro.princeton.edu

Mike Hobson Jr Research Fellow, Cavendish Astrophysics, Univ. of Cambridge, Cambridge, CB3OHE, UK
mph@mrao.cam.ac.uk; http://www.mrao.cam.ac.uk/

Martin Kunz Lecturer, Institute for Theoretical Physics, Univ. of Geneva, Switz Univ. of Geneva, 24, Quai Ernest-Ansermet, CH-1211 Genève 4
martin.kunz@unige.ch; http://theory.physics.unige.ch/~kunz/

Thomas J. Loredo Sr. Research Assoc, Dept. of Astronomy, 620 Space Sciences Bldg., Cornell Univ, Ithaca, NY, 14853-6801
loredo@astro.cornell.edu; http://www.astro.cornell.edu/staff/loredo/bayes/

Marisa Cristina March Astronomy Centre, Postdoctoral Research Fellow (Physics and Astronomy); Postdoctral Research Fellow in Cosmology, Univ. of Sussex, East Sussex, UK
marisa.march06@imperial.ac.uk

Saptarshi Mondal Dept. of Statistics, Calcutta Univ, Kolkata
saptarshi.stat@gmail.com

Contributors

Daniel J. Mortlock Lecturer in Astrostatistics, Astrophysics Group, Dept. of Physics, and Statistics Group, Dept. of Mathematics, Imperial College, London, S Kensington Campus, London, SW7 2AZ
d.mortlock@imperial.ac.uk; mortlock@ic.ac.uk

James Newling Dept. of Mathematics and Applied Mathematics, Univ. of Cape Town, SA
james.newling@gmail.com

Roberto Ruiz de Austri Instituto de F´ýsica Corpuscular, IFIC-UV/CSIC, Univ of Valencia, Valencia, Spain
rruiz@ific.uv.es

Mathew Smith Dept of Physics, Univ. of Western Cape, Bellville 7535, Cape Town, SA; Astrophysics, Cosmology and Gravity Centre (ACGC), Dept. of Mathematics and Applied Mathematics, Univ. of Cape Town, Rondebosch 7700, SA
mathew.smith@uct.ac.za

Glenn Starkman Prof of Physics and Astronomy, and Director, Institute for the Science of Origins, and Director, Center for Education and Research in Cosmology and Astrophysics, Case Western Reserve Univ, Cleveland, OH
glenn.starkman@case.edu

Roberto Trotta Lecturer, Theoretical Cosmology, Astrophysics Group, Dept. of Physics, Imperial College London, Blackett Laboratory, Prince Consort Rd, London SW7 2AZ, UK
r.trotta@imperial.ac.uk; www.robertotrotta.com

Pascal Vaudrevange Fellow, Particle Physics group, Deutsches Elektronen-Synchrotron (DESY), Hamburg, Germany
pascal@vaudrevange.com, pascal.vaudrevange@desy.de

Melvin Varughese Dept. of Statistical Sciences, Univ. of Cape Town, SA
melvin.varughese@uct.ac.za; http://web.uct.ac.za/depts/stats/varughese.htm

Benjamin Wandelt Prof of Theoretical Cosmology and International Chair, Institut d'Astrophysique de Paris, Université Pierre et Marie Curie, 98 bis, boulevard Arago, 75014 Paris
e-mail: wandelt@iap.fr; cosmos.astro.illinois.edu

Chapter 1
Astrostatistics: A Brief History and View to the Future

Joseph M. Hilbe

1.1 A Brief History of Astrostatisics

Humans have long sought to understand astronomical events through the use of mathematical relationships. Ancient monuments were constructed to be aligned with stellar positions at given times of the year, and eclipses were assiduously recorded and predicted for a variety of reasons. In pre-scientific times priests and astrologers attempted to predict astronomical relationships for the purpose of understanding the minds of the gods. It was not until the 5th century BCE that a small group of philosophers living in the area of the Aegean Sea began to value understanding nature for its own sake. Many historians argue that science in fact began during this period. The event that some signal as the beginning of science was the prediction of a total solar eclipse by Thales of Miletus in 585 BCE. By using his knowledge of past eclipses and their locations, Thales predicted that the eclipse would occur in central Lydia close to May 28th of that year. As it happened, the Lydians were in battle against the Medes. The celestial event resulted in a cessation of the war, and a widespread recognition of Thales as a Sage.

Thales's successes inspired later philosophers and mathematicians to investigate the natural world by compiling records of past observations related to celestial events with the aim of predicting future eclipses, lunar and solar positions, and the like. Aristarchus of Samos (ca. 310 – ca. 230 BCE, Eratosthenes (276–197 BCE), Apollonius of Perga (ca. 262 – ca. 190 BCE), and Hipparchus (190–120 BCE) were the most noteworthy astronomers of this period. However, it was Hipparchus, living in various cities along the present-day western Turkish shore, who first clearly applied statistical methods to astronomical data.

Hipparchus is generally acclaimed as one of the leading astronomers and mathematicians of antiquity. He is credited as being the first to develop trigonometry,

---

Joseph M. Hilbe
Jet Propulsion Laboratory, Pasedena, CA 91109, U.S.A.
Arizona State University, Tempe, AZ 85287, U.S.A.
e-mail: hilbe@asu.edu

spherical trigonometry, and trigonometric tables, applying these to the relative motions of both the Sun and Moon. With respect to statistics, however, it is sometimes overlooked to what extent Hipparchus employed statistical measures of central tendency in reconciling observations. Using median reported values for the percent of the sun covered by the Earth's shadow at various sites in the Greek and Roman world at the time, and from the size of the moon's parallax, he calculated the distance from the Earth to both the moon and sun in terms of Earth radii, as well as their comparative volumes. Hipparchus realized however, that the distance varies as the Moon circles the spherical Earth, so preferred giving its median value, plus the minimum and maximum ranges. Hipparchus is said to have calculated the median value as 60.5 Earth radii. The true value is 60.3.

Hipparchus is also well known for calculating the precession of the equinoxes, and for calculating the length of the tropical year. Based on his calculations, Hipparchus argued that the tropical year is 365 days, 5 hours, 55.2 minutes, and 12 seconds in length. Current technology puts the figure at 365 days, 5 hours, 49 minutes, and 19 seconds. The difference is less than 6 minutes per year.

Problems related to the variability and errors of measurement concerned later astronomers as well; e.g. Ptolemy and Galileo. Galileo is perhaps most well known in astronomy for the use of a telescope in discovering the four major moons of Jupiter and the phases of Venus. He also contributed to the understanding of gravity, determining that objects of differing weights fall to the Earth at an equal rate in a vacuum.

With respect to statistics, however, Galileo argued that all measurements come with error, including observations of astronomical bodies. He proposed that measurement errors are symmetrically distributed about the mean of the calculated measurements, and that the mean of this error distribution is zero. In addition, he was aware that smaller errors in measurement occur less frequently than larger errors. Galileo thus came close to discovering the normal distribution, and I suspect that had his interests been more purely mathematical he would have discovered it.

Isaac Newton (1642–1727) of Cambridge University was a truly seminal figure in science, finalizing the break with the Aristotelian worldview that pervaded Western science from the age of Constantine to the beginning of the 18th century. Newton's *Principia Mathematica* was the foundation of modern mechanics, and was the basis of a new form of mathematics called The Calculus, which he co-invented separately with the German Gottfried Leibniz (1646–1716). His mechanics lead to the notion of universal gravitation and of his 3 laws of motion. Of central importance to astronomy was his argument that the motions of astronomical objects were governed by same set of laws as are terrestrial objects.

Newtonian mechanics led to the later discovery of the planets Uranus and Neptune – particularly Neptune. Newton's theories of gravitation and of motion combined to provide a deterministic view of astronomical events. If there were a variation from what was expected of a body moving about the Sun under the Newtonian framework, then there must be a cause. Measuring the disturbance in Uranus's orbit led astronomers in the mid-nineteenth century to conclude that there was another planet external to Uranus in the Solar System, and to predict with near exactness where in the sky the planet could be located.

Actually, Galileo observed Neptune in 1612 and 1613, believing it to be a star. He knew that it had moved relative to other fixed stars, but did not follow up on his discovery. In 1821 Alexis Bouvard published tables of Uranus's orbit based on Newtonian principles. He noticed that its actual orbit varied from what was predicted based on the tables. The conclusion – another planet influencing Uranus's orbit. Several noted astronomers of the time tried to define a position for the "missing" planet. John Adams (1819–1892), a Cambridge University mathematician, actually identified the location and mass of the new planet in 1845, when only 26 years of age, passing along this information to James Challis (1803–1882), director of the Cambridge University observatory, to look for it. Challis saw the planet in 1846, but failed to clearly identify it as such. French mathematician Urbain LeVerrier (1811–1877) finally predicted the location of Neptune, and requested his friend, Johann Galle (1812–1910) of the Berlin Observatory to search where he indicated. Galle was the first to view Neptune, knowing what he was looking at. Adams and LeVerrier appropriately received the actual credit for the discovery.

It is commonly accepted that German mathematician Carl Gauss (1777–1855) first developed the statistical technique of least squares regression when he was but eighteen years of age in 1794. The method was used for the first time by Hungarian astronomer Franz Xaver von Zach (1754–1832) in 1801 for determining the position of Ceres as it came into view from its orbit behind the Sun. This inaugural inferential statistical routine successfully predicted its actual location. However, although Adrien-Marie Legendre (1752–1833) described Gauss's least squares methodology in a 1804 publication, Gauss himself did not fully describe its use until 1809 in a text on Celestial Mechanics.

Mention should be made of Pierre-Simon Laplace (1749–1827) who was clearly one of the leading mathematicians of his age. Like many of his mathematically inclined contemporaries, he was also interested in understanding astronomical observations. In fact, he stated that astronomy was his foremost love. His primary statistical contributions, however, have proved to be of considerable importance to both statistics and astronomy today. Laplace is credited with formulating the central limit theorem, and of independently devising a theory of inverse probability, upon which Bayesian analysis is based. He was also the first mathematician who was interested in dealing with large amounts of complex data as well as being the first statistician to present a formal proof of least squares regression. Laplace also constructed what is known as the *Laplace transform*, which is used to assist in solving differential equations by converting them to algebraic equations. This method has been of considerable use to astronomers and astrophysicists. His contributions to astrostatistics are substantial.

Laplace's manner of presenting inverse probability and of formulating Bayes Rule is fundamental to how current Bayesian statisticians view the methodology. Bayes Theorem or Rule was in fact formulated by Laplace, together with its theoretical justification. He stimulated the later widespread use of the method throughout

Europe for nearly a century. Although Laplace had successfully applied the method to astronomical events, others later applied it to areas other than astronomy.

What is many times not realized however is that after Laplace formulated the Central Limit Theorem in 1810, he returned to using frequentist-based statistics until his death in 1827. The reason was due to his realization that when dealing with large amounts of data the two methodologies frequently gave the same results. He used Bayesian analysis, or what he termed the *probability of causes* approach, when dealing with uncertain data, or with singular event situations. But he in practice all but abandoned the method he formulated. On the other hand, his discovery of the Central Limit Theorem and its consequences in evaluating data led to major advances in frequency-based statistics. It should be noted, however, that Laplace's proof of the Central Limit Theorem entailed various flaws that were later corrected by Simeon Poisson (1781–1840) and Fredrick Bessell (1784–1846). Bessel was the first astronomer to actually use parallax for calculating the distance to a star. In 1838 he determined that 61 Cygni was 10.4 light years distant. Today's calculations give a distance of 11.4 light years.

Many have considered Laplace to be the true founder of Bayesian statistics, but Laplace himself acknowledged Bayes' prior development of the approach, even though Laplace gave it its modern format, including the formula we know and use. It would be more accurate I suspect to refer to *Bayes–Laplace analysis* rather than simply to *Bayes analysis*, or to the *Laplace–Bayes Rule* instead of *Bayes Rule*, but the latter appellation now seems firmly established in statistical nomenclature. Perhaps the foremost reason for this overlook is due to the highly negative charges made against Laplace by a few of the world's leading statisticians from the mid 1900s until a full century later. The charges have since been proved false, but they nevertheless affected Laplace's reputation, and may well have been a reason that Bayesian analysis, which was largely thought of as *Laplace Analysis* from the eighteenth through the mid-twentieth centuries, found so little support during much of the twentieth century. See [1] for a well researched discussion of this situation. In any case, it is unfortunate that Laplace does not get the credit he deserves, and that his name is not attached to what is now known as Bayesian analysis. Laplacean–Bayesian methodology is currently becoming the foremost statistical method used by astrostatisticians, even though traditional frequentist methods still appear to be favored by a number of statistically erudite astronomers. Loredo examines these trends in Chapter 2 of this volume.

Gauss continued to develop least squares methodology after the publication of his 1809 work. Together with von Zach's early success in implementing least squares regression and Laplace's later formal justification of it, it would appear that the method would have been widely used in subsequent astronomical analysis. However, the majority of astronomers showed relatively little interest in applying it to astronomical data. Certainly there were exceptions, as we shall later discuss, and advances occurred in describing astronomical events by employing underlying probability distributions; but astronomers in general turned to non-statistical quantitative methods including spectroscopy and differential equations for the understanding of astronomical data throughout the majority of the 19th and 20th centuries. Astronomers who did employ inferential statistical methods to their data typically relied on some form of least squares algorithm. In large part this was a result of astronomy, and later astrophysics, moving ahead faster in discoveries than statistics. During this period statisticians were predominantly interested in the analysis of tables, of fit statistics, correlation analysis, and the analysis of probability distributions.

Regression analysis failed to substantially advance during the 19th century until the work of Francis Galton (1822–1911). Galton developed the statistical methods of correlation analysis, standard deviation, regression to the mean, the regression least-squares line, histograms, and other now well known statistical techniques. Although he applied his methods to such diverse fields as psychology, agriculture, heredity, linguistics, and even fingerprint analysis, he is likely most known for his work in eugenics. The statistical methods he advanced have become commonplace in the discipline, but they were rarely applied to astronomical analysis.

Karl Pearson (1857–1936) was another leading statistician during this period, who also had an overriding interest in eugenics. His contributions to statistics entailed developing the method of moments, constructing the now well-regarded Pearson correlation coefficient and Pearson *Chi2* test. He also advanced the theory of hypothesis testing, which later served as a key statistical operation in frequentist methodology Pearson's statistical methods became quite influential in social statistics and early econometrics. However, early in the twentieth century he also applied correlation analysis to various properties of stars. Unfortunately, he did not continue with this research, but several astronomers followed in his footsteps.

Least squares regression also had some well known advocates within the astronomical community during first part of the twentieth century. Edwin Hubble (1889–1953), for example, used least squares regression when analyzing the relationship of galactic distance and redshift, leading to the notion of an expanding universe.

Major advances in statistics occurred with the work of British statistician and evolutionary biologist Ronald Fisher (1890–1962). Fisher is foremost responsible for the development of maximum likelihood estimation, likelihood theory in general, design of experiments and ANOVA, exact tests, and of the modern theory of hypothesis testing. His contributions to statistics have left him regarded as the father of modern statistics as well as the founder of modern quantitative genetics.

Fisher's interests were in agriculture and biology, as were many of the statisticians of the early/mid 20th century. Probability theory was also advanced by various statisticians from the insurance industry. Application of these new methods simply did not find their way into mainstream astronomy.

Fisher's view of statistics was based on a frequency interpretation of statistical inference in which probability is thought of in terms of the long range frequency of events. To the frequentist, the mean of a large population has a true value. However, we can typically only obtain a sample from this distribution, which may exist as both past and future events as well as present events. Statistical methods are designed to obtain a sample from a population in order to abstract an unbiased estimate of the true mean, together with a confidence interval. This view of statistics has been predominant in the discipline until recent years.

Frequentist statistics took various turns during the early-mid part of the twentieth century. For example Jerzy Neyman (1894–1981) and Egon Pearson (1895–1980), Karl's son, advanced an alternative view to Fisher's notion of evaluating hypothesis, which is now referred to as the Neyman–Pearson theory of hypothesis testing. Egon and Fisher later developed their own feud, albeit a different one from the feud between Karl Pearson and Fisher, as did Neyman and Fisher. But all three joined in a vigorous opposition to the views of Cambridge geophysicist and mathematician, Harold Jeffreys (1891–1989), who was perhaps the foremost proponent of the Bayes–Laplacean theory of statistics during the majority of the twentieth century. Several other statisticians had advocated Bayesian–Laplacean views before him; e.g. Italian actuary Bruno de Finetti (1906–1985) and Cambridge University mathematics professor Frank Ramsey (1903–1929). Even Egon Pearson had experimented with Bayesian methods during the 1920s before turning to a strict frequentist approach. Interestingly Bayesian methods were employed in cryptology during World War II, but the practitioners were generally careful not to characterize the methods as Bayesian.

Jeffreys attempted to develop a full-fledged Bayesian theory of probability, publishing articles and a book on the subject, *Theory of Probability* (1939). His efforts were largely ignored, but now they are classics in the field, and serve as the basis of the contemporary revival of Bayes–Laplace statistics.

It is important to note that a growing number of current astrostatisticians employ a Bayesian approach to the analysis of astronomical data. Prior to the beginning of this new millennium most astronomers used the more traditional frequentist approach when engaged in inferential statistical analysis – the approach founded in large part with the efforts of Fisher and Neyman-Pearson. The majority of astronomers, as well as statisticians having an interest in astronomy, model astronomical data using maximum likelihood, by some variety of EM algorithm, or by employing quadrature techniques. I believe that the use of finite mixture models and non-parametric quantile count models can be of considerable use in astronomical analysis, but have thus far not observed their implementation in astrostatistical modeling.

In 1996 Jogesh Babu and Eric Feigelson of Pennslyvania State University and the Penn State Center for Astrostatistics authored a brief text titled *Astrostatistics* [2], the first text specifically devoted to the discipline. All of the statistical procedures discussed in the text were based on traditional statistical methods. However, the use of Bayesian methods for the analysis of astronomical data was already in its infancy. In 1990, Thomas Loredo, then a graduate student at the University of Chicago, appears to have been the first researcher to employ Bayesian methods for the analysis of astronomical observations in his 1990 PhD dissertation. His dissertation was based on observing 18 neutrinos from Supernova 1987A and using that information to discern the inner structure of the star's interior. As a single event, frequentist methods were inappropriate. The success of his study stimulated other astronomers to use Bayesian methods when evaluating astronomical events. Now, as reflected in the monographs in this text, many leading astrostatisticians are implementing Bayesian techniques for the analysis of astronomical data.

It is sometimes forgotten that there is a symbiotic relationship between statistics and computer technology [3]. For example, in 1949 Columbia economist Wassily Leontief sought to analyze a problem related to the material and services sectors of the U.S economy. It was determined that he needed to invert a 24x24 matrix in order to determine the estimates of the related model parameters. With the help of Jerome Cornford, a young researcher with the Bureau of Labor Statistics, who had spent a week inverting a 12x12 matrix by hand for a preliminary version of the model, it was determined that such an inversion would take a minimum of over one hundred years, working 12 hours a day – not counting time to correct errors. Leontief and Cornford instead turned to Harvard's Mark II computer, the state of the art machine at the time, to invert the matrix. It took only three days of processing to invert the matrix and estimate the maximum likelihood parameters rather than the hundred plus years needed to invert it by hand. The effort led to Leontief receiving the Noble Prize in economics in 1973 [1]. It was also one of the first uses of a computer to solve a problem in statistical modeling. Now a 24x24 matrix can be inverted in microseconds by any laptop home computer. Much of the advances made in statistics during the twentieth century has been constrained, if not defined, by the available computing power required to execute a procedure. It was simply not possible to perform the highly iterative simulations needed in Bayesian analysis twenty years ago unless one had access to a sizeable mainframe system. A history of the relationship of logistic-based and count model-based regression, software availability, and computing power is given in [4] and [5], respectively.

As mentioned above, astronomers in general paid comparatively little attention to inferential statistical methods throughout most of the twentieth century. The reason for this is likely due to the fact that the computing power required to perform meaningful statistical analysis of astronomical data was not in general available. A brief overview of the relationship of computers and statistics in astronomy follows.

Statistical analyses were executed by hand and on primitive punch-card machines, first developed in 1911, until the development of the Harvard-IBM Mark 1 in 1944, which has been held as the first true computer. In 1952 the IBM 701 was developed, which was the first production computer, and which used a magnetic drum to store memory. Nineteen 701 computers were constructed, only three of which went to research agencies. They were not available for astronomical use. However, the 701 spurred the creation of FORTRAN which was later used by scientists as its favored programming language. IBM switched back to card punch technology two years later when it began mass producing the 650 series of mainframes, which were marketed to universities and research institutes as well as business organizations. In 1955 floating point arithmetic instructions were introduced, which greatly assisted the accuracy of mathematical operations, such as matrix inversion. A few astronomers took advantage of these computers, which steadily grew in power and speed throughout the series, which ended in 1962.

The IBM 360 series of mainframe computers were released in 1964. This was the first series of computers with compatible hardware and software. The use of virtual memory came in 1970 with the 370 series. Speech recognition was developed in 1971 and networking in 1974. Finally, a Reduced Instruction Set Computer (RISC) was created in 1981 which greatly enhanced computer speed. Astronomers working for the government and larger universities and observatories generally had access to mainframe technology from the mid 1950s, but usage cost was high and interactive capabilities were simply not available. Cards or instructions were submitted at a terminal, sent to the university's or agency's computing center, with results returned the following day. Given that statisticians had not yet developed reliable multivariable nonlinear regression software or other inferential procedures that could be helpful to astronomers, the latter generally employed only descriptive statistics and rather simple inferential methods when describing astronomical data. Except for basic statistical methods, astronomers had little use for statistics.

Since needed computing power was not available to most astronomers, even in the first three decades of the second half of the twentieth century, statistical software was not developed to meet the requirements of the discipline. Except for astronomers who created their own statistical routines in FORTRAN and in a few alternative languages, commercial statistical software was not available to the general astronomical community until the late 1960s and early 1970s. Applications such as SAS, SPSS, Systat, GLIM, and some other less known statistical applications were not developed until that period. Recall that the first personal computers had little power compared to their mainframe counterparts, and it was not until August 1981 that the first 4.77MHz 8088 processor IBM personal computer (PC) was released. Like the Apple, Commodore, Adam, and other similar machines of the period, the PC had no hard drive. The IBM XT was released in March 1983, coming with a 10MB hard drive, 128K of RAM, and ran on a 8088 processor. The IBM AT was the second generation PC, released in late 1984 with a 20MB hard drive, a new 6MHz 80286 processor, and 512K RAM. A user could update the RAM up to 640K maximum by installing it themselves. However, the AT was not inexpensive. The list price for the basic machine was over US$8,000. Academics could purchase the machine for half price through their University, but $4,000 was nevertheless a substantial sum at the time for a personal computer. Most computer work resided with department mini computers and observatory main frames. I shall focus on PC and PC software development though since this appears to had more impact on the subsequent relationship of astronomers and statisticians.

Enhancements to the IBM PC and the many clones which came into the market became available to the user on a regular basis. A 80386 processor came out in October 1985, the 80486 in 1989, and a 32-bit 80586 computer called the Pentium in 1993. Pentiums contained 3.3 million transistors, nearly three times the power of the 80486 machines. The Pentium II computer (1997) later provided users with 7.5 million transistors and 333MHz speed. A Pentium III (1999) with 1000MHz or 1GHz speed was introduced in May 2000. The Pentium 4 came out later that same year, with a full 2GHz version sold in August of 2001, exactly twenty years after the 4.77K 8088 machine was introduced to the world. Fortunately prices dropped as fast as computer speed and memory increased, so computers with sufficient memory to use for astronomical research were becoming more and more feasible.

The Intel Core architecture of microprocessors were released in 2006. It is the current state of PC operation. In 2010 the Core i7 computer was released. As of this writing a 2GHz 64-bit laptop having 6Ghz of RAM and running on a Windows 7 operating system can be purchased for under a thousand dollars. The speed with which operations may be calculated, and the amount of data that can be analyzed at one time has grown from the earliest computers to the point that astronomers can engage in serious statistical and mathematical analysis on a laptop computer. The turning point when astronomers determined, albeit rather slowly, that meaningful inferential statistical analysis of astronomical data was a viable enterprise occurred in the late 1980s and early 1990s. It was at this time that conferences between astronomers and statisticians began to be held at various sites. Now nearly all astronomers see a value in astrostatistics. We shall look at this conversion in more detail in the following section. An excellent overview of the modern and contemporary eras of the history of astrostatistics can be found in Feigelson & Babu, 2012.

1.2 Recent Developments in Astrostatistics

Beginning in the mid 1980s, astronomers began to organize small conferences devoted to what we may now call astrostatistics. One of the first was the "Statistical Methods in Astronomy" conference held in Strasbourg in 1983. A number of other conferences and series of conferences began at that time, but only one has maintained a regular timetable over a decade or more – the "Statistical Challenges in Modern Astronomy" conference, which has been held every 5 years since its inception in 1991. Under the direction of Jogesh Babu and Eric Feigelson of the Pennsylvania State University Center for Astrostatistics, the conference has brought together both astronomers and statisticians from around the world for week-long series of discussions.

During the 1990s several groups were organized consisting of astronomers and statisticians having a common interest in developing new statistical tools for understanding astronomical data. Two of the foremost groups are the California/Boston/Smithsonian Astrostatistics Collaboration (CHASC), headed by David van Dyk of the University of California, Irvine, and the International Computational Astrostatistics (InCA) Group, which is primarily comprised of researchers from Carnegie Mellon University and the University of Pittsburgh. CHASC, InCA, and the Pennsylvania State University all belong to Large Synoptic Survey Telescope (LSST) Project, which will provide huge amounts of data for analysis. The 8.4 meter LSST is currently scheduled to begin surveying activities in 2014.

Several sites in various parts of the world are presently engaged in developing astrostatistics programs and collaborations. An astrostatistics program is being developed by the joint efforts of the departments of Statistics and Astronomy/Astrophysics at Imperial College in London. Conferences on astrostatistics and specializations in the discipline are also being developed at the University of Calcutta, and at Pennsylvania State University, the University of Pittsburgh, Carnegie Mellon, Harvard University, University of Florida, University of Birmingham, and at other sites.

When astronomers again began to utilize inferential statistical methods into their published research, many of the articles employed inappropriate statistical analyses, or if correct methodology was employed, the analyses generally failed to account for possible violations of the assumptions upon which the research models were based. That is, they did not fully appreciate the statistical theory underlying their analyses. It was certainly not that astronomers lacked mathematical expertise to understand these assumptions; rather it was that many had no special training in statistical estimation. Moreover, many astronomers tended to use only a limited number of statistical procedures. They had not become aware of the vast range of statistical capabilities that had become available to professional statisticians and other researchers [6]. Of course, there were noted exceptions, but it became readily apparent in the late twentieth and during the first decade of the twenty-first centuries that astronomers in general needed to enhance their statistical knowledge. Those astronomers who took up this challenge believed that the best way to address the problem was to conduct conferences and organize collaborative research groups consisting of both astronomers and statisticians. I earlier mentioned some of these groups and conferences.

As of 2009, a relative handful of astronomers and statisticians with an interest in the statistical analysis of astronomical data were associated with collaborative organizations such as CHASC and InCa. Perhaps 40 to 50 attended the quint-annual *Statistical Challenges* conference at Pennsylvania State University. Other conferences have also been ongoing in Europe; e.g. *Cosmostat*. The remaining astrostatisticians have established collaborative associations within their own universities, or within a small group of universities. Some excellent work was being done in the area of astrostatistics, but communication between astrostatisticians on a global basis has been rather haphazard. Until recently there has been no overall organization or association for the discipline as a whole.

A requisite of my association with NASA's Jet Propulsion Laboratory entails that I participate on conference calls with the directors of various NASA and JPL projects and missions. Since 2007 I repeatedly heard from principal investigators that statistical issues were going to be a problem in the analysis of their data. This in turn encouraged me to explore the possibility of forming an association of astrostatisticians that would foster a global collaboration of statisticians and astronomers with the aim of effecting better statistical research. In early 2008 I formed an exploratory astrostatistics interest group within the fold of the *International Statistical Institute* (ISI), the world association of statisticians, with headquarters in The Netherlands. A meeting of the interest group, comprised of both statisticians and astronomers, was held in conjunction with the biannual ISI World Statistics Congress held in Durban, South Africa in August 2009. The members of the interest group agreed to propose that an ISI astrostatistics committee be formed and recognized as part of the Institute. By this time there were some fifty members in the interest group.

In December 2009 the ISI Council approved the existence of astrostatistics as a full standing committee of the ISI. However, ISI committees consist of no more than twelve to fifteen members. In order to extend the committee to incorporate anyone having an interest in the statistical analysis of astronomical data, I formed what is now known as the *ISI International Astrostatistics Network* as a separate body of researchers, with the ISI astrostatistics committee serving as the Network executive board. Membership has grown to over 200 members from some 30 nations and all populated continents.

After two years of existence, the Network has established solid relationships with both the ISI and *International Astronomical Union* (IAU) whose leadership has supported the Network and its goals. Network members were awarded an invited papers session and two special topics sessions at the 2011 World Statistics Congress in Dublin, Ireland. In addition, discussions have been underway with several publishing houses regarding a *Journal of Astrostatistics*. The Network will only proceed with such a venture if it is assured that there will be a rather steady long-term stream of quality submissions made to the journal's editorial board. As of this time we are not convinced that this will be the case in the immediate future, but do plan for such a journal in the future.

As a consequence of the initial successes of the Network, in December 2010 Springer Science and Business Media begun a *Springer Series on Astrostatistics*, on which Network members hold the editorial board positions. The series will publish texts and monographs on a wide variety of astrostatistical and astrophysical subjects. The Series will also publish Proceedings papers. Series books will be published in both print and e-book format. The first two books in the new Series include *Astrostatistics and Data Mining* (2012) [7] and this volume.

It is clear that many in the astrostatistics community believe that the existence of a global association of astrostatisticians is a worthwhile body to support. However, the Network is not aimed to be a governing organization, but rather an association that seeks to augment and support the ongoing efforts of established astrostatistics groups and conferences. It is an association of researchers with a common interest and a resource to help disseminate information regarding astrostatistically related literature, conferences, and research. Most importantly, it can also serve as the professional society for those identifying themselves as astrostatisticians. Astrostatistics as a profession is but in its infancy at this time, but it is hoped that a viable profession will be established within the next twenty years.

Astrostatistics faces some formidable challenges. The National Virtual Observatory (NVO) is now being constructed which will link archival astronomical databases and catalogues from the many ongoing surveys now being maintained, including LSST. The goal is to make all gathered astronomical data available to astronomers and astrostatisticians for analysis. However, this will involve many petabytes of information. In a relatively short time the amount of data may exceed an Exabyte, or a thousand petabytes. This is a truly huge amount of data. Even when dealing with terabytes, current statistical software is not capable of handling such an amount of information. A regression of a billion observations with 10 predictors results in a matrix inversion that far exceeds current and realistically foreseeable capabilities. New methods of statistical analysis will need to be developed to deal with these large datasets. And new statistical methods will need to be created that can evaluate such large amounts of data. There are a host of statistical and data mining problems related to evaluating huge masses of data in the attempt to determine the probability of some proposed outcome or event.

Another major step in the advance of the discipline of astrostatistics occurred in March of 2012. The *Astrostatistics and Astroinformatics Portal* <http://asaip.psu.edu> hosted by Pennsylvania State University, was established, providing astrostatisticians and others in related areas with an interactive web site including discussion pages or forums, a component where selected journal articles and pre-publication manuscripts can be downloaded, conferences announced, and so forth. Members of the Network automatically became Portal members. Others apply through application.

On August 28, 2012 the ISI astrostatistics Network approved becoming an independent professional organization for astronomers, statisticians and information scientists having an interest in the statistical analysis of astronomical data. It is named the *International Astrostatistics Association* (IAA), with the Portal as its host website. Soon following the creation of the IAA, both the IAU and *American Astronomical Society* (AAS) approved Astrostatistics Working Groups, the first time for any astronomical organization. The Portal was amended to have three components: the IAA, other international, national and astrostatistical specialty area organizations, and the LSST project. The Portal has in effect become the web site for the discipline.

I believe that astrostatistics will only develop into a mature discipline, capable of handing the looming data and analytic problems, by becoming a profession. This can be done by developing joint programs in the discipline, sponsored and maintained by the mutual efforts of the departments of statistics and astronomy/astrophysics at leading universities. Graduates will be awarded MS and PhD degrees in astrostatistics, and be trained in statistical analysis, astrophysics, and computer and computational logic. With a new generation of astrostatisticians engaged in handling the problems I have previously mentioned, there is more likelihood that the foremost questions we have of the early universe, as well a host of other queries, can be answered. Astrostatistics has a deep heritage within the disciplines of both statistics and astronomy, reaching back to their very beginnings, but its value rests in the future. With continually increased computer computation speed and enhanced memory available to astrostatisticians in the future, it is likely that answers may be found to many of the fundamental questions that are of concern to present-day astronomy and cosmology. I also suspect that many of the methods of analysis discovered and used by astrostatisticians will be of considerable value to researchers in other disciplines as well.

References

1. McGrayne, S.B.: The theory that would not die. Yale University Press, New Haven, CT (2011)
2. Babu, G.J., Feigelson, E.D.: Astrostatistics. Chapman & Hall, Baton Rouge, FL (1996)
3. Hilbe, J.M.: The Co-evolution of Statistics and Hz. In: Sawilowsky, S. (ed.), Real Data Analysis, pp 3–20, Information Age Publishing, Charlotte, NC (2007)
4. Hilbe, J.M.: Logistic Regression Models. Chapman & Hall/CRC, Boca Raton, FL (2009)
5. Hilbe, J.M.: Negative Binomial Regression, 2nd edn. Cambridge University Press, Cambridge, UK (2011)

6. Babu, G.J., Feigelson, E.D.: Statistical Challenges in Modern Astronomy, arXiv:astro-ph/0401404v1, Cornell University Library Astrophysical Archives (2004)
7. Baro, L.M.S, Eyer, L., O'Mullane, W., de Ridder, J. (eds.): Astrostatistics and Data Mining, New York: Springer (2004)
8. Feigelson, E.D. and Babu, G.J.: Modern Statistical Methods for Astronomy, Cambidge University Press, Cambridge, UK (2012)

Chapter 2
Bayesian Astrostatistics: A Backward Look to the Future

Thomas J. Loredo

Abstract This "perspective" chapter briefly surveys (1) past growth in the use of Bayesian methods in astrophysics; (2) current misconceptions about both frequentist and Bayesian statistical inference that hinder wider adoption of Bayesian methods by astronomers; and (3) multilevel (hierarchical) Bayesian modeling as a major future direction for research in Bayesian astrostatistics.

This volume contains presentations from the first invited session on astrostatistics to be held at an International Statistical Institute (ISI) World Statistics Congress. This session was a major milestone for astrostatistics as an emerging cross-disciplinary research area. It was the first such session organized by the ISI Astrostatistics Committee, whose formation in 2010 marked formal international recognition of the importance and potential of astrostatistics by one of its information science parent disciplines. It was also a significant milestone for Bayesian astrostatistics, as this research area was chosen as a (non-exclusive) focus for the session.

As an early (and elder!) proponent of Bayesian methods in astronomy, I have been asked to provide a "perspective piece" on the history and status of Bayesian astrostatistics. I begin by briefly documenting the rapid rise in use of the Bayesian approach by astrostatistics researchers over the past two decades. Next, I describe three misconceptions about both frequentist and Bayesian methods that hinder wider adoption of the Bayesian approach across the broader community of astronomer data analysts. Then I highlight the emerging role of multilevel (hierarchical) Bayesian modeling in astrostatistics as a major future direction for research in Bayesian astrostatistics. I end with a provocative recommendation for survey data reporting, motivated by the multilevel Bayesian perspective on modeling cosmic populations.

Thomas J. Loredo
Center for Radiophysics and Space Research, Cornell University, Ithaca, NY 14853-6801
e-mail: loredo@astro.cornell.edu

2.1 Looking back

Bayesian ideas entered modern astronomical data analysis in the late 1970s, when Gull and Daniell [1, 2] framed astronomical image deconvolution in Bayesian terms.[1] Motivated by Harold Jeffreys' Bayesian *Theory of Probability* [4], and Edwin Jaynes's introduction of Bayesian and information theory methods into statistical mechanics and experimental physics [5], they addressed image estimation by writing down Bayes's theorem for the posterior probability for candidate images, adopting an entropy-based prior distribution for images. They focused on finding a single "best" image estimate based on the posterior: the maximum entropy (MaxEnt) image (maximizing the entropic prior probability subject to a somewhat ad hoc likelihood-based constraint expressing goodness of fit to the data). Such estimates could also be found using frequentist penalized likelihood or regularization approaches. The Bayesian underpinnings of MaxEnt image deconvolution thus seemed more of a curiosity than the mark of a major methodological shift.

By about 1990, genuinely Bayesian data analysis — in the sense of reporting Bayesian probabilities for statistical hypotheses, or samples from Bayesian posterior distributions — began appearing in astronomy. The Cambridge MaxEnt group of astronomers and physicists, led by Steve Gull and John Skilling, began developing "quantified MaxEnt" methods to quantify uncertainty in image deconvolution (and other inverse problems), rather than merely reporting a single best-fit image. On the statistics side, Brian Ripley's group began using Gibbs sampling to sample from posterior distributions for astronomical images based on Markov random field priors [6]. My PhD thesis (defended in 1990) introduced parametric Bayesian modeling of Poisson counting and point processes (including processes with truncation or thinning, and measurement error) to high-energy astrophysics (X-ray and gamma-ray astronomy) and to particle astrophysics (neutrino astronomy). Bayesian methods were just beginning to be used for parametric modeling of ground- and space-based cosmic microwave background (CMB) data [e.g., 7].

It was in this context that the first session on Bayesian methods to be held at an astronomy conference (to my knowledge) took place, at the first *Statistical Challenges in Modern Astronomy* conference (SCMA I), hosted by statistician G. Jogesh Babu and astronomer Eric Feigelson at Pennsylvania State University in August 1991. Bayesian methods were not only new, but also controversial in astronomy at that time. Of the 22 papers published in the SCMA I proceedings volume [8], only two were devoted to Bayesian methods ([6] and [9]; see also the unabridged version of the latter, [10]).[2] Both papers had a strong pedagogical component (and a bit of polemic). Of the 131 SCMA I participants (about 60% astronomers and

---

[1] Notably, Sturrock [3] earlier introduced astronomers to the use of Bayesian probabilities for "bookkeeping" of subjective beliefs about astrophysical hypotheses, but he did not discuss statistical modeling of measurements per se.

[2] A third paper [11] had some Bayesian content but focused on frequentist evaluation criteria, even for the one Bayesian procedure considered; these three presentations, with discussion, comprised the Bayesian session.

40% statisticians), only two were astronomers whose research prominently featured Bayesian methods (Steve Gull and me).

Twenty years later, the role of Bayesian methods in astrostatistics research is dramatically different. The 2008 Joint Statistical Meetings included two sessions on astrostatistics predominantly devoted to Bayesian research. At SCMA V, held in June 2011, two sessions were devoted entirely to Bayesian methods in astronomy: "Bayesian analysis across astronomy" (BAA), with eight papers and two commentaries, and "Bayesian cosmology," including three papers with individual commentaries. Overall, 14 of 32 invited presentations (not counting commentaries) featured Bayesian methods, and the focus was on calculations and results rather than on pedagogy and polemic. About two months later, the ISI World Congress session on astrostatistics commemorated in this volume was held; as already noted, its focus was Bayesian astrostatistics.

On the face of it, these events seem to indicate that Bayesian methods are not only no longer controversial, but are in fact now widely used, even favored for some applications (most notably for parametric modeling in cosmology). But how representative are the conference presentations of broader astrostatistical practice?

Fig. 2.1 shows my amateur attempt at bibliometric measurement of the growing adoption of Bayesian methods in both astronomy and physics, based on queries of publication data in the NASA Astrophysics Data System (ADS). Publication counts indicate significant and rapidly growing use of Bayesian methods in both astronomy and physics.[3] Cursory examination of the publications reveals that Bayesian methods are being developed across a wide range of astronomical subdisciplines.

It is tempting to conclude from the conference and bibliometric indicators that Bayesian methods are now well-established and well-understood across astronomy. But the conference metrics reflect the role of Bayesian methods in the astrostatistics research community, not in bread-and-butter astronomical data analysis. And as impressive as the trends in the bibliometric metrics may be, the absolute numbers remain small in comparison to all astronomy and physics publications, even limiting consideration to data-based studies. Although their impact is growing, Bayesian methods are not yet in wide use by astronomers.

My interactions with colleagues indicate that significant misconceptions persist about fundamental aspects of both frequentist and Bayesian statistical inference, clouding understanding of how these rival approaches to data analysis differ and relate to one another. I believe these misconceptions play no small role in hindering broader adoption of Bayesian methods in routine data analysis. In the following section I highlight a few misconceptions I frequently encounter. I present them here as a challenge to the Bayesian astrostatistics community; addressing them may accelerate the penetration of sound Bayesian analysis into routine astronomical data analysis.

---

[3] Roberto Trotta and Martin Hendry have shown similar plots in various venues, helpfully noting that the recent rate of growth apparent in Fig. 2.1 is much greater than the rate of growth in the number of *all* publications; i.e., not just the amount but also the prevalence of Bayesian work is rapidly rising.

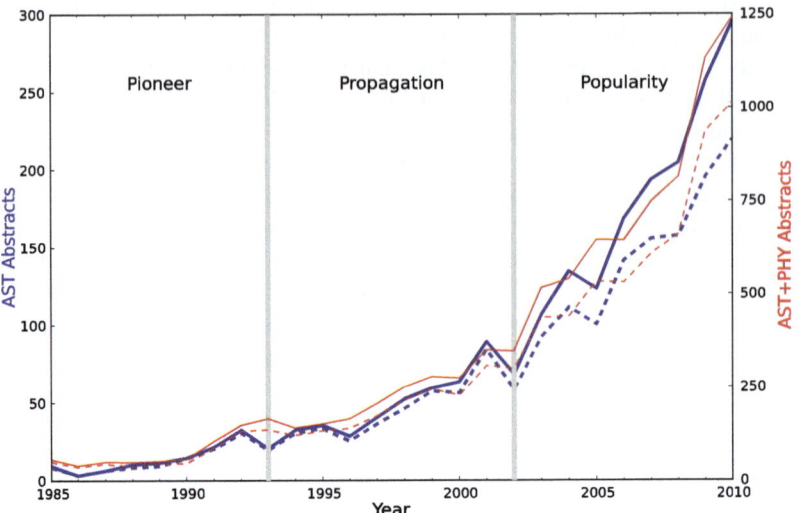

Fig. 2.1 Simple bibliometrics measuring the growing use of Bayesian methods in astronomy and physics, based on queries of the NASA ADS database in October 2011. Thick (blue) curves (against the left axis) are from queries of the astronomy database; thin (red) curves (against the right axis) are from joint queries of the astronomy and physics databases. For each case the dashed lower curve indicates the number of papers each year that include "Bayes" or "Bayesian" in the title or abstract. The upper curve is based on the same query, but also counting papers that use characteristically Bayesian terminology in the abstract (e.g., the phrase "posterior distribution" or the acronym "MCMC"); it is meant to capture Bayesian usage in areas where the methods are well-established, with the "Bayesian" appellation no longer deemed necessary or notable.

2.2 Misconceptions

For brevity, I will focus on just three important misconceptions I repeatedly encounter about Bayesian and frequentist methods, posed as (incorrect!) "conceptual equations." They are:

- **Variability = Uncertainty**: This is a misconception about frequentist statistics that leads analysts to think that good frequentist statistics is easier to do than it really is.
- **Bayesian computation = Hard**: This is a misconception about Bayesian methods that leads analysts to think that implementing them is harder than it really is, in particular, that it is harder than implementing a frequentist analysis with comparable capability.
- **Bayesian = Frequentist + Priors**: This is a misconception about the role of prior probabilities in Bayesian methods that distracts analysts from more essential features of Bayesian inference.

I will elaborate on these faulty equations in turn.

Variability = Uncertainty: Frequentist statistics gets its name from its reliance on the long-term frequency conception of probability: frequentist probabilities describe the long-run variability of outcomes in repeated experimentation. Astronomers who work with data learn early in their careers how to quantify the variability of data analysis procedures in quite complicated settings using Monte Carlo methods that simulate data. How to get from quantification of *variability* in the outcome of a procedure applied to an ensemble of simulated data, to a meaningful and useful statement about the *uncertainty* in the conclusions found by applying the procedure to the one actually observed data set, is a subtle problem that has occupied the minds of statisticians for over a century. My impression is that many astronomers fail to recognize the distinction between variability and uncertainty, and thus fail to appreciate the achievements of frequentist statistics and their relevance to data analysis practice in astronomy. The result can be reliance on overly simplistic "home-brew" analyses that at best may be suboptimal, but that sometimes can be downright misleading. A further consequence is a failure to recognize fundamental differences between frequentist and Bayesian approaches to quantification of uncertainty (e.g., that Bayesian probabilities for hypotheses are not statements about variability of results in repeated experiments).

To illustrate the issue, consider estimation of the parameters of a model being fit to astronomical data, say, the parameters of a spectral model. It is often straightforward to find best-fit parameters with an optimization algorithm, e.g., minimizing a χ^2 measure of misfit, or maximizing a likelihood function. A harder but arguably more important task is quantifying uncertainty in the parameters. For abundant data, an asymptotic Gaussian approximation may be valid, justifying use of the Hessian matrix returned by many optimization codes to calculate an approximate covariance matrix for defining confidence regions. But when uncertainties are significant and models are nonlinear, we must use more complicated procedures to find accurate confidence regions.

Bootstrap resampling is a powerful framework statisticians use to develop methods to accomplish this. There is a vast literature on applying the bootstrap idea in various settings; much of it is devoted to the nontrivial problem of devising algorithms that enable useful and accurate uncertainty statements to be derived from simple bootstrap variability calculations. Unfortunately, this literature is little-known in the astronomical community, and too often astronomers misuse bootstrap ideas. The variability-equals-uncertainty misconception appears to be at the root of the problem.

As a cartoon example, suppose we have spectral data from a source that we wish to fit with a simple thermal spectral model with two parameters, an amplitude, A (e.g., proportional to the source area and inversely proportional to its distance squared), and a temperature, T (determining the shape of the spectrum as a function of energy or wavelength); we denote the parameters jointly by $\mathcal{P} = (A, T)$. Fig. 2.2 depicts the two-dimensional (A, T) parameter space, with the best-fit parameters, $\hat{\mathcal{P}}(D_{\text{obs}})$, indicated by the blue four-pointed star. We can use simulated data to find the variability of the estimator (i.e., of the function $\hat{\mathcal{P}}(D)$ defined by the optimizer). But how should we simulate data when we do not know the true nature of the signal

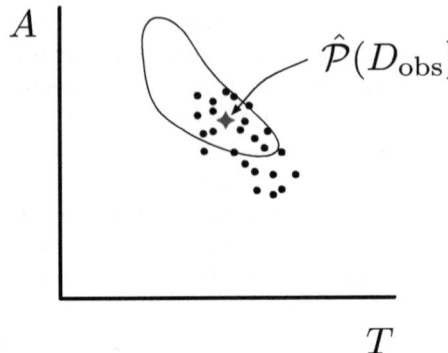

Fig. 2.2 Illustration of the nontrivial relationship between variability of an estimator, and uncertainty of an estimate as quantified by a frequentist confidence region. Shown is a two-dimensional parameter space with a best-fit estimate to the observed data (blue 4-pointed star), best-fit estimates to boostrapped data (black dots) showing variability of the estimator, and a contour bounding a parametric bootstrap confidence region quantifying uncertainty in the estimate.

(and perhaps of noise and instrumental distortions)? And how should the variability of simulation results be used to quantify the uncertainty in results based on the observed data?

The underlying idea of the bootstrap is to use the observed data set to define the ensemble of hypothetical data to use in variability calculations, and to find functions of the data (statistics) whose variability can be simply used to quantify uncertainty (e.g., via moments or a histogram). Normally using the observed data to define the ensemble for simulations would be cheating and would invalidate one's inferences; all frequentist probabilities formally must be "pre-observation" calculations. A major achievement of the bootstrap literature is showing how to use the observed data in a way that gives approximately valid results (hopefully with a rate of convergence better than the $O(1/\sqrt{N})$ rate achieved by simple Gaussian approximations, for sample size N).

One way to proceed is to use the full model we are fitting (both the signal model and the instrument and noise model) to simulate data, with the parameters fixed at $\hat{\mathcal{P}}(D_{\text{obs}})$ as a surrogate for the true values. Statisticians call this the *parametric bootstrap*; it was popularized to astronomers in a well-known paper by Lampton, Margon and Bowyer ([12], hereafter LMB76; statisticians introduced the "parametric bootstrap" terminology later). Alternatively, if some probabilistic aspects of the model are not trusted (e.g., the error distribution is considered unrealistic), an alternative approach is the *nonparametric bootstrap*, which "recycles" the observed data to generate simulated data (in some simple cases, this may be done by sampling from the observed data with replacement to generate each simulated data set). Whichever approach we adopt, we will generate a set of simulated data, $\{D_i\}$, to which we can apply our fitting procedure to generate a set of best-fit parameter points $\{\hat{\mathcal{P}}(D_i)\}$ that together quantify the variability of our estimator. The black dots in Fig. 2.2 show a scatterplot or "point cloud" of such parameter estimates.

What does the point cloud tell us about the uncertainty we should associate with the estimate from the observed data (the blue star)? The practice I have seen in too many papers is to interpret the points as samples from a probability distribution in parameter space.[4] The cloud itself might be shown, with a contour enclosing a specified fraction of the points offered as a joint confidence region. One-dimensional histograms may be used to find "1σ" (i.e., 68.3%) confidence regions for particular parameters of interest. Such procedures naively equate variability with uncertainty: the uncertainty of the estimate is identified with the variability of the estimator. Regions created this way are *wrong*, plainly and simply. They will not cover the true parameters with the claimed probability, and they are skewed in the wrong direction. This is apparent from the figure; the black points are skewed *down* and to the *right* from the parameter values used to produce the simulated data; the parameters that produced the observed data are thus likely to be *up* and to the *left* of the star.

In a correct parametric bootstrapping calculation (i.e., with a trusted model), one can use simulated data to calibrate χ^2 or likelihood contours for defining confidence regions. The procedure is not very complicated, and produces regions like the one shown by the contour in Fig. 2.2, skewed in just the right way; LMB76 described the construction. But frequently investigators are attempting a *non*parametric bootstrap, corresponding to a trusted signal model but an untrusted noise model (often this is done without explicit justification, as if nonparametric bootstrapping is the only type of bootstrapping). In this case devising a sound bootstrap confidence interval algorithm is not so simple. Indeed, there is a large statistics literature on nonparametric bootstrap confidence intervals, its size speaking to nontrivial challenges in making the bootstrap idea work. In particular, no simple procedure is currently known for finding accurate joint confidence regions for multiple parameters in nonlinear models using the nonparametric bootstrap; yet results purporting to come from such a procedure appear in a number of astronomical publications. Strangely, the most cited reference on bootstrap confidence regions in the astronomical literature is a book on numerical methods, authored by astronomers, with an extremely brief and misleadingly simplistic discussion of the nonparametric bootstrap (and no explicit discussion of parametric bootstrapping). Sadly, this is not the only area where our community appears content disregarding a large body of relevant statistics research (frequentist or otherwise).

What explains such disregard of relevant and nontrivial expertise? I am sure there are multiple factors, but I suspect an important one is the misconception that variability may be generically identified with uncertainty. If one believes this, then doing (frequentist) statistics appears to be a simple matter of simulating data to quantify the variability of procedures. For linear models with normally-distributed errors, the identification is practically harmless; it is conceptually flawed but leads to correct

---

[4] I am not providing references to publications exhibiting the problem for diplomatic reasons and for a more pragmatic and frustrating reason: In the field where I have repeatedly encountered the problem—analysis of infrared exoplanet transit data—authors routinely fail to describe their analysis methods with sufficient detail to know what was done, let alone to enable readers to verify or duplicate the analysis. While there are clear signs of statistical impropriety in many of the papers, I only know the details from personal communications with exoplanet transit scientists.

results by accident. But more generally, determining how to use variability to quantify uncertainty can be subtle and challenging. Statisticians have worked for a century to establish how to map variability to uncertainty; when we seek frequentist quantifications of uncertainty in nontrivial settings, we need to mine their expertise.

Why devote so much space to a misconception about frequentist statistics in a commentary on Bayesian methods? The variability-equals-uncertainty misconception leads data analysts to think that frequentist statistics is easier than it really is, in fact, to think that they already know what they need to know about it. If astronomers realize that sound statistical practice is nontrivial and requires study, they may be more likely to study Bayesian methods, and more likely to come to understand the differences between the frequentist and Bayesian approaches. Also, for the example just described, the way some astronomers misuse the bootstrap is to try to use it to define a probability distribution over parameter space. Frequentist statistics denies such a concept is meaningful, but it is exactly what Bayesian methods aim to provide, e.g., with point clouds produced via Markov chain Monte Carlo (MCMC) posterior sampling algorithms. This brings us to the topic of Bayesian computation.

Bayesian computation = Hard: A too-common (and largely unjustified) complaint about Bayesian methods is that their computational implementation is difficult, or more to the point, that Bayesian computation is harder than frequentist computation. Analysts wanting a quick and usable result are thus dissuaded from considering a Bayesian approach. It is certainly true that posterior sampling via MCMC—generating pseudo-random parameter values distributed according to the posterior—is harder to do well than is generating pseudo-random data sets from a best-fit model. (In fact, I would argue that our community may not appreciate how hard it can be to do MCMC well.) But this is an apples-to-oranges comparison between methods making very different types of approximations. Fair comparisons, between Bayesian and frequentist methods *with comparable capabilities and approximations*, tell a different story. Advocates of Bayesian methods need to give this complaint a proper and public burial.

Consider first basic "textbook" problems that are analytically accessible. Examples include estimating the mean of a normal distribution, estimating the coefficients of a linear model (for data with additive Gaussian noise), similar estimation when the noise variance is unknown (leading to Student's t distribution), estimating the intensity of a Poisson counting process, etc.. In such problems, the Bayesian calculation is typically *easier* than its frequentist counterpart, sometimes significantly so. Nowhere is this more dramatically demonstrated than in Jeffreys's classic text, *Theory of Probability* [4]. In chapter after chapter, Jeffreys solves well-known statistics problems with arguments significantly more straightforward and mathematics significantly more accessible than are used in the corresponding frequentist treatments. The analytical tractability of such foundational problems is an aid to developing sound statistical intuition, so this is not a trivial virtue of the Bayesian approach. (Of course, ease of use is no virtue at all if you reject an approach—Bayesian, frequentist, or otherwise—on philosophical grounds.)

Turn now to problems of more realistic complexity, where approximate numerical methods are necessary. The main computational challenge for both frequentist and Bayesian inference is multidimensional integration, over the sample space for frequentist methods, and over parameter space for Bayesian methods. In high dimensions, both approaches tend to rely on Monte Carlo methods for their computational implementation. The most celebrated approach to Bayesian computation is MCMC, which builds a multivariate pseudo-random number generator producing dependent samples from a posterior distribution. Frequentist calculation instead uses Monte Carlo methods to simulate data (i.e., draw samples from the sampling distribution). Building an MCMC posterior sampler, and properly analyzing its output, is certainly more challenging than simulating data from a model, largely because data are usually independently distributed; nonparametric bootstrap resampling of data may sometimes be simpler still. But the comparison is not fair. In typical frequentist Monte Carlo calculations, one simulates data from the best-fit model (or a bootstrap surrogate), not from the true model (or other plausible models). The resulting frequentist quantities are approximate, not just because of Monte Carlo error, but because the *integrand* (or family of integrands) that would appear in an exact frequentist calculation is being approximated (asymptotically). In some cases, an exact finite-sample formulation of the problem may not even be known. In contrast, MCMC posterior sampling makes no approximation of integrands; results are approximate only due to Monte Carlo sampling error. No large-sample asymptotic approximations need be invoked.

This point deserves amplification. Although the main computational challenge for frequentist statistics is integration over sample space, there are additionally serious *theoretical* challenges for finite-sample inference in many realistically complex settings. Such challenges do not typically arise for Bayesian inference. These theoretical challenges, and the analytical approximations that are adopted to address them, are ignored in comparisons that pit simple Monte Carlo simulation of data or bootstrapping against MCMC or other nontrivial Bayesian computation techniques. Arguably, accurate finite-sample parametric inference is often computationally *simpler* for Bayesian methods, because an accurate frequentist calculation is simply impossible, and an approximate calculation can quantify only the rate of convergence of the approximation, not the actual accuracy of the specific calculation being performed.

If one is content with asymptotic approximations, the fairer comparison is between asymptotic frequentist and asymptotic Bayesian methods. At the lowest order, asymptotic Bayesian computation using the Laplace approximation is not significantly harder than asymptotic frequentist calculation. It uses the same basic numerical quantities—point estimates and Hessian matrices from an optimizer—but in different ways. It provides users with nontrivial new capabilities, such as the ability to marginalize over nuisance parameters, or to compare models using Bayes factors that include an "Ockham's razor" penalty not present in frequentist significance tests, and that enable straightforward comparison of rival models that need not be nested (with one being a special case of the other). And the results are sometimes accurate to one order higher (in $1/\sqrt{N}$) than corresponding frequentist results (making them potentially competitive with some bootstrap methods seeking similar asymptotic approximation rates).

Some elaboration of these issues is in [13], including references to literature on Bayesian computation up to that time. A more recent discussion of this misconception about Bayesian methods (and other misconceptions) is in an insightful essay by the Bayesian economist Christopher Sims, winner of the 2011 Nobel Prize in economics [14].

Bayesian = Frequentist + Priors: I recently helped organize and present two days of tutorial lectures on Bayesian computation, as a prelude to the SCMA V conference mentioned above. As I was photocopying lecture notes for the tutorials, a colleague walked into the copy room and had a look at the table of contents for the tutorials. "Why aren't all of the talks about priors?" he asked. In response to my puzzled look, he continued, "Isn't that what Bayesian statistics is about, accounting for prior probabilities?"

Bayesian statistics gets its name from Bayes's theorem, establishing that the posterior probability for a hypothesis is proportional to the product of its prior probability, and the probability for the data given the hypothesis (i.e., the sampling distribution for the data). The latter factor is the likelihood function when considered as a function of the hypotheses being considered (with the data fixed to their observed values). Frequentist methods that directly use the likelihood function, and their least-squares cousins (e.g., χ^2 minimization), are intuitively appealing to astronomers and widely used. On the face of it, Bayes's theorem appears merely to add modulation by a prior to likelihood methods. By name, *Bayesian* statistics is evidently about using *Bayes's theorem*, so it would seem it must be about how frequentist results should be altered to account for prior probabilities.

It would be hard to overstate how wrong this conception of Bayesian statistics is.

The name is unfortunate; Bayesian statistics uses *all* of probability theory, not just Bayes's theorem, and not even primarily Bayes's theorem. What most fundamentally distinguishes Bayesian calculations from frequentist calculations is not modulation by priors, but the key role of probability distributions *over parameter (hypothesis) space* in the former, and the complete absence of such distributions in the latter. Via Bayes's theorem, a prior enables one to use the likelihood function—describing a family of measures over the *sample* space—to build the posterior distribution—a measure over the *parameter* (hypothesis) space (where by "measure" I mean an additive function over sets in the specified space). This construction is just one step in inference. Once it happens, the rest of probability theory kicks in, enabling one to assess scientific arguments directly by calculating probabilities quantifying the strengths of those arguments, rather than indirectly, by having to devise a way that variability of a cleverly chosen statistic across hypothetical data might quantify uncertainty across possible choices of parameters or models for the observed data.

Perhaps the most important theorem for doing Bayesian calculations is the *law of total probability* (LTP) that relates marginal probabilities to joint and conditional probabilities. To describe its role, suppose we are analyzing some observed data,

D_{obs}, using a parametric model with parameters θ. Let M denote all the modeling assumptions—definitions of the sample and parameter spaces, description of the connection between the model and the data, and summaries of any relevant prior information (parameter constraints or results of other measurements). Now consider some of the common uses of LTP in Bayesian analysis:

- Calculating the probability in a *credible region*, R, for θ:

$$p(\theta \in R | D_{\text{obs}}) = \int_R d\theta \, p(\theta \in R|\theta) \, p(\theta|D_{\text{obs}}) \qquad \| \, M. \qquad (2.1)$$

Here I have introduced a convenient shorthand due to John Skilling: "$\| M$" indicates that M is conditioning information common to all displayed probabilities.

- Calculating a *marginal posterior distribution* when a vector parameter θ has both an interesting subset of parameters, ψ, and nuisance parameters, η:

$$p(\psi|D_{\text{obs}}) = \int d\eta \, p(\psi, \eta|D_{\text{obs}}) \qquad \| \, M. \qquad (2.2)$$

- Comparing rival parametric models M_i (each with parameters θ_i) via posterior odds or Bayes factors, which requires computation of the *marginal likelihood* for each model given by

$$p(D_{\text{obs}}|M_i) = \int d\theta_i \, p(\theta_i|M_i) \, p(D_{\text{obs}}|\theta_i, M_i) \qquad \| \, M_1 \vee M_2 \dots . \qquad (2.3)$$

In words, this says that the likelihood for a model is the average of the likelihood function for that model's parameters.

- Predicting future data, D', with the posterior predictive distribution,

$$p(D'|D_{\text{obs}}) = \int d\theta \, p(D'|\theta) \, p(\theta|D_{\text{obs}}) \qquad \| \, M. \qquad (2.4)$$

Arguably, if this approach to inference is to be named for a theorem, "total probability inference" would be a more appropriate appellation than "Bayesian statistics." It is probably too late to change the name. But it is not too late to change the emphasis.

In axiomatic developments of Bayesian inference, priors play no fundamental role; rather, they *emerge* as a required ingredient when one seeks a consistent or coherent calculus for the strengths of arguments that reason from data to hypotheses. Sometimes priors are eminently useful, as when one wants to account for a positivity constraint on a physical parameter, or to combine information from different experiments or observations. Other times they are frankly a nuisance, but alas still a necessity.

A physical analogy I find helpful for elucidating the role of priors in Bayesian inference appeals to the distinction between intensive and extensive quantities in thermodynamics. Temperature is an intensive property; in a volume of space it is meaningful to talk about the temperature $T(x)$ at a *point* x, but not about the "total temperature" of the *volume*; temperature does not add or integrate across space. In contrast, heat is an extensive property, an additive property of volumes; in mathematical parlance, it may be described by a measure (a mapping from regions, rather than points, to a real number). Temperature and heat are related; the heat in a volume V is given by $Q = \int_V dx\, [\rho(x)c(x)]T(x)$, where $\rho(x)$ is the density and $c(x)$ is the specific heat capacity. The product ρc is extensive, and serves to convert the intensive temperature to its extensive relative, heat. In Bayesian inference, the prior plays an analogous role, not just "modulating" likelihood, but converting intensive likelihood to extensive probability. In thermodynamics, a region with a high temperature may have a small amount of heat if its volume is small, or if, despite having a large volume, the value of ρc is small. In Bayesian statistics, a region of parameter space with high likelihood may have a small probability if its volume is small, or if the prior assigns low probability to the region.

This *accounting for volume in parameter space* is a key feature of Bayesian methods. What makes it possible is having a measure over parameter space. Priors are important, not so much as modulators of likelihoods, but as converters from intensities (likelihoods) to measures (probabilities). With poetic license, one might say that frequentist statistics focuses on the "hottest" (highest likelihood) hypotheses, while Bayesian inference focuses on hypotheses with the most "heat" (probability).

Incommensurability: The growth in the use of Bayesian methods in recent decades has sometimes been described as a "revolution," presumably alluding to Thomas Kuhn's concept of scientific revolutions [15]. Although adoption of Bayesian methods in many disciplines has been growing steadily and sometimes dramatically, Bayesian methods have yet to completely or even substantially replace frequentist methods in any broad discipline I am aware of (although this has happened in some subdisciplines). I doubt the pace and extent of change qualifies for a Kuhnian revolution. Also, the Bayesian and frequentist approaches are not rival scientific theories, but rather rival paradigms for a part of the scientific method itself (how to build and assess arguments from data to scientific hypotheses). Nevertheless, the competition between Bayesian and frequentist approaches to inference does bear one hallmark of a Kuhnian revolution: *incommensurability*. I believe Bayesian-frequentist incommensurability is not well-appreciated, and that it underlies multiple misconceptions about the approaches.

Kuhn insisted that there could be no neutral or objective measure allowing comparison of competing paradigms in a budding scientific revolution. He based this claim on several features he found common to scientific revolutions, including the following: (1) Competing paradigms often adopt different meanings for the same term or statement, making it very difficult to effectively communicate across paradigms (a standard illustration is the term "mass," which takes on different meanings in Newtonian and relativistic physics). (2) Competing paradigms adopt different standards of evaluation; each paradigm typically "works" when judged by its own standards, but the standards themselves are of limited use in comparing across paradigms.

These aspects of Kuhnian incommensurability are evident in the frequentist and Bayesian approaches to statistical inference. (1) The term "probability" takes different meanings in frequentist and Bayesian approaches to uncertainty quantification, inviting misunderstanding when comparing frequentist and Bayesian answers to a particular inference problem. (2) Long-run performance is the gold standard for frequentist statistics; frequentist methods aim for specified performance across repeated experiments by construction, but make no probabilistic claims about the result of application of a procedure to a particular observed dataset. Bayesian methods adopt more abstract standards, such as coherence or internal consistency, that apply to inference for the case-at-hand, with no fundamental role for frequency-based long-run performance. A frequentist method with good long-run performance can violate Bayesian coherence or consistency requirements so strongly as to be obviously unacceptable for inference in particular cases.[5] On the other hand, Bayesian algorithms do not have guaranteed frequentist performance; if it is of interest, it must be separately evaluated, and priors may need adjustment to improve frequentist performance.[6]

Kuhn considered rival paradigms to be so "incommensurable" that "proponents of competing paradigms practice their trades in different worlds." He argued that incommensurability is often so stark that for an individual scientist to adopt a new paradigm requires a psychological shift that could be termed a "conversion experience." Following Kuhn, philosophers of science have debated how extreme incommensurability really is between rival paradigms, but the concept is widely considered important for understanding significant changes in science. In this context, it is notable that statisticians, and scientists more generally, often adopt a particular almost-religious terminology in frequentist vs. Bayesian discussions: rather than a *method* being described as frequentist or Bayesian, the *investigator* is so described. This seems to me to be an unfortunate tradition that should be abandoned. Nevertheless, it does highlight the fundamental incommensurability between these rival paradigms for uncertainty quantification. Advocates of one approach or the other (or of a nuanced combination) need to more explicitly note and discuss this incommensurability, especially with non experts seeking to choose between approaches.

The fact that both paradigms remain in broad use suggests that ideas from both approaches may be relevant to inference; perhaps they are each suited to addressing different types of scientific questions. For example, my discussion of misconceptions has been largely from the perspective of parametric modeling (parameter estimation and model comparison). *Non*parametric inference raises more subtle issues

---

[5] Efron [16] describes some such cases by saying the frequentist result can be *accurate* but not *correct*. Put another way, the performance claim is *valid*, but the long-run performance can be *irrelevant* to the case-at-hand, e.g., due to the existence of so-called recognizable subsets in the sample space (see [9] and [16] for elaboration of this notion). This is a further example of how nontrivial the relationship between variability and uncertainty can be.

[6] There are theorems linking single-case Bayesian probabilities and long-run performance in some general settings, e.g., establishing that, for fixed-dimension parametric inference, Bayesian credible regions with probability P have frequentist coverage close to P (the rate of convergence is $O(1/\sqrt{N})$ for flat priors, and faster for so-called reference priors). But the theorems do not apply in some interesting classes of problems, e.g., nonparametric problems.

regarding both computation and the role of long-term performance in Bayesian inference; see [17] and [14] for insightful discussions of some of these issues. Also, Bayesian model checking (assessing the adequacy of a parametric model without an explicitly specified class of alternatives) typically invokes criteria based on predictive frequencies [17, 49, 19]. A virtue of the Bayesian approach is that one may predict or estimate frequencies when they are deemed relevant; explicitly distinguishing probability (as degree of strength of an argument) from frequency (in finite or hypothetical infinite samples) enables this. This suggests some kind of unification of approaches may be easier to achieve from the Bayesian direction. This is a worthwhile task for research; see [20] for a brief overview of some recent work on the Bayesian/frequentist interface.

No one would claim that the Bayesian approach is a data analysis panacea, providing the best way to address all data analysis questions. But among astronomers outside of the community of astrostatistics researchers, Bayesian methods are significantly underutilized. Clearing up misconceptions should go a long way toward helping astronomers appreciate what both frequentist and Bayesian methods have to offer for both routine and research-level data analysis tasks.

2.3 Looking forward

Having looked at the past growth of interest in Bayesian methods and present misconceptions, I will now turn to the future. As inspiration, I cite Mike West's commentary on my SCMA I paper [21]. In his closing remarks he pointed to an especially promising direction for future Bayesian work in astrostatistics:

> On possible future directions, it is clear that Bayesian developments during recent years have much to offer—I would identify prior modeling developments in *hierarchical* models as particularly noteworthy. Applications of such models have grown tremendously in biomedical and social sciences, but this has yet to be paralleled in the physical sciences. Investigations involving repeat experimentation on similar, related systems provide the archetype logical structure for hierarchical modeling... There are clear opportunities for exploitation of these (and other) developments by astronomical investigators....

However clear the opportunities may have appeared to West, for over a decade following SCMA I, few astronomers pursued hierarchical Bayesian modeling. A particularly promising application area is modeling of populations of astronomical sources, where hierarchical models can naturally account for measurement error, selection effects, and "scatter" of properties across a population. I discussed this at some length at SCMA IV in 2006 [22], but even as of that time there was relatively little work in astronomy using hierarchical Bayesian methods, and for the most part only the simplest such models were used.

The last few years mark a change-point in this respect, and evidence of the change is apparent in the contributions to the summer 2011 Bayesian sessions at both SCMA V and the ISI World Congress. Several presentations in both forums described recent and ongoing research developing sophisticated hierarchical models for complex astronomical data. Other papers raised issues that may be addressed with hierarchical models. Together, these papers point to hierarchical Bayesian modeling as an important emerging research direction for astrostatistics.

To illustrate the notion of a hierarchical model—also known as a *multilevel model* (MLM) — we start with a simple parametric density estimation problem, and then promote it to a MLM by adding measurement error.

Suppose we would like to estimate parameters θ defining a probability density function $f(x;\theta)$ for an observable x. A concrete example might be estimation of a galaxy luminosity function, where x would be two-dimensional, $x = (L, z)$ for luminosity L and redshift z, and $f(x;\theta)$ would be the normalized luminosity function (i.e., a probability density rather than a galaxy number density). Consider first the case where we have a set of precise measurements of the observables, $\{x_i\}$ (and no selection effects). Panel (a) in Fig. 2.3 depicts this simple setting. The likelihood function for θ is $\mathcal{L}(\theta) \equiv p(\{x_i\}|\theta, M) = \prod_i f(x_i;\theta)$. Bayesian estimation of θ requires a prior density, $\pi(\theta)$, leading to a posterior density $p(\theta|\{x_i\}, M) \propto \pi(\theta)\mathcal{L}(\theta)$.

An alternative way to write Bayes's theorem expresses the posterior in terms of the joint distribution for parameters and data:

$$p(\theta|\{x_i\}, M) = \frac{p(\theta, \{x_i\}|M)}{p(\{x_i\}|M)}. \tag{2.5}$$

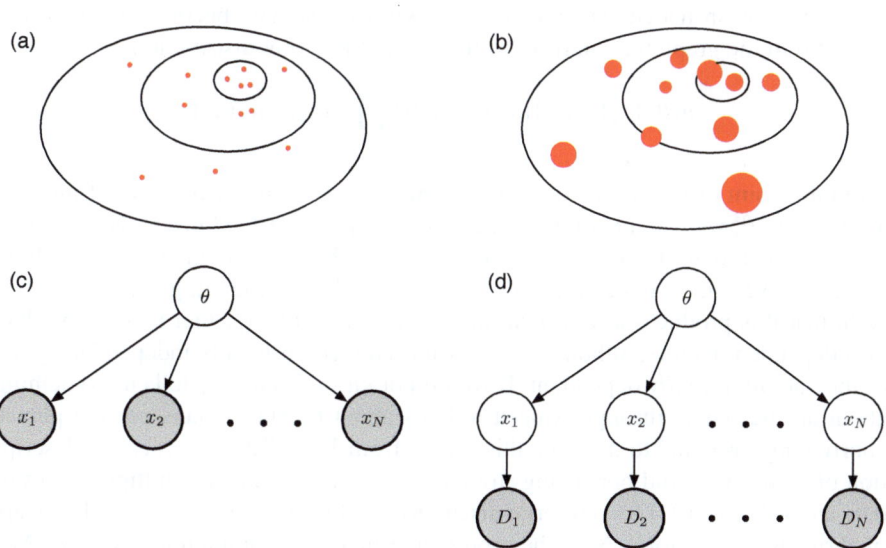

Fig. 2.3 Illustration of multilevel model approach to handling measurement error. (a) and (b) (top row): Measurements of a two-dimensional observable and its probability distribution (contours); in (a) the measurements are precise (points); in (b) they are noisy (filled circles depict uncertainties). (c) and (d): Graphical models corresponding to Bayesian estimation of the density in (a) and (b), respectively.

This "probability for everything" version of Bayes's theorem changes the process of modeling from separate specification of a prior and likelihood, to specification of the joint distribution for everything; this proves helpful for building models with complex dependencies. Panel (c) depicts the dependencies in the joint distribution with a graph—a collection of nodes connected by edges—where each node represents a probability distribution for the indicated variable, and the directed edges indicate dependences between variables. Shaded nodes indicate variables whose values are known (here, the data); we may manipulate the joint to condition on these quantities. The graph structure visually displays how the joint distribution may be factored as a sequence of independent and conditional distributions: the θ node represents the prior, and the x_i nodes represent $f(x_i; \theta)$ factors, dependent on θ but independent of other x_i values when θ is given (i.e., conditionally independent). The joint distribution is thus $p(\theta, \{x_i\}|M) = \pi(\theta) \prod_i f(x_i; \theta)$. In a sense, the most important edges in the graph are the *missing* edges; they indicate independence that makes factors simpler than they might otherwise be.

Now suppose that, instead of precise x_i measurements, for each observation we get noisy data, D_i, producing a measurement likelihood function $\ell_i(x_i) \equiv p(D_i|x_i, M)$ describing the uncertainties in x_i (we might summarize it with the mean and standard deviation of a Gaussian). Panel (b) depicts the situation; instead of points in x space, we now have likelihood functions (depicted as "1σ" error circles). Panel (d) shows a graph describing this measurement error problem, which adds a $\{D_i\}$ level to the previous graph; we now have a multilevel model.[7] The x_i nodes are now unshaded; they are no longer known, and have become *latent parameters*. From the graph we can read off the form of the joint distribution:

$$p(\theta, \{x_i\}, \{D_i\}|M) = \pi(\theta) \prod_i f(x_i; \theta)\ell_i(x_i). \qquad (2.6)$$

From this joint distribution we can make inferences about any quantity of interest. To estimate θ, we use the joint to calculate $p(\theta, \{x_i\}|\{D_i\}, M)$ (i.e., we condition on the known data using Bayes's theorem), and then we marginalize over all x_i variables. We can estimate all the x_i values jointly by instead marginalizing over θ. Note that this produces a joint marginal distribution for $\{x_i\}$ that is not a product of independent factors; although the x_i values are conditionally independent given θ, they are *marginally* dependent. If we do not know θ, each x_i tells us something about all the others through what it tells us about θ. Statisticians use the phrase "borrowing strength" to describe this effect, from John Tukey's evocative description of "mustering and borrowing strength" from related data in multiple stages of data analysis (see [23] for a tutorial discussion of this effect and the related concept of shrinkage estimators). Note the prominent role of LTP in inference with MLMs, where inference at one level requires marginalization over unknowns at other levels.

---

[7] The convention is to reserve the term for models with three or more levels of nodes, i.e., two or more levels of edges, or two or more levels of nodes for *uncertain variables* (i.e., unshaded nodes). The model depicted in panel (d) would be called a two-level model.

The few Bayesian MLMs used by astronomers through the 1990s and early 2000s did not go much beyond this simplest hierarchical structure. For example, unbeknownst to West, at the time of his writing my thesis work had already developed a MLM for analyzing the arrival times and energies of neutrinos detected from SN 1987A; the multilevel structure was needed to handle measurement error in the energies (an expanded version of this work appears in [24]). Panel (a) of Fig. 2.4 shows a graph describing the model. The rectangles are "plates" indicating substructures that are repeated; the integer variable in the corner indicates the number of repeats. There are two plates because neutrino detectors have a limited (and energy-dependent) detection efficiency. The plate with a known repeat count, N, corresponds to the N detected neutrinos with times t and energies ϵ; the plate with an unknown repeat count, $\overline{N}$, corresponds to undetected neutrinos, which must be considered in order to constrain the total signal rate; $\overline{D}$ denotes the nondetection data, i.e., reports of zero events in time intervals between detections.

Other problems tackled by astronomers with two-level MLMs include: modeling of number-size distributions ("log N–log S" or "number counts") of gamma-ray bursts and trans-Neptunian objects [e.g., 25, 26]; performing linear regression with measurement error along both axes, e.g., for correlating quasar hardness and luminosity ([27]; see Kelly's contribution in [28] for an introduction to MLMs for measurement error); accounting for Eddington and Malmquist biases in cosmology [23]; statistical assessment of directional coincidences with gamma-ray bursts [29, 30]

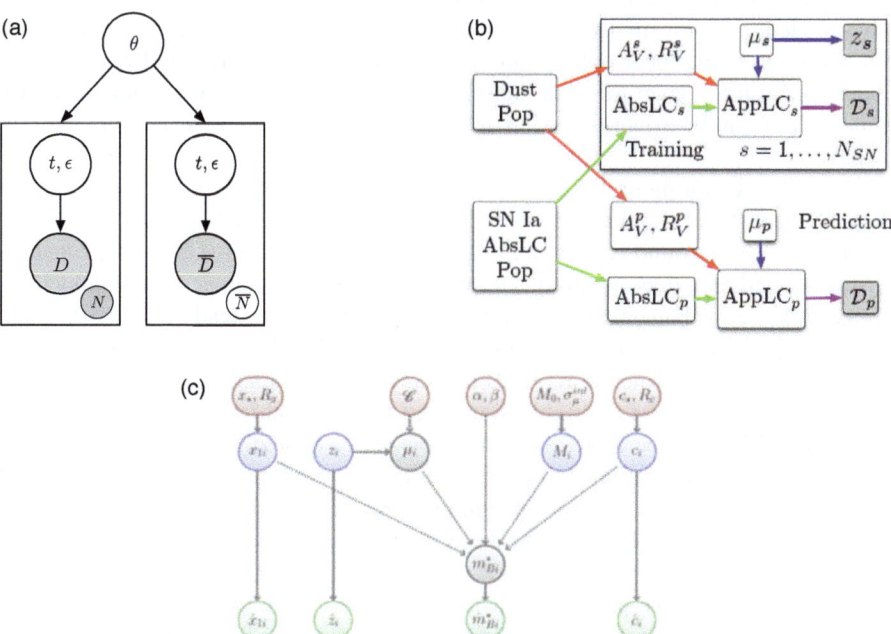

Fig. 2.4 Graphs describing multilevel models used in astronomy, as described in the text.

and cross-matching catalogs produced by large star and galaxy surveys ([31]; see [32] for discussion of the underlying MLM); and handling multivariate measurement error when estimating stellar velocity distributions from proper motion survey data [33].

Beginning around 2000, interdisciplinary teams of astronomers and information scientists began developing significantly more sophisticated MLMs for astronomical data. The most advanced such work has come from a collaboration of astronomers and statisticians associated with the Harvard-Smithsonian Center for Astrophysics (CfA). Much of their work has been motivated by analysis of data from the *Chandra* X-ray observatory satellite, whose science operations are managed by CfA. van Dyk et al. [34] developed a many-level MLM for fitting *Chandra* X-ray spectral data; a host of latent parameters enable accurate accounting for uncertain backgrounds and instrumental effects such as pulse pile-up. Esch et al. [35] developed a Bayesian image reconstruction algorithm for *Chandra* imaging data that uses a multiscale hierarchical prior to build spatially-adaptive smoothing into image estimation and uncertainty quantification. van Dyk et al. [36] showed how to analyze stellar cluster color-magnitude diagrams (CMDs) using finite mixture models (FMMs) to account for contamination of the data from stars not lying in the targeted cluster. In FMMs, categorical class membership variables appear as latent parameters; the mixture model effectively averages over many possible graphs (corresponding to different partitions of the data into classes). FMMs have long been used to handle outliers and contamination in Bayesian regression and density estimation. This work showed how to implement it with computationally expensive models and informative class membership priors. In the area of time series, the astronomer-engineer collaboration of Dobigeon, Tourneret, and Scargle [37] developed a three-level MLM to tackle joint segmentation of astronomical arrival time series (a multivariate extension of Scargle's well-known Bayesian Blocks algorithm).

Cosmology is a natural arena for multilevel modeling, because of the indirect link between theory and observables. For example, in modeling both the cosmic microwave background (CMB) and the large scale structure (LSS) of the galaxy distribution, theory does not predict a specific temperature map or set of galaxy locations (these depend on unknowable initial conditions), but instead predicts statistical quantities, such as angular or spatial power spectra. Modeling observables given theoretical parameters typically requires introducing these quantities as latent parameters. In [38] I described a highly simplified hierarchical treatment of CMB data, with noisy CMB temperature difference time series data at the lowest level, $l = 2$ spherical harmonic coefficients in the middle, and a single physical parameter of interest, the cosmological quadrupole moment Q, at the top. While a useful illustration of the MLM approach, I noted there were enormous computational challenges facing a more realistic implementation. It took a decade for such an implementation to be developed, in the pioneering work of Wandelt et al. [39]. And only recently have explicit hierarchical models been implemented for LSS modeling [e.g., 40].

This brings us to the present. Contributions in this volume and in the SCMA V proceedings [28] document burgeoning interest in Bayesian MLMs among astrostatisticians. I discuss the role of MLMs in the SCMA V contributions elsewhere [41]. Four of the contributions in this volume rely on MLMs. Two address complex problems and help mark an era of new complexity and sophistication in astrophysical MLMs. Two highlight the potential of basic MLMs for addressing common astronomical data analysis problems that defy accurate analysis with conventional methods taught to astronomers. I will highlight the MLM aspects of these contributions in turn.

Some of the most impressive new Bayesian multilevel modeling in astronomy addresses the analysis of measurements of multicolor light curves (brightness vs. time in various wavebands) from Type Ia supernovae (SNe Ia). In the late 1990s, astronomers discovered that these enormous stellar thermonuclear explosions are "standardizable candles;" the *shapes* of their light curves are strongly correlated with their *luminosities* (the intrinsic amplitudes of the light curves). This enables use of SNe Ia to measure cosmic distances (via a generalized inverse-square law) and to trace the history of cosmic expansion. The 2011 Nobel Prize in physics went to three astronomers who used this capability to show that the expansion rate of the universe is *growing* with time ("cosmic acceleration"), indicating the presence of "dark energy" that somehow prevents the deceleration one would expect from gravitational attraction.

A high-water mark in astronomical Bayesian multilevel modeling was set by Mandel et al. [42], who address the problem of estimating supernova luminosities from light curve data. Their model has three levels, complex connections between latent variables (some of them random *functions*—light curves—rather than scalars), and jointly describes three different types of data (light curves, spectra, and host galaxy spectroscopic redshifts). Panel (b) of Fig. 2.4 shows the graph for their MLM (the reader will have to consult [42], or Mandel's more tutorial overview in [28], for a description of the variables and the model). In this volume, March et al. tackle a subsequent SNe Ia problem: how to use the output of light curve models to estimate cosmological parameters, including the density of dark energy, and an "equation of state" parameter aiming to capture how the dark energy density may be evolving. Their framework can fuse information from SN Ia with information from other sources, such as the power spectrum of CMB fluctuations, and characterization of the baryon acoustic oscillation (BAO) seen in the large scale spatial distribution of galaxies. Panel (c) of Fig. 2.4 shows the graph for their model, also impressive for its complexity (see their contribution in this volume for an explanation of this graph). I display the graphs (without much explanation) to show how much emerging MLM research in astronomy is leapfrogging the simple models of the recent past, exemplified by Panel (a) in the figure.

Equally impressive is Wandelt's contribution, describing a hierarchical Bayes approach (albeit without explicit MLM language) for reconstructing the galaxy density field from noisy photometric redshift data measuring the line-of-sight velocities of galaxies (which includes a component from cosmic expansion and a "peculiar velocity" component from gravitational attraction of nearby galaxies and dark matter). His team's framework (described in detail by [43]) includes nonparametric estimation of the density field at an upper level, which adaptively influences estimation of galaxy distances and peculiar velocities at a lower level, borrowing strength in the manner described above. The nonparametric upper level, and the size of the calculation, make this work groundbreaking.

The contributions by Andreon and by Kunz et al. describe simpler MLMs, but with potentially broad applicability. Andreon describes regression with measurement error (fitting lines and curves using data with errors in both the abscissa and ordinate) using basic MLMs, including testing model adequacy with a relatively simple predictive distribution test resembling the tail-area p-value tests familiar to astronomers. Such problems are common in astronomy. Kelly ([27]; see also Kelly's more introductory treatment in [28]) provided a more formal account of the Bayesian treatment of such problems, drawing on the statistics literature on measurement error problems [e.g., 44]. The complementary emphasis of Andreon's account is on how straightforward—indeed, almost automatic—implementation can be using modern software packages such as BUGS and JAGS.[8] Kunz et al. further develop the Bayesian estimation applied to multiple species (BEAMS) framework first described by [45]. BEAMS aims to improve parameter estimation in nonlinear regression when the data may come from different types of sources ("species" or classes), with different error or population distributions, but with uncertainty in the type of each datum. The classification labels for the data become discrete latent parameters in a MLM; marginalizing over them (and possibly estimating parameters in the various error distributions) can greatly improve inferences. They apply the approach to estimating cosmological parameters using SNe Ia data, and show that accounting for uncertainty in supernova classification has the potential to significantly improve the precision and accuracy of estimates. In the context of this volume, one cannot help but wonder what would come of integrating something like BEAMS into the MLM framework of March et al.

In Section 2 I noted how the "variability = uncertainty" misconception leads many astronomers to ignore relevant frequentist statistics literature; it gives the analyst the impression that the ability to quantify variability is all that is needed to devise a sound frequentist procedure. There is a similar danger for Bayesian inference. Once one has specified a model for the data (embodied in the likelihood function), Bayesian inference appears to be automatic in principle; one just follows the rules of probability theory to calculate probabilities for hypotheses of interest (after assigning priors). But despite the apparent simplicity of the sum and product rules, probability theory can exhibit a sophisticated subtlety, with apparently innocuous assumptions and calculations sometimes producing surprising results. The huge literature on Bayesian methods is more than mere crank turning; it amasses significant experience with this sophisticated subtlety that astronomers should mine to guide development of new Bayesian methods, or refine existing ones. For this non-statistician reader, it is the simpler of the contributions in this volume that point to opportunities for such "borrowing and mustering of strength" from earlier work by statisticians.

---

[8] http://www.mrc-bsu.cam.ac.uk/bugs/

Andreon's "de-TeXing" approach to Bayesian modeling (i.e., transcribing model equations to BUGS or JAGS code, and assigning simple default priors) is both appealing and relatively safe for simple models, such as standard regression (parametric curve fitting with errors only in the ordinate) or density estimation from precise point measurements in low dimensions. But the implications of modeling assumptions become increasingly subtle as model complexity grows, particularly when one starts increasing the number of levels or the dimensions of model components (or both). This makes me a little wary of an automated style of Bayesian modeling. Let us focus here on some MLM subtlety.

Information gain from the data tends to weaken as one considers parameters at increasingly high levels in a multilevel model [46]. On the one hand, if one is interested in quantities at lower levels, this weakens dependence on assumptions made at high levels. On the other hand, if one is interested in high-level quantities, sensitivity to the prior becomes an issue. The weakened impact of data on high levels has the effect that improper (i.e., non-normalized) priors that are safe to use in simple models (because the likelihood makes the posterior proper) can be dangerous in MLMs; nearly improper "vague" default priors may hide the problem without ameliorating it [47, 18]. Paradoxically, in some settings one may need to assign very informative upper-level priors to allow lower level distributions to adapt to the data (see [35] for an astronomical example). Also, the impact of the graph structure on a model's predictive ability becomes less intuitively accessible as complexity grows, making predictive tests of MLMs important, but also nontrivial; simple posterior predictive tests may be insensitive to significant discrepancies [48, 49, 50]. An exemplary feature of the SNe Ia MLM work of Mandel et al. is the use of careful predictive checks, implemented via a frequentist cross-validation procedure, to quantitatively assess the adequacy of various aspects of the model (notably, Mandel audited graduate-level statistics courses to learn the ins and outs of MLMs for this work, comprising his PhD thesis). In the context of nonlinear regression with measurement error, Carroll et al. [44] provides a useful entry point to both Bayesian and frequentist literature, incidentally also describing a number of frequentist approaches to such problems that would be more fair competitors to Bayesian MLMs than the "χ^2" approach that serves as Andreon's straw man competitor.

The problem addressed by the BEAMS framework—essentially the problem of data contamination, or mixed data—is not unique to astronomy, and there is significant statistics literature addressing similar problems with lessons to offer astronomers. BEAMS is a form of Bayesian regression using FMM error distributions. Statisticians first developed such models a few decades ago, to treat outliers (points evidently not obeying the assumed error distribution) using simple two-component mixtures (e.g., normal distributions with two different variances). More sophisticated versions have since been developed for diverse applications. An astronomical example building on some of this expertise is the finite mixture modeling of stellar populations by [36], mentioned above. The most immediate lessons astronomers may draw from this literature are probably computational; for example, algorithms using data augmentation (which involves a kind of guided, iterative

Monte Carlo sampling of the class labels and weights) may be more effective for implementing BEAMS than the weight-stepping approach currently used.

2.4 Provocation

I will close this commentary with a provocative recommendation I have offered at meetings since 2005 but not yet in print, born of my experience using multilevel models for astronomical populations. It is that astronomers cease producing catalogs of estimated fluxes and other source properties from surveys. This warrants explanation and elaboration.

As noted above, a consequence of the hierarchical structure of MLMs is that the values of latent parameters at low levels cannot be estimated independently of each other. In a survey context, this means that the flux (and potentially other properties) of a source cannot be accurately or optimally estimated considering only the data for that source. This may initially seem surprising, but at some level astronomers already know this to be true. We know — from Eddington, Malmquist, and Lutz and Kelker — that simple estimates of source properties will be misleading if we do not take into account information besides the measurements and selection effects, i.e., specification of the population distribution of the property. The standard Malmquist and Lutz-Kelker corrections adopt an a priori fixed (e.g., spatially homogeneous) population distribution, and produce an independent corrected estimate for each object. What the fully Bayesian MLM approach adds to the picture is the ability to handle uncertainty in the population distribution. After all, a prime reason for performing surveys is to learn about populations. When the population distribution is not well-known a priori, each source property measurement bears on estimation of the population distribution, and thus indirectly, each measurement bears on the estimation of the properties of every other source, via a kind of adaptive bias correction.[9] This is Tukey's "mustering and borrowing of strength" at work again.

To enable this mustering and borrowing, we have to stop thinking of a catalog entry as providing all the information needed to infer a particular source's properties (even in the absence of auxiliary information from outside a particular survey). Such a complete summary of information is provided by the marginal posterior distribution for that source, which depends on the data from *all* sources—and on population-level modeling assumptions. However, in the MLM structure (e.g., panel (d) of Fig. 2.3), the *likelihood function* for the properties of a particular source may be independent of information about other sources. The simplest output of a survey that would enable accurate and optimal subsequent analysis is thus a *catalog of likelihood functions* (or possibly marginal likelihood functions when there are uncertain

---

[9] It is worth pointing out that this is not a uniquely Bayesian insight. Eddington, Malmquist, and Lutz and Kelker used frequentist arguments to justify their corrections; Eddington even offered adaptive corrections. The large and influential statistics literature on shrinkage estimators leads to similar conclusions; see [22] for further discussion and references.

survey-specific backgrounds or other "nuisance" effects the surveyor must account for).

For a well-measured source, the likelihood function may be well-approximated by a Gaussian that can be easily summarized with a mean and standard deviation. But these should not be presented as point estimates and uncertainties.[10] For sources near the "detection limit," more complicated summaries may be justified. Counterpart surveys should cease reporting upper limits when a known source is not securely detected; instead they should report a more informative non-Gaussian likelihood summary. Discovery surveys (aiming to detect new sources rather than counterparts) could potentially devise likelihood summaries that communicate information about sources with fluxes *below* a nominal detection limit, and about uncertain source multiplicty in crowded fields. Recent work on maximum-likelihood fitting of "pixel histograms" (also known as "probability of deflection" or $P(D)$ distributions), which contain information about undetected sources, hints at the science such summaries might enable in a MLM setting [e.g., 51].

In this approach to survey reporting, the notion of a detection limit as a decision boundary identifying sources disappears. In its place there will be decision boundaries, driven by both computational and scientific considerations, that determine what type of likelihood summary is associated with each possible candidate source location.

Coming at this issue from another direction, Hogg and Lang [52] have recently made similar suggestions, including some specific ideas for how likelihoods may be summarized. Multilevel models provide a principled framework, both for motivating such a thoroughgoing revision of current practice, and for guiding its detailed development. Perhaps by the 2015 ISI World Congress in Brazil we will hear reports of analyses of the first survey catalogs providing such more optimal, MLM-ready summaries.

But even in the absence of so revolutionary a development, I think one can place high odds in favor of a bet that Bayesian multilevel modeling will become increasingly prevalent (and well-understood) in forthcoming astrostatistics research. Whether Bayesian methods (multilevel and otherwise) will start flourishing *outside* the astrostatistics research community is another matter, dependent on how effectively astrostatisticians can rise to the challenge of correcting misconceptions about both frequentist and Bayesian statistics, such as those outlined above. The abundance of young astronomers with enthusiasm for astrostatistics makes me optimistic.

Acknowledgements I gratefully acknowledge NSF and NASA for support of current research underlying this commentary, via grants AST-0908439, NNX09AK60G and NNX09AD03G. I thank Martin Weinberg for helpful discussions on information propagation within multilevel models. Students of Ed Jaynes's writings on probability theory in physics may recognize the last part of my title, borrowed from a commentary by Jaynes on the history of Bayesian and maximum entropy

---

[10] I am tempted to recommend that, even in this regime, the likelihood summary be chosen so as to deter misuse as an estimate, say by tabulating the $+1\sigma$ and -2σ points rather than means and standard deviations. I am only partly facetious about this!

ideas in the physical sciences [53]. This bit of plagiarism is intended as a homage to Jaynes's influence on this area — and on my own research and thinking.

References

1. Gull, S.F., Daniell, G.J.: Image reconstruction from incomplete and noisy data. Nature **272**, 686–690 (1978). DOI 10.1038/272686a0
2. Gull, S.F., Daniell, G.J.: The Maximum Entropy Method (invited Paper). In: van Schooneveld, C. (ed.) IAU Colloq. 49: Image Formation from Coherence Functions in Astronomy, *Astrophysics and Space Science Library*, D. Reidel Publishing Company, Dordrecht vol. 76, p. 219 (1979)
3. Sturrock, P.A.: Evaluation of Astrophysical Hypotheses. Astrophysical Journal **182**, 569–580 (1973). DOI 10.1086/152165
4. Jeffreys, H.: Theory of probability. Third edition. Clarendon Press, Oxford (1961)
5. Jaynes, E.T.: Probability theory in science and engineering. Colloquium lectures in pure and applied science. Socony Mobil Oil Co. Field Research Laboratory (1959). URL http://books.google.com/books?id=Ft4-AAAAIAAJ
6. Ripley, B.D.: Bayesian methods of deconvolution and shape classification. In: Feigelson, E. D., Babu, G. J. (eds.) Statistical Challenges in Modern Astronomy, Springer, New York pp. 329–346 (1992)
7. Readhead, A.C.S., Lawrence, C.R.: Observations of the isotropy of the cosmic microwave background radiation. Annual Review of Astronomy and Astrophysics **30**, 653–703 (1992). DOI 10.1146/annurev.aa.30.090192.003253
8. Feigelson, E.D., Babu, G.J. (eds.): Statistical Challenges in Modern Astronomy. Springer (1992)
9. Loredo, T.J.: Promise of Bayesian inference for astrophysics. In: Feigelson, E. D., Babu, G. J. (eds.) Statistical Challenges in Modern Astronomy, pp. 275–306 (1992)
10. Loredo, T.J.: The promise of bayesian inference for astrophysics (unabridged). Tech. rep., Department of Astronomy, Cornell University (1992). http://citeseerx.ist.psu.edu/viewdoc/summary?doi=10.1.1.56.1842CiteSeer DOI 10.1.1.56.1842
11. Nousek, J.A.: Source existence and parameter fitting when few counts are available. In: Feigelson, E. D., Babu, G. J. (eds.) Statistical Challenges in Modern Astronomy, pp. 307–327 (1992)
12. Lampton, M., Margon, B., Bowyer, S.: Parameter estimation in X-ray astronomy. Astrophysical Journal **208**, 177–190 (1976). DOI 10.1086/154592
13. Loredo, T.J.: Computational Technology for Bayesian Inference. In: Mehringer, D. M., Plante, R. L., Roberts, D. A. (eds.) Astronomical Data Analysis Software and Systems VIII, *Astronomical Society of the Pacific Conference Series*, Astronomical Society of the Pacific, San Francisco vol. 172, p. 297 (1999)
14. Sims, C.: Understanding non-bayesians. Unpublished chapter, Department of Economics, Princeton University (2010). URL http://www.princeton.edu/~sims/#UndstndngNnBsns
15. Kuhn, T.S.: The structure of scientific revolutions. Second edition. University of Chicago Press, Chicago (1970)
16. Efron, B.: Bayesians, frequentists, and physicists. In: L. Lyons (ed.) PHYSTAT2003: Statistical Problems in Particle Physics, Astrophysics, and Cosmology, SLAC, Stanford CA, pp. 17–24 (2003)
17. Bayarri, M.J., Berger, J.O.: The interplay of Bayesian and frequentist analysis. Statist. Sci. **19**(1), 58–80 (2004). DOI 10.1214/088342304000000116. URL http://dx.doi.org/10.1214/088342304000000116
18. Gelman, A., Carlin, J.B., Stern, H.S., Rubin, D.B.: Bayesian data analysis, second edn. Texts in Statistical Science Series. Chapman and Hall/CRC, Boca Raton, FL (2004)
19. Little, R.J.: Calibrated Bayes: a Bayes/frequentist roadmap. Amer. Statist. **60**(3), 213–223 (2006). DOI 10.1198/000313006X117837. URL http://dx.doi.org/10.1198/000313006X117837

20. Loredo, T.J.: Statistical foundations and statistical practice (contribution to a panel discussion on the future of astrostatistics). In: Feigelson, E. D., Babu, G. J. (eds.) Statistical Challenges in Modern Astronomy, p. 7pp. Springer (2012) (in press)
21. West, M.: Commentary. In: Feigelson, E. D., Babu, G. J. (eds.) Statistical Challenges in Modern Astronomy, p. 328ff (1992)
22. Loredo, T.J.: Analyzing Data from Astronomical Surveys: Issues and Directions. In: Babu, G. J., Feigelson, E. D. (eds.) Statistical Challenges in Modern Astronomy IV, *Astronomical Society of the Pacific Conference Series*, Astronomical Society of the Pacific, San Francisco vol. 371, p. 121 (2007)
23. Loredo, T.J., Hendry, M.A.: Bayesian multilevel modelling of cosmological populations. In: Hobson, M. P., Jaffe, A. H., Liddle, A. R., Mukeherjee, P., Parkinson, D. (eds.) Bayesian Methods in Cosmology, p. 245. Cambridge University Press, Cambridge (2010)
24. Loredo, T.J., Lamb, D.Q.: Bayesian analysis of neutrinos observed from supernova SN 1987A. Physical Review D **65**(6), 063002 (2002). DOI 10.1103/PhysRevD.65.063002
25. Loredo, T.J., Wasserman, I.M.: Inferring the Spatial and Energy Distribution of Gamma-Ray Burst Sources. II. Isotropic Models. Astrophysical Journal **502**, 75 (1998). DOI 10.1086/305870
26. Petit, J.M., Kavelaars, J.J., Gladman, B., Loredo, T.: Size Distribution of Multikilometer Transneptunian Objects. In: Barucci, M. A., Boehnhardt, H., Cruikshank, D. P., Morbidelli, A., Dotson, R. (eds.) The Solar System Beyond Neptune, pp. 71–87. University of Arizona Press Tucson (2008)
27. Kelly, B.C.: Some Aspects of Measurement Error in Linear Regression of Astronomical Data. Astrophysical Journal **665**, 1489–1506 (2007). DOI 10.1086/519947
28. Feigelson, E.D., Babu, G.J. (eds.): Statistical Challenges in Modern Astronomy V. Springer (2012)
29. Luo, S., Loredo, T., Wasserman, I.: Likelihood analysis of GRB repetition. In: Kouveliotou, C., Briggs, M. F., Fishman, G. J. (eds.) American Institute of Physics Conference Series, *American Institute of Physics Conference Series*, AIP, Woodbury vol. 384, pp. 477–481 (1996). DOI 10.1063/1.51706
30. Graziani, C., Lamb, D.Q.: Likelihood methods and classical burster repetition. In: Rothschild, R. E., Lingenfelter, R. E. (eds.) High Velocity Neutron Stars, *American Institute of Physics Conference Series*, AIP Woodbury vol. 366, pp. 196–200 (1996). DOI 10.1063/1.50246
31. Budavári, T., Szalay, A.S.: Probabilistic Cross-Identification of Astronomical Sources. Astrophysical Journal **679**, 301–309 (2008). DOI 10.1086/587156
32. Loredo, T.J.: Commentary on Bayesian coincidence assessment (cross-matching). In: Feigelson, E. D., Babu, G. J. (eds.) Statistical Challenges in Modern Astronomy, p. 6pp. Springer (2012) (in press)
33. Bovy, J., Hogg, D.W., Roweis, S.T.: Extreme deconvolution: Inferring complete distribution functions from noisy, heterogeneous and incomplete observations. Ann. Appl. Stat. **5**(2B), 1657–1677 (2011)
34. van Dyk, D.A., Connors, A., Kashyap, V.L., Siemiginowska, A.: Analysis of Energy Spectra with Low Photon Counts via Bayesian Posterior Simulation. Astrophysical Journal **548**, 224–243 (2001). DOI 10.1086/318656
35. Esch, D.N., Connors, A., Karovska, M., van Dyk, D.A.: An Image Restoration Technique with Error Estimates. Astrophysical Journal **610**, 1213–1227 (2004). DOI 10.1086/421761
36. van Dyk, D.A., DeGennaro, S., Stein, N., Jefferys, W.H., von Hippel, T.: Statistical analysis of stellar evolution. Ann. Appl. Stat. **3**(1), 117–143 (2009). DOI 10.1214/08-AOAS219. URL http://dx.doi.org/10.1214/08-AOAS219
37. Dobigeon, N., Tourneret, J.Y., Scargle, J.D.: Joint segmentation of multivariate astronomical time series: Bayesian sampling with a hierarchical model. IEEE Trans. Signal Process. **55**(2), 414–423 (2007). DOI 10.1109/TSP.2006.885768.
38. Loredo, T.J.: The return of the prodigal: Bayesian inference For astrophysics. In: Bernardo, J. M. Berger, J. O., Dawid, A. P., Smith, A. F. M. (eds.) *Bayesian Statistics 5 Preliminary Proceedings*, volume distributed to participants of the 5th Valencia Meeting on Bayesian

Statistics (1995). http://citeseerx.ist.psu.edu/viewdoc/summary?doi=10.1.1.55.3616. CiteSeer DOI 10.1.1.55.3616
39. Wandelt, B.D., Larson, D.L., Lakshminarayanan, A.: Global, exact cosmic microwave background data analysis using Gibbs sampling. Physical Review D **70**(8), 083511 (2004). DOI 10.1103/PhysRevD.70.083511
40. Kitaura, F.S., Enßlin, T.A.: Bayesian reconstruction of the cosmological large-scale structure: methodology, inverse algorithms and numerical optimization. Mon. Not. Roy. Astron. Soc. **389**, 497–544 (2008). DOI 10.1111/j.1365-2966.2008.13341.x
41. Loredo, T.J.: Commentary on Bayesian analysis across astronomy. In: Feigelson, E. D., Babu, G. J. (eds.) Statistical Challenges in Modern Astronomy, p. 12pp. Springer (2012) (in press)
42. Mandel, K.S., Narayan, G., Kirshner, R.P.: Type Ia Supernova Light Curve Inference: Hierarchical Models in the Optical and Near-infrared. Astrophysical Journal **731**, 120 (2011). DOI 10.1088/0004-637X/731/2/120
43. Jasche, J., Wandelt, B.D.: Bayesian inference from photometric redshift surveys. ArXiv/1106.2757 (2011)
44. Carroll, R.J., Ruppert, D., Stefanski, L.A., Crainiceanu, C.M.: Measurement error in nonlinear models, *Monographs on Statistics and Applied Probability*, vol. 105, second edn. Chapman and Hall/CRC, Boca Raton, FL (2006). DOI 10.1201/9781420010138. URL http://dx.doi.org/10.1201/9781420010138. A modern perspective
45. Kunz, M., Bassett, B.A., Hlozek, R.A.: Bayesian estimation applied to multiple species. Physical Review D **75**(10), 103508 (2007). DOI 10.1103/PhysRevD.75.103508
46. Goel, P.K., DeGroot, M.H.: Information about hyperparameters in hierarchical models. J. Amer. Statist. Assoc. **76**(373), 140–147 (1981).
47. Hadjicostas, P., Berry, S.M.: Improper and proper posteriors with improper priors in a Poisson-gamma hierarchical model. Test **8**(1), 147–166 (1999). DOI 10.1007/BF02595867.
48. Sinharay, S., Stern, H.S.: Posterior predictive model checking in hierarchical models. J. Statist. Plann. Inference **111**(1-2), 209–221 (2003). DOI 10.1016/S0378-3758(02)00303-8.
49. Gelman, A.: Prior distributions for variance parameters in hierarchical models (comment on article by Browne and Draper). Bayesian Anal. **1**(3), 515–533 (electronic) (2006)
50. Bayarri, M.J., Castellanos, M.E.: Bayesian checking of the second levels of hierarchical models. Statist. Sci. **22**(3), 322–343 (2007). DOI 10.1214/07-STS235.
51. Patanchon, G., et al.: Submillimeter Number Counts from Statistical Analysis of BLAST Maps. Astrophysical Journal **707**, 1750–1765 (2009). DOI 10.1088/0004-637X/707/2/1750
52. Hogg, D.W., Lang, D.: Telescopes don't make catalogues! In: EAS Publications Series, *EAS Publications Series*, EDP Sciences, Les Ulis vol. 45, pp. 351–358 (2011).DOI 10.1051/eas/1045059
53. Jaynes, E.T.: A Backward Look to the Future. In: Grandy, W. T., Jr., Milonni, P. W. (eds.) Physics and Probability, Cambridge University Press, Cambridge pp. 261–276 (1993)

Chapter 3
Understanding Better (Some) Astronomical Data Using Bayesian Methods

S. Andreon

Abstract Current analysis of astronomical data is confronted with the daunting task of modelling the awkward features of astronomical data, among which are heteroscedastic (point-dependent) errors, intrinsic scatter, non-ignorable data collection (selection effects), data structure, non-uniform populations (often called Malmquist bias), non-Gaussian data, and upper/lower limits. This chapter shows, by examples, how to model all these features using Bayesian methods. In short, one just needs to formalise, using maths, the logical link between the involved quantities, how the data arise and what we already know on the quantities we want to study. The posterior probability distribution summarises what we know on the studied quantities after analysing the data, and numerical computation is left to (special) Monte Carlo programs such as JAGS. As examples, we show to predict the mass of a new object disposing of a calibrating sample, how to constraint comological parameters from supernovae data and how to check if the fitted data are in tension with the adopted fitting model. Examples are given with their coding. These examples can be easily used as template for completely different analysis, on totally unrelated astronomical objects, requiring to model the same awkward data features.

3.1 Introduction

Astronomical data present a number of quite common awkward features (see [1] for a review):

- **Heteroscedastic errors**: Error sizes vary from point to point.

S. Andreon
INAF–Osservatorio Astronomico di Brera, Milano, Italy
e-mail: stefano.andreon@brera.inaf.it

- **non-Gaussian data**: the likelihood is asymmetric and thus errors are, e.g. $3.4^{+2.5}_{-1.2}$. Upper/lower limits, as 2.3 at 90% probability, are (perhaps extreme) examples of asymmetric likelihood.
- **non uniform populations or data structure**: the number of objects per unit parameter(s) is non-uniform. This is the source of the Malmquist- or Eddington-like bias that affect most astronomical quantities, as parallaxes, star and galaxy counts, mass and luminosity, galaxy cluster velocity dispersions, supernovae luminosity corrections, etc.
- **intrinsic scatter**: data often scatter more than allowed by the errors. The extra-scatter can be due to unidentified sources of errors, often called systematic errors, or indicates an intrinsic spread of the population under study, i.e. the fact that astronomical objects are not identically equal.
- **noisy estimates of the errors**: as every measurement, errors are known with a finite degree of precision. This is even more true when one is measuring complex, and somewhat model dependent, quantities like mass.
- **non-random sampling**: in simple terms, the objects in the sample are not a random sampling of those in the Universe. In some rare occasions in astronomy, sampling is planned to be non-random on purpose, but most of the times non-random sampling is due to selection effects: harder-to-observe objects are very often missed in samples.
- **mixtures**: very often, large samples include the population under interest, but also contaminating objects. Mixtures also arise when one measure the flux of a source in presence of a background (i.e. always).
- **prior**: we often known from past data or from theory that some values of the parameters are more likely than other. In order terms, we have prior knowledge about the parameter being investigated. If we known anything, not even the order of magnitude, about a parameter, it is difficult even to choose which instrument, or sample, should be used to measure the parameter.
- **non-linear**: laws of Nature can be more complicated than $y = ax + b$.

Bayesian methods allow to deal with these features (and also other ones), even all at the same time, as we illustrate in Section 3.3 and 3.4 with two research examples, it is just matter of stating in mathematical terms our wordy statements about the nature of the measurement and of the objects being measured. The probabilistic (Bayesian) approach returns the whole (posterior) probability distribution of the parameters, very often in form of a Monte Carlo sampling of it.

In this paper we make an attempt to be clear at the cost of being non-rigorous. We defer the reader looking for rigour to general textbooks, as [2], and, to [3] for our first research example.

3.2 Parameter Estimation in Bayesian Inference

Before addressing a research example, let's consider an idealised applied problem to explain the basics of the Bayesian approach.

Suppose one is interested in estimating the (log) mass of a galaxy cluster, *lgM*. Before collecting any data, we may have certain beliefs and expectations about the values of *lgM*. In fact, these thoughts are often used in deciding which instrument will be used to gather data and how this instrument may be configured. For example, if we plan to measure the mass of a poor cluster via the virial theorem, we will select a spectroscopic set-up with adequate resolution, in order to avoid velocity errors that are comparable to, or larger than, the likely low velocity dispersion of poor clusters. Crystallising these thoughts in the form of a probability distribution for *lgM* provides the prior $p(lgM)$, on which relies the feasibility section of the telescope time proposal, where instrument, configuration and exposure time are set.

For instance, one may believe (e.g. from the cluster being somewhat poor) that the log of the cluster mass is probably not far from 13, plus or minus 1; this might be modelled by saying that the (prior) probability distribution of the log mass, here denoted *lgM*, is a Gaussian centred on 13 and with σ, the standard deviation, equal to 0.5, i.e. $lgM \sim \mathcal{N}(13, 0.5^2)$.

Once the appropriate instrument and its set-up have been selected, data can be collected. In our example, this means we record a measurement of log mass, say *obslgM*200, via, for example, a virial theorem analysis, i.e. measuring distances and velocities.

The likelihood describes how the noisy observation *obslgM*200 arises given a value of *lgM*. For example, we may find that the measurement technique allows us to measure masses in an unbiased way but with a standard error of 0.1 and that the error structure is Gaussian, i.e. $obslgM200 \sim \mathcal{N}(lgM, 0.1^2)$, where the tilde symbol reads "is drawn from" or "is distributed as". If we observe $obslgM200 = 13.3$, we usually summarise the above by writing $lgM = 13.3 \pm 0.1$.

How do we update our beliefs about the unobserved log mass *lgM* in light of the observed measurement *obslgM*200? Expressing this probabilistically, what is the posterior distribution of *lgM* given *obslgM*200, i.e. $p(lgM|obslgM200)$? Bayes Theorem [4, 5] tells us that

$$p(lgM|obslgM200) \propto p(obslgM200|lgM) p(lgM)$$

i.e. the posterior (the left-hand side) is equal to the product of likelihood and prior (the right-hand side) times a proportionality constant of no importance in parameter estimation.

Simple algebra shows that in our example the posterior distribution of *lgM*|*obslgM*200 is Gaussian, with mean

$$\mu = \frac{13.0/0.5^2 + 13.3/0.1^2}{1/0.5^2 + 1/0.1^2} = 13.29$$

and

$$\sigma^2 = \frac{1}{1/0.5^2 + 1/0.1^2} = 0.0096.$$

μ is just the usual weighted average of two "input" values, the prior and the observation, with weights given by prior and observation σ's.

From a computational point of view, only simple examples such as the one described above can generally be tackled analytically, realistic analysis should be instead tacked numerically by special (Markov Chain) Monte Carlo methods. These are included in BUGS-like programs [6] such as JAGS [7], allowing scientists to focus on formalising in mathematical terms our wordy statements about the quantities under investigation without worrying about the numerical implementation. In the idealised example, we just need to write in an ASCII file the symbolic expression of the prior, $lgM \sim \mathcal{N}(13, 0.5^2)$, and of likelihood, $obslgM200 \sim \mathcal{N}(lgM, 0.1^2)$ to get the posterior in form of samplings. From the Monte Carlo sampling one may directly derive mean values, standard deviations, and confidence regions. For example, for a 90% interval it is sufficient to peak up the interval that contain 90% of the samplings.

3.3 First Example: Predicting Mass from a Mass Proxy

Mass estimates are one of the holy grails of astronomy. Since these are observationally expensive to measure, or even unmeasurable with existing facilities, astronomers use mass proxies, far less expensive to acquire: from a (usually small) sample of objects, the researcher measures masses, y and the mass proxy, x. Then, he regress x vs y and infer y for those objects having only x. This is the way most of the times galaxy cluster masses are estimated, for example using the X-ray luminosity, X-ray temperature, Y_X, Y_{SZ}, the cluster richness or the total optical luminosity. Here we use the cluster richness, i.e. the number of member galaxies, but with minor changes this example can be adapted for other cases.

Andreon and Hurn [1] show that predicted y using the Bayesian approach illustrated here are more precise than any other method and that the Bayesian approach does not show the large systematics of other approaches. This means, in the case of masses, that more precise masses can be derived for the same input data, i.e. at the same telescope time cost.

3.3.1 Step 1: Put in Formulae What You Know

3.3.1.1 Heteroscedasticity

Clusters have widely different richnesses, and thus widely different errors. Some clusters have better determined masses than other. Heteroscedasticity means that errors have an index i, because they differ from point to point.

3.3.1.2 Contamination (Mixtures), Non-Gaussian Data and Upper Limits

Galaxies in the cluster direction are both cluster members and galaxies on the line of sight (fore/background). The contamination may be estimated by observing a reference line of sight (fore/background), perhaps with a C_i times larger solid angle (to improve the quality of the determination).

The mathematical translation of our words, when counts are modelled as Poisson, is:

$$obsbkg_i \sim \mathcal{P}(nbkg_i) \qquad \text{\# Poisson with intensity } nbkg_i \tag{3.1}$$
$$obstot_i \sim \mathcal{P}(nbkg_i/C_i + n200_i) \qquad \text{\# Poisson with intensity } (nbkg_i/C_i + n200_i) \tag{3.2}$$

The variables $n200_i$ and $nbkg_i$ represent the true richness and the true background galaxy counts in the studied solid angles, whereas we add a prefix "obs" to indicate the observed values.

Upper limits are automatically accounted for. Suppose, for exposing simplicity, that we observed five galaxies, $obstot_i = 5$, in the cluster direction and that in the control field direction (with $C_i = 1$ for exposing simplicity) we observe four background galaxies, $obsbkg = 4$. With one net galaxy and Poisson fluctuations of a few, $n200_i$ is poorly determined at best, and the conscientious researcher would probably report upper limits of a few. To use the information contained in the upper limit in our regression analysis, we only need to list in the data file the raw measurements ($C_i = 1$, $obstot_i = 5$, $obsbkg_i = 4$), as for the other clusters. These data will be treated independently on whether an astronomer decides to report a measurement or an upper limit, because Nature does not care about astronomer decisions.

3.3.1.3 Non-linearity and Extra-Scatter

The relation between mass, $M200$, and proxy, $n200$, (richness) is usually parametrised as a power-law:

$$M200_i \propto n200_i^\beta$$

Allowing for a Gaussian intrinsic scatter σ_{scat} (clusters of galaxies of a given richness may not all have the very same mass) and taking the log, the previous equation becomes:

$$lgM200_i \sim \mathcal{N}(\alpha + \beta \log(n200_i), \sigma_{scat}^2)$$
$$\text{\# Gaussian scatter around}(M200_i \propto n200_i^\beta) \tag{3.3}$$

where the intercept is α and the slope is β.

3.3.1.4 Noisy Errors

Once logged, mass has Gaussian errors. In formulae:

$$obslgM200_i \sim \mathcal{N}(lgM200_i, \sigma_i^2) \quad \text{\# Gauss errors on lg mass} \quad (3.4)$$

However, errors (as everything) is measured with a finite degree of precision. We assume that the measured error, $obserrlgM200_i$, is not biased (i.e. it is not systematically larger or smaller that the true error, σ_i) but somewhat noisy. If a χ^2 distribution is adopted, it satisfies both our request of unbiasedness and noisiness. In formulae:

$$obserrlgM200_i^2 \sim \sigma_i^2 \chi_\nu^2/\nu \quad \text{\# Unbiased errors} \quad (3.5)$$

where the parameter ν regulates the width of the distribution, i.e. how precise measured errors are. Since we are 95% confident that quoted errors are correct up to a factor of 2,

$$\nu = 6 \quad \text{\# 95\% confident within a factor 2} \quad (3.6)$$

3.3.1.5 Prior Knowledge and Population Structure

The data used in this investigation are of quality good enough to determine all parameters, but one, to a sufficient degree of accuracy that we should not care about priors and we can safely take weak (almost uniform) priors, zeroed for un-physical values of parameters (to avoid, for example, negative richnesses). The exception is given by the prior on the errors (i.e. σ_i), for which there is only one measurement per datum. The adopted prior (eq. 3.11) is supported by statistical considerations (see [3] for details). The same prior is also used for the intrinsic scatter term, although any weak prior would return the same result, because this term is well determined by the data.

$$\alpha \sim \mathcal{N}(0.0, 10^4) \quad \text{\# Almost uniform prior on intercept} \quad (3.7)$$

$$\beta \sim t_1 \quad \text{\# Uniform prior on angle} \quad (3.8)$$

$$n200_i \sim \mathcal{U}(0, \infty) \quad \text{\# Uniform, but positive, cluster richness} \quad (3.9)$$

$$nbkg_i \sim \mathcal{U}(0, \infty) \quad \text{\# Uniform, but positive, background rate} \quad (3.10)$$

$$1/\sigma_i^2 \sim \Gamma(\epsilon, \epsilon) \quad \text{\# Weak prior on error} \quad (3.11)$$

$$1/\sigma_{scat}^2 \sim \Gamma(\epsilon, \epsilon) \quad \text{\# Weak prior on intrinsic scatter} \quad (3.12)$$

Richer clusters are rarer. Therefore, the prior on the cluster richness is, for sure, not uniform, contrary to our assumption (eq. 3.9). Modelling the population structure is un-necessary for the data used in [3] and here, but is essential if noisier richnesses were used. Indeed, [3] shows that a previous published richness-mass calibration, which uses richnesses as low as $obsn200 = 3$ and neglects the $n200$ structure, shows a slope biased by five times the quoted uncertainty. Therefore, the population structure cannot be overlooked in general.

3.3.2 Step 2: Remove TeXing, Perform Stochastic Computations and Publish

At this point, we have described, using mathematical symbols, the link between the quantities that matter for our problem, and we only need to compute the posterior probability distribution of the parameters by some sort of sampling (the readers with exquisite mathematical skills may instead attempt an analytical computation).

Just Another Gibb Sampler (JAGS,[1] [7]) can return it at the minor cost of de-TeXing Eqs. 3.1 to 3.12 (compare them to the JAGS code below). Poisson, normal and uniform distributions become `dpois`, `dnorm dunif`, respectively. JAGS, following BUGS [6], uses precisions, $prec = 1/\sigma^2$, in place of variances σ^2. Furthermore, it uses neperian logarithms, instead of decimal ones. Equation 3.5 has been rewritten using the property that the χ^2 is a particular form of the Gamma distribution. Equation 3.3 is split in two JAGS lines for a better reading. The arrow symbol reads "take the value of". `obsvarlgM200` is the square of *obserrlgM*200. For computational advantages, $\log(n200)$ is centred at an average value of 1.5 and α is centred at -14.5. Finally, we replaced infinity with a large number.

The model above, when inserted in JAGS, reads:

```
model
{
for (i in 1:length(obstot)) {
  obsbkg[i] ~ dpois(nbkg[i])                              # eq 3.1
  obstot[i] ~ dpois(nbkg[i]/C[i]+n200[i])                 # eq 3.2
  n200[i] ~ dunif(0,3000)                                 # eq 3.9
  nbkg[i] ~ dunif(0,3000)                                 # eq 3.10

  precy[i] ~ dgamma(1.0E-5,1.0E-5)                        # eq 3.12
  obslgM200[i] ~ dnorm(lgM200[i],precy[i])                # eq 3.4
  obsvarlgM200[i] ~ dgamma(0.5*nu,0.5*nu*precy[i])        # eq 3.5

  z[i] <- alpha+14.5+beta*(log(n200[i])/2.30258-1.5)      # eq 3.3
  lgM200[i] ~ dnorm(z[i], prec.intrscat)                  # eq 3.3
  }
intrscat <- 1/sqrt(prec.intrscat)
prec.intrscat ~ dgamma(1.0E-5,1.0E-5)                     # eq 3.11
alpha ~ dnorm(0.0,1.0E-4)                                 # eq 3.7
beta ~ dt(0,1,1)                                          # eq 3.8
nu <-6                                                    # eq 3.6
}
```

JAGS samples the posterior distribution of all quantity of interests, such as intercept, slope and intrinsic scatter by Gibb sampling (a sort of Monte Carlo). From

---
[1] http://calvin.iarc.fr/~martyn/software/jags/

these samplings, it is straightforward to compute (posterior) mean and standard deviations (by computing the average and the standard deviation!), to plot posterior marginals (by ignoring the values of the other parameters) and confidence contours, data and mean model, etc. Therefore, our effort is over, we only need to produce nice plots and summaries of our results.

Figure 3.1 shows the data used in this analysis (see [3] for details), the mean scaling (solid line) and its 68% uncertainty (shaded yellow region) and the mean intrinsic scatter (dashed lines) around the mean relation. The ±1 intrinsic scatter band is not expected to contain 68% of the data points, because of the presence of measurement errors.

Figure 3.2 shows the posterior marginals for the intercept, slope and intrinsic scatter σ_{scat}. These marginals are reasonably well approximated by Gaussians. The intrinsic mass scatter at a given richness, $\sigma_{scat} = \sigma_{lgM200|\log n200}$, is small, 0.19 ± 0.03. (Unless otherwise stated, results of the statistical computations are quoted in the form $x \pm y$ where x is the posterior mean and y is the posterior standard deviation.)

The found relation is:

$$lgM200 = (0.96 \pm 0.15)(\log n200 - 1.5) + 14.36 \pm 0.04 \qquad (3.13)$$

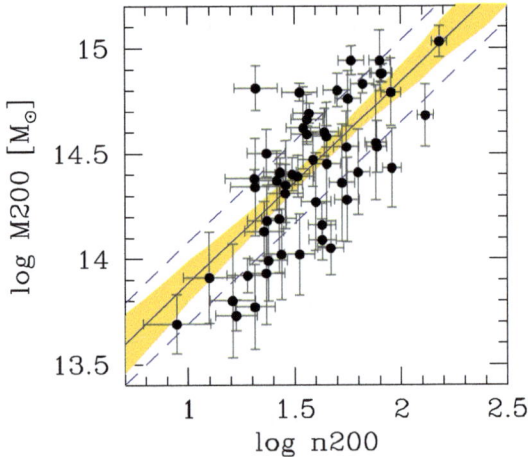

Fig. 3.1: Richness-mass scaling. The solid line marks the mean fitted regression line, while the dashed line shows this mean plus or minus the intrinsic scatter σ_{scat}. The shaded region marks the 68% highest posterior credible interval for the regression. Error bars on the data points represent observed errors for both variables. The distances between the data and the regression line is due in part to the measurement error and in part to the intrinsic scatter. From [3], reproduced with permission.

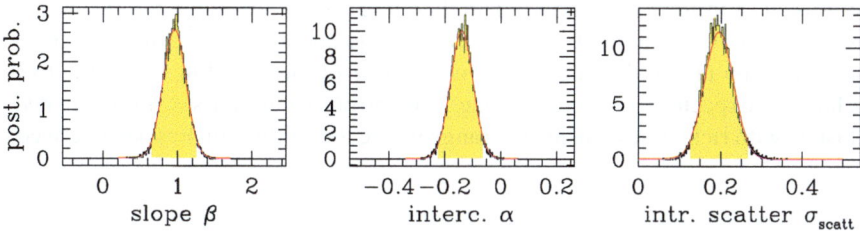

Fig. 3.2 Posterior probability distribution for the parameters of the richness-mass scaling. The *black jagged histogram* shows the posterior as computed by MCMC, marginalised over the other parameters. The *red curve* is a Gauss approximation of it. The *shaded (yellow) range* shows the 95% highest posterior credible interval. (From [3], reproduced with permission.)

3.3.3 Predicting Masses

As mentioned, one of the reasons why astronomers regress a quantity x against another one, y, is to predict the latter when a direct measurement is missing (usually because it is observationally expensive to acquire). It is clear that the uncertainty on the predicted y, called $\tilde{y}$ hereafter, should account for: (a) the intrinsic scatter between y and x (a larger scatter implies a lower quality $\tilde{y}$ estimate); (b) the uncertainty of the x (the larger it is, the noisier will be the $\tilde{y}$); (c) the quality of the calibration between y and x (better determined relations should return more precise estimates of $\tilde{y}$); and (d) extrapolation errors, i.e. should penalise attempts to infer $\tilde{y}$ values corresponding to x values absent from the calibrator sample (e.g. outside the range sampled by it). All these requirements are satisfied using the posterior predictive distribution

$$p(\tilde{y}) = \int p(\tilde{y}|\theta) p(\theta|y) d\theta \qquad (3.14)$$

where θ are the regression parameters (intercept, slope intrinsic scatter). This apparently complicated expression is easy to understand: one should combine (multiply) the uncertainties of the calibrating relation, $p(\theta|y)$, to the uncertainty of predicting new data if the calibrating relation were perfectly known, $p(\tilde{y}|\theta)$. Since we are now interested in predicted values only, we get rid of non-interesting parameters (θ) by marginalisation (integration).

Posterior predictive distributions are very basic to be introduced at page 8 of the > 700 pages "Bayesian Data Analysis" book [2] and to be offered as a standard output of JAGS. Of course, we need to list the x values (richnesses), and errors of the clusters for which we want to infer $\tilde{y}$ (predicted mass) in the data file, listing e.g., in the data file *obsbkg* = 12, *obtot* = 32, C = 5 and mass *obslgM200* = NA ("not available"), to indicate that these quantities should be estimated using the regression computed from the points with available masses and galaxy counts. JAGS returns $p(\widetilde{y})$ in form of sampling and therefore, as for any other parameter, a point estimate may be obtained by taking the average and a 68% (credible) interval can be derived by taking the interval including 68% of the samplings. Returned values behave as expected, and indeed have large errors when masses are estimated for clusters with richnesses outside the range where the calibration has been derived [8].

3.4 Second example: Cosmological Parameters from SNIa

Supernovae (SNIa) are very bright objects with very similar luminosities. The luminosity spread can be made even smaller by accounting for the correlation with colour and stretch parameter (the latter is a measurement of how slowly SNIa fade), as illustrated in Figure 3.3 for the sample in [9]. These features make SNIa very useful for cosmology: they can be observed far away and the relation between flux (basically the rate of photons received) and luminosity (the rate of photons emitted)

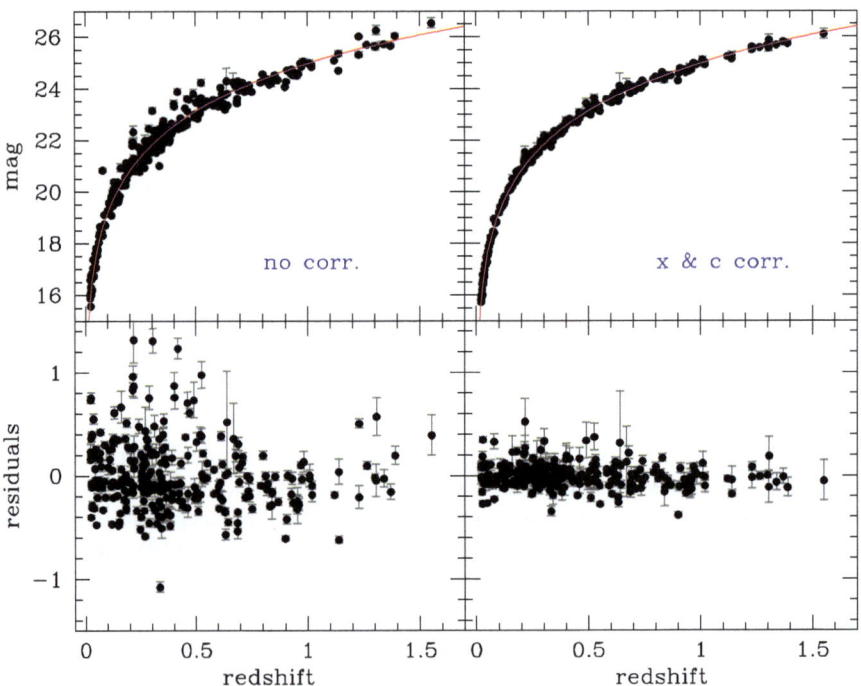

Fig. 3.3: Apparent magnitude vs redshift of the SNIa sample (upper panels), and their residual from a $\Omega_M = 0.3$, $\Omega_\Lambda = 0.7$ cosmological model (bottom panels) before (left panels) or after (right panels) correcting for stretching and colour parameter.

3 Understanding Better (Some) Astronomical Data using Bayesian Methods

is modulated by the luminosity distance (to the square), which in turn is a function of the cosmological parameters. Therefore, measuring SNIa fluxes (and redshift) allows us to put constraints on cosmological parameters. The only minor complication is that SNIa luminosities are functions of their colour and stretch parameters, and these parameters have an intrinsic scatter too, which in turn has to be determined from the data at the same time as the other parameters.

March et al. [10] show that the Bayesian approach delivers tighter statistical constraints on the cosmological parameters over 90% of the times, that it reduces the statistical bias typically by a factor $\sim$2–3 and that it has better coverage properties than the usual chi-squared approach.

In this second example, we can proceed a bit faster in illustrating this non-linear regression with heteroscedastic errors, non-uniform data structure and intrinsic scatter. In this example, we also briefly discuss the prior sensitivity, i.e. how much the results are affected by the chosen prior, and we also check the quality of the model fit.

3.4.1 Step 1: Put in Formulae What You Know

We observe SNIa magnitudes $obsm_i$ ($= -2.5 log(flux) + c$) with Gaussian errors $\sigma_{m,i}$, i.e.

$$obsm_i \sim \mathcal{N}(m_i, \sigma_{m,i}^2) \qquad (3.15)$$

These m_i are related to the distance modulus $distmod_i$, via

$$m_i = M + distmod_i - \alpha x_i + \beta c_i$$

with a Gaussian intrinsic scatter σ_{scat}. More precisely:

$$m_i \sim \mathcal{N}(M + distmod_i - \alpha x_i + \beta c_i, \sigma_{scat}^2) \qquad (3.16)$$

where M is the (unknown) mean absolute magnitude of SNIa, and α and β allow to reduce the SNIa luminosity scatter by accounting for the correlation with the stretch and colour parameters.

Similarly to [10], the M, α, β and $\log \sigma_{scat}$ priors are taken uniform in a wide range:

$$\log_{10} \sigma_{scat} \sim \mathcal{U}(-3, 0) \qquad (3.17)$$
$$\alpha \sim \mathcal{U}(-2, 2) \qquad (3.18)$$
$$\beta \sim \mathcal{U}(-4, 4) \qquad (3.19)$$
$$M \sim \mathcal{U}(-20.3, -18.3) \qquad (3.20)$$

x_i and c_i are the true value of the stretch and colour parameters, of which we observe (the noisy) $obsx_i$ and $obsc_i$ with errors $\sigma_{x,i}$ and $\sigma_{c,i}$. In formulae:

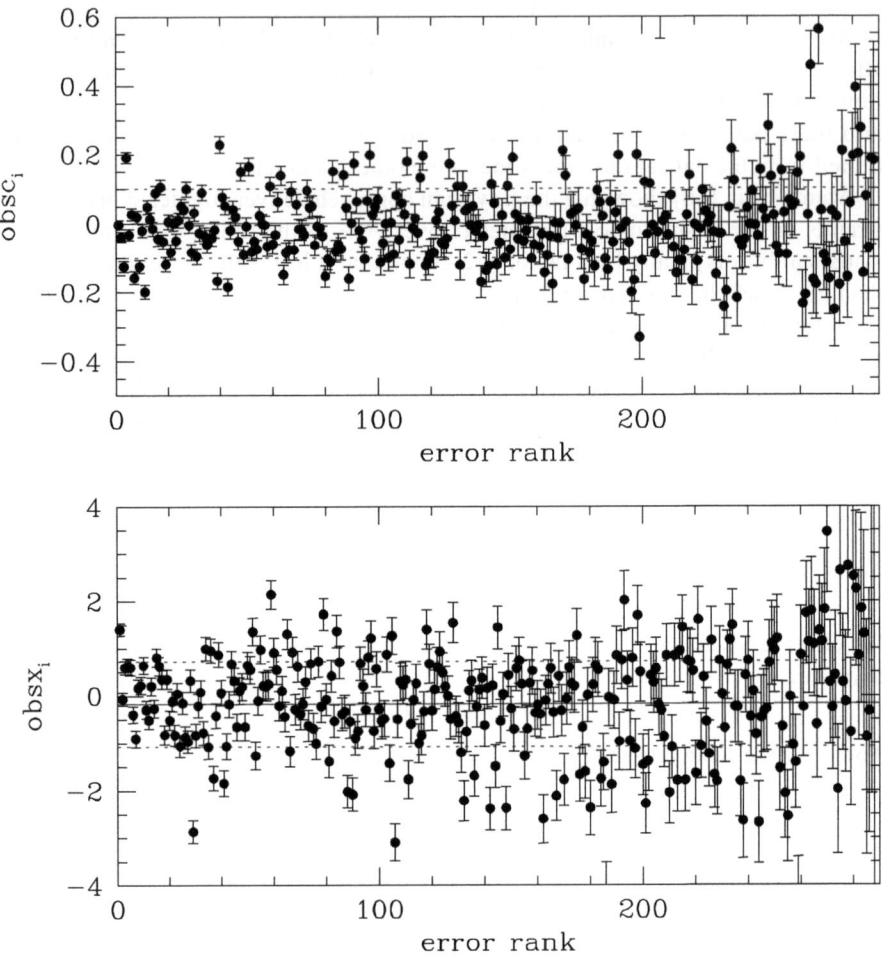

Fig. 3.4: Observed values of the stretch parameters, $obsx_i$, and of the colour parameter, $obsc_i$ ranked by error. Points scatter more than the error bars (see the left side of the figure). The dashed lines indicate the size of the intrinsic scatter as determined by our analysis.

$$obsx_i \sim \mathcal{N}(x_i, \sigma_{x,i}^2) \qquad (3.21)$$
$$obsc_i \sim \mathcal{N}(c_i, \sigma_{c,i}^2) \qquad (3.22)$$

The key point of the modelling is that the $obsx_i$ and $obsc_i$ values scatter more than their errors, but not immensely so, see Figure 3.4. The presence of a non-uniform distribution induces a Malmquist-like bias if not accounted for (e.g. large $obsx_i$ values are more likely low x_i values scattered to large values than vice versa, because of the larger abundance of low x_i values). Therefore, we model, as [10] do, the individual x_i and c_i as drawn from independent normal distributions centred on xm and cm with standard deviation R_x and R_c, respectively. In formulae:

$$x_i \sim \mathcal{N}(xm, R_x^2) \tag{3.23}$$

$$c_i \sim \mathcal{N}(cm, R_c^2) \tag{3.24}$$

We take uniform priors for xm and sc, and uniform priors on $\log R_x$ and on $\log R_c$, between the indicated boundaries:

$$xm \sim \mathcal{U}(-10, +10) \tag{3.25}$$
$$cm \sim \mathcal{U}(-3, +3) \tag{3.26}$$
$$\log_{10} R_x \sim \mathcal{U}(-5, +2) \tag{3.27}$$
$$\log_{10} R_c \sim \mathcal{U}(-5, +2) \tag{3.28}$$

That is almost all: we need to remember the definition of distance modulus:

$$distmod_i = 25 + 5 \log_{10} dl_i \tag{3.29}$$

where the luminosity distance, dl is a complicate expression, involving integrals, of the redshift z_i and the cosmological parameters $\Omega_\Lambda, \Omega_M, w, H_0$ (see any recent cosmology textbook for the mathematical expression).

Redshift, in the considered sample, has heteroscedastic Gaussian errors $\sigma_{z,i}$:

$$obsz_i \sim \mathcal{N}(z_i, \sigma_{z,i}^2) \tag{3.30}$$

The redshift prior is assumed to be uniform

$$z_i \sim \mathcal{U}(0, 2) \tag{3.31}$$

Supernovae alone do not allow to determine all cosmological parameters, so we need external prior on them, notably on H_0, taken from [11] to be

$$H_0 \sim \mathcal{N}(72, 8^2) \tag{3.32}$$

At this point, we may decide which sets of cosmological models we want to investigate using SNIa, for e.g. a flat universe with a possible $w \neq 0$ with the following priors:

$$\Omega_M \sim \mathcal{U}(0, 1) \tag{3.33}$$
$$\Omega_k = 0 \tag{3.34}$$
$$w \sim \mathcal{U}(-4, 0) \tag{3.35}$$

or a curved universe with $w = -1$:

$$\Omega_M \sim \mathcal{U}(0,1) \qquad (3.36)$$
$$\Omega_k \sim \mathcal{U}(-1,0) \qquad (3.37)$$
$$w = -1 \qquad (3.38)$$

or any other set.

Both considered cosmologies have:

$$\Omega_k = 1 - \Omega_m - \Omega_\Lambda \qquad (3.39)$$

Finally, one may want to use some data. As shortly mentioned, we use the compilation of 288 SNIa in [9].

3.4.2 Step 2: remove TeXing, perform stochastic computations and publish

Most of the distributions above are Normal, and the posterior distribution can be almost completely analytically computed [10]. However, numerical evaluation of the stochastic part of the model on an (obsolete) laptop takes about one minute, therefore there is no need for speed up. Instead, the evaluation of the luminosity distance is CPU intensive (it takes $\approx 10^3$ more times, unless approximate analytic formulae for the luminosity distance are used), because an integral has to be evaluated a number of times equal to the number of supernovae times the number of target posterior samplings, i.e. about four millions times in our numerical computation. The JAGS implementation of the luminosity distance integral is implemented as a sum over a tightly packed grid on redshift.

As the previous example, eq 3.15 to 3.32 can be de-TeXed and used in JAGS, adding one of the two set of priors, 3.33–3.35 or 3.36–3.38, depending on which problem one is interested in.

```
data {
# JAGS like precisions
 precmag <-1/errmag/errmag
 precobsc <- 1/errobsc/errobsc
 precobsx <- 1/errobsx/errobsx
 precz <- 1/errz/errz
# grid for distance modulus integral evaluation
 for (k in 1:1500){
   grid.z[k] <- (k-0.5)/1000.
 }
 step.grid.z <-grid.z[2]-grid.z[1]
}
model {
for (i in 1:length(obsz)) {
obsm[i] ~ dnorm(m[i],precmag[i])             # eq 15
m[i] ~ dnorm(Mm+distmod[i]- alpha* x[i] + beta*c[i], precM)   # eq 16
obsc[i] ~ dnorm(c[i], precobsc[i])           # eq 22
c[i] ~ dnorm(cm,precC)                       # eq 24
obsx[i] ~ dnorm(x[i], precobsx[i])           # eq 21
```

3 Understanding Better (Some) Astronomical Data using Bayesian Methods

```
x[i] ~ dnorm(xm, precx)                                        # eq 3.23
# distmod definition & H0 term
distmod[i] <- 25 + 5/2.3026 * log(dl[i]) -5/2.3026* log(H0/300000)
z[i] ~ dunif(0,2)                                              # eq 3.31
obsz[i] ~ dnorm(z[i],precz[i])                                 # eq 3.30
######### dl computation (slow and tedious)
tmp2[i] <- sum(step(z[i]-grid.z) * (1+w) / (1+grid.z)) * step.grid.z
omegade[i] <- omegal * exp(3 * tmp2[i])
xx[i] <- sum(pow((1+grid.z)^3*omegam + omegade[i] + (1+grid.z)^2*omegak,-0.5)*
            *step.grid.z * step(z[i]-grid.z))
# implementing if, to avoid diving by 0 added 1e-7 to omegak
zz[1,i] <- sin(xx[i]*sqrt(abs(omegak))) * (1+z[i])/sqrt(abs(omegak+1e-7))
zz[2,i] <- xx[i]    * (1+z[i])
zz[3,i] <- (exp(xx[i]*sqrt(abs(omegak)))-exp(-xx[i]*sqrt(abs(omegak))))/2 *
            *(1+z[i])/sqrt(abs(omegak+1e-7))
dl[i] <- zz[b,i]
}
b <- 1 + (omegak==0) + 2*(omegak > 0)
########## end dl computation

# JAGS uses precisions
precM <- 1/ intrscatM /intrscatM
precC <- 1/ intrscatC /intrscatC
precx <- 1/ intrscatx /intrscatx
# priors
Mm~ dunif(-20.3, -18.3)                                        # eq 3.20
alpha ~ dunif(-2,2.0)                                          # eq 3.18
beta ~ dunif(-4,4.0)                                           # eq 3.19
cm ~ dunif(-3,3)                                               # eq 3.26
xm ~ dunif(-10,10)                                             # eq 3.25
# uniform prior on logged quantities
intrscatM <- pow(10,lgintrscatM)                               # eq 3.17
lgintrscatM ~ dunif(-3,0)                                      # eq 3.17
intrscatx <- pow(10,lgintrscatx)                               # eq 3.27
lgintrscatx ~ dunif(-5,2)                                      # eq 3.27
intrscatC <- pow(10,lgintrscatC)                               # eq 3.28
lgintrscatC ~ dunif(-5,2)                                      # eq 3.28
#cosmo priors
H0 ~ dnorm(72,1/8./8.)                                         # eq 3.32
omegal<-1-omegam-omegak                                        # eq 3.39
# cosmo priors 1st set LCDM
#omegam~dunif(0,1)                                             # eq 3.36
#omegak~dunif(-1,1)                                            # eq 3.37
#w <- -1                                                       # eq 3.38
# cosmo priors 2nd set: wCDM
omegam~dunif(0,1)                                              # eq 3.33
omegak <-0                                                     # eq 3.34
w ~ dunif(-4,0)                                                # eq 3.35
}
```

Figure 3.5 shows the prior (dashed blue line) and posterior (histogram) probability distribution for the three intrinsic scatter terms present in the cosmological parameter estimation: the scatter in absolute luminosity after colour and stretch corrections, (σ_{scat}), and the intrinsic scatter in the distribution of the colour and stretch terms (R_c and R_x). This plot shows that the posterior probability at intrinsic scatters near zero is approximately zero and, thus, that the three intrinsic scatter terms are necessary parameters for the modelling of SNIa, and not useless complications. The three posteriors are dominated by the data, being the prior quite flat in the range where the

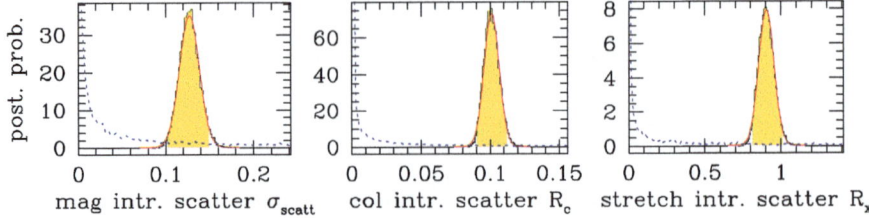

Fig. 3.5: Prior and posterior probability distribution for the three intrinsic scatter terms in the SNIa problem. The black jagged histogram shows the posterior as computed by MCMC, marginalised over the other parameters. The red curve is a Gauss approximation of it. The shaded (yellow) range shows the 95% highest posterior credible interval. The adopted priors are indicated by the blue dotted curve.

posterior is appreciably not zero (Figure 3.5). Therefore, any other prior choice, as long as smooth and shallow over the shown parameter range, would have returned indistinguishable results.

Not only SNIa have luminosities that depend on colour and stretch terms, but these in turns have their own probability distribution (taken Gaussian for simplicity) with a well determined width. Figure 3.6 depicts the Malmquist- like bias one should incur if the spread of the distribution of colour and stretch parameters is ignored: it reports the observed values (as in Figure 3.4), $obsx_i$ and $obsc_i$ as well as the true values x_i and c_i (posterior means). The effect of equations 3.23 and 3.24 is to pull values toward the mean, and more so those with large errors, to compensate the systematic shift (Malmquist-like bias) toward larger observed values.

Figure 3.7 shows the probability distribution of the two the colour and stretch slopes: $\alpha = 0.12 \pm 0.02$ and $\beta = 2.70 \pm 0.14$ respectively. As for the intrinsic scatter terms, the posterior is dominated by the data and therefore any other prior, smooth and shallow, would have returned indistinguishable results.

Finally, Figure 3.8 reports perhaps the most wanted result: contours of equal probability for the cosmological parameters Ω_M and w.

For one dimensional marginals, we found: $\Omega_M = 0.40 \pm 0.10$ and $w = -1.2 \pm 0.2$, but with non-Gaussian probability distributions.

3.4.3 Model Checking

The work of the careful researcher does not end by finding the parameter set that best describe the data, (s)he also checks whether the adopted model is a good description of the data, or it is misspecified, i.e. in tension with the fitted data. In the non-Bayesian paradigm this is often achieved by computing a p-value, i.e. the probability to obtain data more discrepant than those in hand once parameters are taken at the best fit value. The Bayesian version of the concept (e.g. [2]) acknowledges that

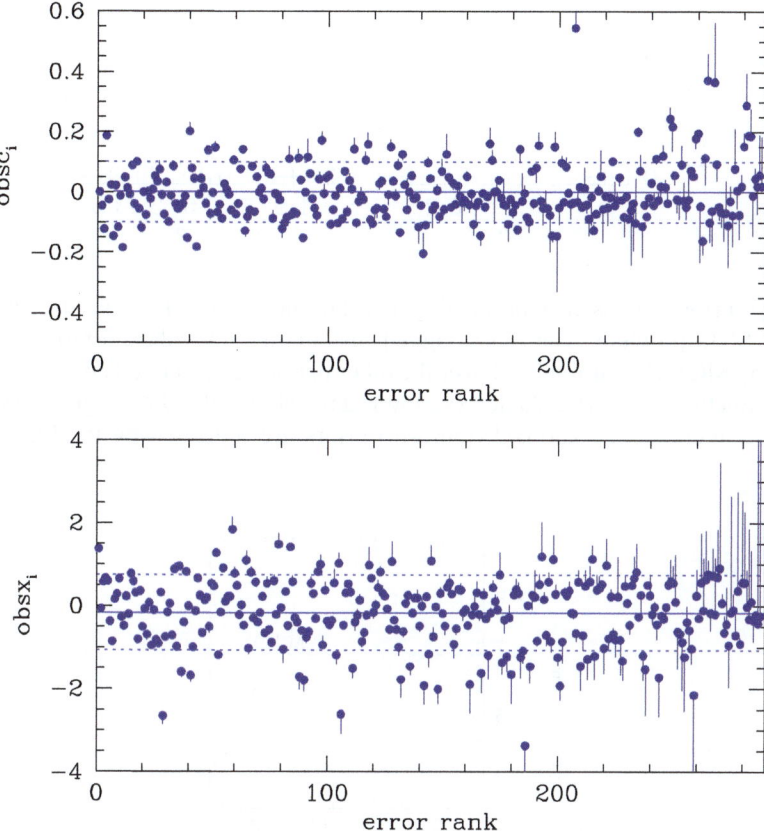

Fig. 3.6 Effect of the population structure for the stretch and the colour parameters. Each tick goes from the observed value to the posterior mean. The population modelling attempt to counterbalance the increased spread (Malmquist-like), especially those with larger error (on the right, in the figure), pulling values towards the mean.

parameters are not perfectly known, and therefore one should also explore, in addition to best fit value, other values of the parameters. Therefore, discrepancy becomes a vector, instead of a scalar, of dimension j, that measures the distance between the data and j models, one per set of parameters considered. Of course, more probable models should occur more frequently in the list to quantify that discrepancy from an unlikely model is less detrimental than discrepancy from a likely model. In practice, if parameters are explored by sampling, it is just a matter of computing the discrepancy of the data in hand for each set j of parameters stored in the chain, instead of relying on one single set of parameter (say, those that maximise the likelihood). Then, one repeats the computation for fake data generated from the model and counts how many times fake data are more extreme of real data.

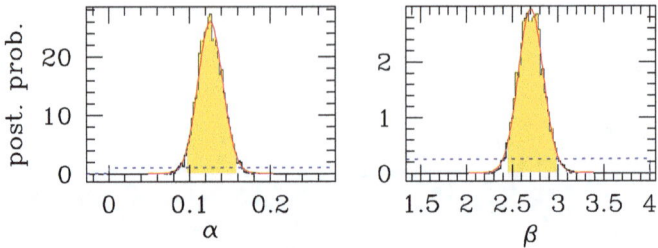

Fig. 3.7: Prior and posterior probability distribution for the colour and stretch slopes of the SNIa problem. The black jagged histogram shows the posterior as computed by MCMC, marginalised over the other parameters. The red curve is a Gauss approximation of it. The shaded (yellow) range shows the 95% highest posterior credible interval. The adopted (uniform) priors are indicated by the blue dotted curve.

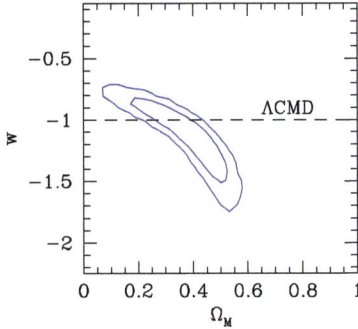

Fig. 3.8: Constraints on the cosmological parameters Ω_M and w. The two contours delimit 68% and 95% constraints.

For example, if we want to test the modelling of the observed spread of magnitude (i.e. equation 3.15 and 3.16), let's define:

$$mcor_i = M + distmod_i - \alpha\, x_i + \beta\, c_i \quad (3.40)$$

We generate fake supernovae mag:

$$m.fake_i \sim \mathcal{N}(mcor_i, \sigma^2_{scat}) \quad (3.41)$$

and fake observed values of them,

$$obsm.fake_i \sim \mathcal{N}(m.fake_i, \sigma^2_{m,i}) \quad (3.42)$$

Then, we adopt a modified χ^2 to quantify discrepancy (or its contrary, agreement). For the real data set, we have:

$$\chi^2_{real,j} = \sum_i \frac{(obsm_i - mcor_{i,j})^2}{\sigma^2_{m,i} + E^2(\sigma_{scat})} \qquad (3.43)$$

where summation is over the data and j refers to the index in the sampling chain.

Apart for the j index, Eq. 3.43 is just the usual χ^2, the difference between observed, $obsm_i$, and true, $mcor_i$, values, weighted by the expected variance, computed as quadrature sum of errors, $\sigma_{m,i}$, and supernovae mag intrinsic scatter σ_{scat}. The χ^2 of the jth fake data set $\chi^2_{fake,j}$ is:

$$\chi^2_{fake,j} = \sum_i \frac{(obsm.fake_{i,j} - mcor_{i,j})^2}{\sigma^2_{m,i} + E^2(\sigma_{scat})} \qquad (3.44)$$

At this point, we only need to compute for which fraction of the simulations $\chi^2_{fake,j} > \chi^2_{real,j}$ and quote the result. If the modelling is appropriate, then the computed fraction (p-value) is not extreme (far from zero or one). If not, our statistically modelling needs to be revised, because the data are in disagreement with the model.

We performed 15,000 simulations,[2] each one generating 288 fake measurements of SNIa. In practice, we added the following three JAGS lines:

```
mcor[i]<-Mm+distmod[i]- alpha* x[i] + beta*c[i]    # eq 3.40
m.fake[i] ~ dnorm(mcor, precM)                     # eq 3.41
obsm.fake[i] ~ dnorm(m.fake[i],precmag[i])         # eq 3.42
```

and we can simplify Eq. 3.16 in

```
m[i] ~ dnorm(mcor[i], precM)                       # eq 3.16
```

We found a p-value of 45%, i.e. that the discrepancy of the data in hand is quite typical (similar to the one of the fake data). Therefore, real data are quite common and the tested part of the model shows no evidence of misspecification. The careful researcher should then move to the other parts of the model, whose detailed exploration is left as exercise.

In such exploration of possible model misfits, it is very useful to visually inspect several data summaries to guide the choice of which discrepancy measure one should adopt (Eq. 3.43 or something different), and, if the adopted model turns out to be unsatisfactory, to guide how to revise the modelling of the tested part of the model. A possible (and common) data summary is the distribution of normalised residuals, e.g. that for $obsx_i$ reads

$$stdresobsx_i = \frac{obsx_i - E(x_i)}{\sqrt{\sigma^2_{x,i} + E^2(R_x)}} \qquad (3.45)$$

---

[2] Skilled readers may note that we are dealing, by large, with Gaussian distributions, and may attempt an analytic computation.

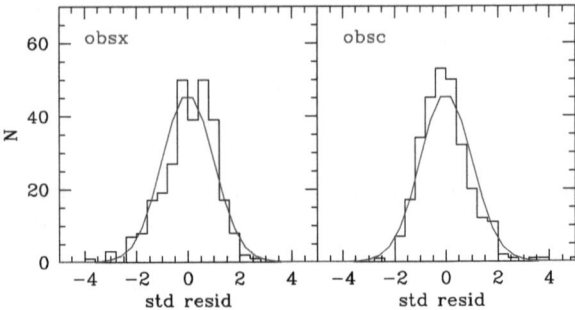

Fig. 3.9: Standardised residuals (histogram) and their expected distribution (blue curve).

i.e. observed minus expected value of x_i divided by their expected spread (the sum in quadrature of errors and intrinsic spread). A similar summary may be built for $obsc_i$ too. To first order (at least), standardised residuals should be normal distributed with standard deviation one (by construction). Figure 3.9 shows the distribution of normalised residuals of both $obsx_i$ and $obsc_i$, with superposed a Gaussian centred in 0 with standard deviation equal to one (in blue). Both distributions show a possible hint of asymmetry. At this point, the careful researcher may want to use a discrepancy measure sensitive to asymmetries, as the skewness index, in addition to the χ^2 during model testing. While leaving the actual computation to the reader, we emphasise that if an extreme Bayesian p-value if found (on $obsx_i$ for exposing simplicity), then one may replace its modelling (eq. (3.23) in the case of $obsx_i$) with a distribution allowing a non-zero asymmetry and this can be easily performed in a Bayesian approach, and easily implemented in JAGS, it is just matter of replacing the adopted Gaussian with an asymmetric distribution. If instead the data exploration gives an hint of double-bumped distribution (again on $obsx_i$ for exposing simplicity), and an extreme Bayesian p-value is found when a measure of discrepancy sensitive to double-bumped distributions is adopted, then one may adopt a mixture of Gaussians, replacing (eq. (3.23) in the case of $obsx_i$) with

$$obsx_i \sim \lambda \mathcal{N}(x_i, \sigma_{x,i}^2) + (1-\lambda)\mathcal{N}(xx_i, \sigma_{xx,i}^2) \qquad (3.46)$$

Even more simple is the (hypothetical) case of possible distribution (again of $obsx_i$ for exposing simplicity) with fat tails: one may adopt a Cauchy distribution. In such case, coding in JAGS is it is just matter of replacing a `dnorm` with `dt` in the JAGS implementation. And so on.

In summary, model checking consists in updating the model until it produces data similar to those in hand. One should start by carefully and attentively inspect the data and their summaries. This inspection should suggest a discrepancy measure to be used to quantify the model misfit and, if one is found, to guide the model updating. The procedure should be iterated until the model fully reproduces the data.

3.5 Summary and Conclusions

These two analyses offer a template for modelling the common awkward features of astronomical data, namely heteroscedastic errors, non-Gaussian likelihood (inclusive of upper/lower limits), non-uniform populations or data structure, intrinsic scatter (either due to unidentified source of errors or population spreads), noisy estimates of the errors, mixtures and prior knowledge.

In a Bayesian framework, learning from the data and the prior, it is just matter of formalising in mathematical terms our wordy statements about the quantities under investigation and how the data arise. The actual numerical computation of the posterior probability distribution of the parameters is left to (special) Monte Carlo programs, which are used to compute the integral of a function.

The great advantage of the Bayesian modelling is its high flexibility: if the data (or theory) call for a more complex modelling, or call for using distributions different from those initially taken, it is just a matter of replacing them in the model, because there are no simplifications forced by the need to reach the finishing line, as, instead, is the case in other modelling approaches. Furthermore, if one is interested in constraining another cosmological model, one with a redshift- dependent dark energy equation of state $w = w_0 + w_1(1+z)/(1+z_p)$, one should just replace just replace w in the JAGS code with the above equation. This flexible, hierarchical modelling is native with Bayesian methods.

In both our examples, the Bayesian approach performs, unsurprisingly, better than non-Bayesian methods obliged to discard part of the available information in order to reach the finishing line: Bayesian methods just use all the provided information.

As a final note, we remember that the careful researcher, whether using a Bayesian modelling or not, before publishing his own result should check that the numerical computation is adequate for his purpose and that the model is appropriate for the data. Therefore, if you, gentle reader, are using our examples as templates, remember to include in your work a sensitivity analysis, by checking that your assumptions (both likelihood and prior) are reasonable. Some prior sensitivity analysis has been performed in both examples to emphasise the importance of showing how much conclusions (the posterior) rely on assumptions (prior). For what concerns the likelihood, i.e. misfitting, the key point consists in updating the model formulation until it produces data similar to those in hand, as we have shown in great detail for the second example.

More in general, we hope that the template modelling shown in these two examples may be useful for any analysis confronted with modelling the awkward features of astronomical data, among which are heteroscedastic (point-dependent) errors, intrinsic scatter, data structure, non-uniform population (often called Malmquist bias) and non-Gaussian data, inclusive of upper/lower limits.

Acknowledgements The first part of this chapter is largely based on papers written in collaboration with Merrilee Hurn. It is a pleasure to thank Merrilee for the fruitful collaboration and her wise suggestions along the years, and for comments on an early draft of this chapter. Errors and inconsistencies remain my own.

References

1. Andreon, S., Hurn, M. A.: Statistical aspects of regressions and scaling relations in astrophysics. Statistical Analysis and Data Mining (2011, submitted)
2. Gelman, A., Carlin, J. B., Stern, H. S., Rubin, D. B.: Bayesian data analysis. Champan and Hall/CRC (2004)
3. Andreon, S., Hurn, M. A.: The scaling relation between richness and mass of galaxy clusters: a Bayesian approach. Monthly Notices of the Royal Astronomical Society **404**(4), 1922–1937 (2010)
4. Bayes, T.: Essay Towards Solving a Problem in the Doctrine of Chances, in Philosophical Transactions of the Royal Society of London (1764)
5. Laplace, P. -S.: Théorie Analytique des Probabilitées (1812)
6. Lunn, D., Spiegelhalter, D., Thomas, A., Best, N.: The BUGS project: Evolution, critique and future directions. Statistics in Medicine **28**(25), 3049–3067 (2009)
7. Plummer, M.: JAGS Version 2.2.0 user manual (2010)
8. Andreon, S., Moretti, A.: Do X-ray dark, or underluminous, galaxy clusters exist? Astronomy & Astrophysics, Arxiv preprint arXiv:1109.4031 (2011, in press)
9. Kessler, R. et al.: First-Year Sloan Digital Sky Survey-II Supernova Results: Hubble Diagram and Cosmological Parameters. The Astronomical Journal Supplement, **185**, 32–84 (2009)
10. March, M. C., Trotta, R., Berkes, P., Starkman, G. D., Vaudrevange, P. M.: Improved constraints on cosmological parameters from SNIa data. *Arxiv preprint arXiv:1102.3237* (2011)
11. Freedman, W. L., Madore, B. F., Gibson, B. K., Ferrarese, L., Kelson, D. D., Sakai, S., Mould, J. R., Kennicutt, R. C., Jr., Ford, H. C., Graham, J. A., Huchra, J. P., Hughes, S. M. G., Illingworth, G. D., Macri, L. M., Stetson, P. B.: Final Results from the Hubble Space Telescope Key Project to Measure the Hubble Constant. The Astrophysical Journal **553**, 47–72 (2001)

Chapter 4
BEAMS: Separating the Wheat from the Chaff in Supernova Analysis

Martin Kunz, Renée Hlozek, Bruce A. Bassett, Mathew Smith, James Newling, and Melvin Varughese

Abstract We present Bayesian Estimation Appliedto Multiple Species (BEAMS), an algorithm designed to deal with parameter estimation when using contaminated data. We introduce the algorithm and demonstrate how it works with the help of a Gaussian simulation. We then apply it to supernova data from the Sloan Digital Sky Survey (SDSS), showing how the resulting confidence contours of the cosmological parameters shrink significantly.

4.1 Introduction

As demonstrated by the 2011 Nobel prize in physics, supernovae of type Ia (SN-Ia) are one of the most important tools to study the expansion history of the universe [1, 2]. Supernovae are exploding stars that can be seen at extremely large distances. The most distant supernova currently known (designated SN 19941, a type IIn supernova) is at a distance of 11 billion light years [3]. Due to the finite speed of light, a signal from a very distant source takes a very long time reach us, in this case some 11 billion years (the universe itself is about 13.7 billion years old [4]). During the time that the light of the explosion travels towards us, the universe expands and the wavelength of the light is redshifted. By looking at spectral lines, it is possible to measure by how much the frequency changed. In the case of SN 19941 the redshift is $z = 2.36$, i.e. the wavelength of the light was stretched by a factor of 2.36. This also implies that the universe has grown by a factor of $(1 + z) = 3.36$ over the last 11 billion years.

In this way it is possible to map the expansion history of the universe and to compare it to predictions from General Relativity (GR). But in GR, space-time is curved, which makes the definition of distances somewhat tricky. One way to do

---

Martin Kunz
Département de Physique Théorique and Center for Astroparticle Physics, Université de Genève, Quai E. Ansermet 24, CH-1211 Genève 4, Switzerland, e-mail: Martin.Kunz@unige.ch

this assumes that we know the intrinsic luminosity of an object. Such an object is called a *standard candle*, and astrophysical research has shown that SN-Ia are fairly good standard candles after some data processing. If we have a standard candle, then the observed brightness can be linked directly to a distance (called luminosity distance d_L): the more distant the standard candle, the dimmer we see it. Astronomers tend to use not the distance but its logarithm $\mu \sim \log_{10} d_L$, for historical reasons (supposedly because the eye uses a roughly logarithmic scale to gauge brightness, apparently the magnitude system dates back to Greek astronomers). The magnitude-redshift diagram, a plot of μ versus z, can then be used to study the way the universe has behaved over most of its lifetime.

Unfortunately not all supernovae are standard candles. There are two main classes that are due to very different mechanisms. Stellar explosions of type Ia are thought to occur when a small dead star (a white dwarf) exists in a binary system with a large, red star and begins to accrete material from that larger star. At some point this cosmic cannibal begins to over-eat, and its stellar structure becomes unstable. At this point, reminiscent somewhat of the unforgettable dinner scene in the Monty Python sketch, the white dwarf explodes in a gigantic conflagration. Although the exact mechanism is not yet fully understood, it is plausible that such events always produce supernovae of comparable luminosity, as the instability always occurs under about the same conditions.

The second class is due to supermassive stars running out of fuel: during most of the life of gigantic stars, the gravitational force of their own mass that wants to crush them is balanced by the radiation pressure generated by burning hydrogen, and later heavier elements. This is possible because it is energetically favorable to fuse light elements together, but once the fusion process reaches lead, the energetics reverse and for heavier elements it is actually favorable to split (this is why fusion plants would use light elements while fission plants use very heavy elements). So stars will eventually run out of fuel, and thus will lose the radiation pressure. If the star is heavy enough then the matter itself also cannot support the self-gravity of the star, and it will collapse under its own weight. The result is called a core-collapse supernova, and much of the gravitational energy liberated in the collapse is radiated away in neutrinos and photons. Such supernovae occur unfortunately with a wide variety of intrinsic luminosities and so are unsuitable for distance measurements.

Luckily the different types can be distinguished with the help of spectral analysis of the supernova light. But measuring a spectrum requires much more light and effort than simply measuring the brightness of an object, as we need to split the light into the different wavelengths. While taking spectra has been feasible for the hundreds supernovae that have been observed to date, we are now seeing a transition where large surveys are finding thousands of supernovae (e.g. the supernova data of the Sloan Digital Sky Survey, SDSS-II SN [5, 6]), which are too many to take all the spectra. Future astronomical projects like the large synoptic survey telescope (LSST [7]) will find tens of thousands to hundreds of thousands of supernovae *per year*. It is impossible to follow up more than a tiny fraction of this data with spectroscopic observations. The "normal" observations will still provide light-curves in several different color bands, of the kind shown in Figure 4.1. Such observations will yield

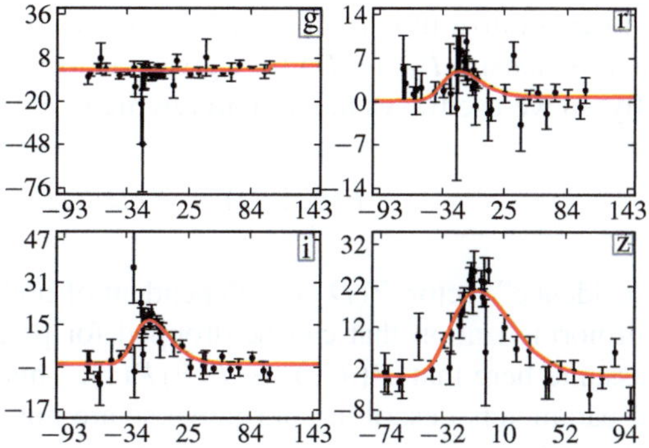

Fig. 4.1 Fitting lightcurves to supernova data: A simulated set of SN-Ia light curves in different bands from the Supernova Photometric Classification Challenge [13], together with interpolating curves from [14].

some idea what kind of supernova we are looking at, but they cannot provide the near-certainty of spectra. We are then faced with a stark choice: either we throw away over 99% of the data, or we develop a statistical method that is robust against mis-identification of supernovae. Here we will make an attempt at providing such a method. The material presented here is based on several publications [8–10] where more details can be found (see also [11, 12]).

In the following section, we introduce the BEAMS formalism and discuss in more detail the role of the probabilities. We then present our choice of likelihood functions for the different types of supernovae in section 4.3, where we also provide some tests of the algorithm itself. In section 4.4 we apply the algorithm to the SDSS-II supernova data. In the final section we summarize the chapter, providing conclusions and an outlook to future work.

4.2 Basic BEAMS

4.2.1 The BEAMS Formalism

Let us first introduce the mathematical formalism (see also [8] for simple examples and basic tests). We normally want to know the posterior distribution $P(\theta|D)$ for parameters θ given data D. Now assume that there is an additional dependence on the type of population that the data has been drawn from. For simplicity, let us assume that there are two kinds of data, type A (corresponding for example to type Ia supernovae) and type B (all other kinds of supernovae). Introducing a type vector τ of the same length N as the data vector D and with entries $\tau_i = A$ or $\tau_i = B$, we can then write

$$P(\theta|D) = \sum_\tau P(\theta, \tau|D) \qquad (4.1)$$

where we marginalized over all 2^N possible τ vectors. It is obviously straightforward to generalize this to an arbitrary number of different populations. The joint probability density $P(\theta, \tau | D)$ for a given vector τ is probably difficult to determine directly, so we use Bayes theorem to rewrite it,

$$P(\theta, \tau | D) = P(D | \theta, \tau) \frac{P(\theta, \tau)}{P(D)}. \tag{4.2}$$

The "evidence" factor $P(D)$ is independent of both the parameters and τ and is an overall normalization that can be dropped for parameter estimation. We will further assume here that $P(\theta, \tau) \approx P(\theta) P(\tau)$. This simplification assumes that the actual parameters describing our universe are not significantly correlated with the probability of a given supernova to be of type Ia or of some other type. Although it is possible that there is some influence, we can safely neglect it for current data as our parameters are describing the large-scale evolution of the universe, while the type of supernova should mainly depend on local astrophysics. In this case $P(\theta)$ is the usual prior parameter probability. We will also assume $P(\tau)$ to separate into independent factors,

$$P(\tau) = \prod_{\tau_i = A} P_i \prod_{\tau_j = B} (1 - P_j), \tag{4.3}$$

for a discussion of this approximation please see [9]. Here the product over "$\tau_i = A$" should be interpreted as a product over those indices i in the vector τ for which $\tau_i = A$. In other words, given a population vector τ with entries "A" for SN-Ia and "B" for other types, the total probability $P(\tau)$ is the product over all entries, with a factor P_j if the j-th entry is "A" and $1 - P_j$ otherwise (if the j-th entry is "B"). In this way P_j is always a probability to find an entry $\tau_j = A$ in the vector τ before using the data. Notice that we discuss here only one given vector τ, the uncertainty is taken care of by the outer sum over all possible such vectors. The full expression is therefore

$$P(\theta | D) \propto P(\theta) \sum_{\tau} P(D | \theta, \tau) \prod_{\tau_i = A} P_i \prod_{\tau_j = B} (1 - P_j). \tag{4.4}$$

The factor $P(D | \theta, \tau)$ is the likelihood, but now conditional on the data types. This means when we write down later on an expression for the likelihood, we can do it assuming that the type of each data point is known.

The price to pay is that we then have to marginalize over all possible vectors τ, evaluating a sum composed of 2^N terms for N data points. The exponential scaling with the number of data points means that we can in general not evaluate the full posterior directly, but have to use a clever approximation. Here we will instead make an additional assumption that the data points are not correlated,

4 BEAMS

$$P(D|\theta, \tau) = \prod_{i=1}^{N} P(D_i|\theta, \tau) \tag{4.5}$$

$$= \prod_{\tau_i = A} P(D_i|\theta, \tau_i = A) \prod_{\tau_j = B} P(D_j|\theta, \tau_j = B). \tag{4.6}$$

In the second line we have made the reasonable assumption that the probability of observation i does not depend on the assumed type of the object $j \neq i$. We have also indicated that the likelihood of each observation naturally splits into two populations, those which have entry A in τ, and those with entry B. In general the form of these two likelihood classes will be different. In toy model applications, we will usually know how they look like, but for actual data they may be unknown and we will have to leave some additional freedom.

The form of Eqs. (4.4) and (4.6) allows for a huge computational simplification: the posterior is the sum over all possible products of the type $A_1 A_2 A_3 \ldots$, $B_1 A_2 A_3 \ldots$, $A_1 B_2 A_3 \ldots$, etc. This sum of 2^N terms can be generated simply by computing the product over N terms $\prod_k (A_k + B_k)$, and the posterior Eq. (4.4) can be written as

$$P(\theta|D) \propto P(\theta) \prod_{i=1}^{N} \{P(D_i|\theta, \tau_i = A) P_i + P(D_i|\theta, \tau_i = B)(1 - P_i)\}. \tag{4.7}$$

This is the form of the BEAMS posterior that we will be using for the rest of this chapter.

4.2.2 BEAMS Probabilities

The probabilities in BEAMS are of central importance, so it is worth to take a closer look by studying a simple toy model: we assume that we are dealing with two populations (let us call them 'red' and 'blue') drawn from two normal distributions with means at $\pm \theta$ and equal variances of $\sigma^2 = 1$, see the top panel of Figure 4.2. In addition to the basic formalism discussed in section 4.2.1 above, we will introduce an extra parameter A that can adjust the relative normalization of the probabilities. As we will see, this parameter allows for automatically adjusting for an unknown relative rate of the two populations. We introduce the parameter as a change of the relative probability B (the Bayes factor) to be of the red or blue kind:

$$B = \frac{P}{1 - P} \rightarrow \tilde{B} = BA = A \frac{P}{1 - P} = \frac{\tilde{P}}{1 - \tilde{P}}. \tag{4.8}$$

The effective, adjusted probability is then

$$\tilde{P} = \frac{AP}{1 - P + AP}, \tag{4.9}$$

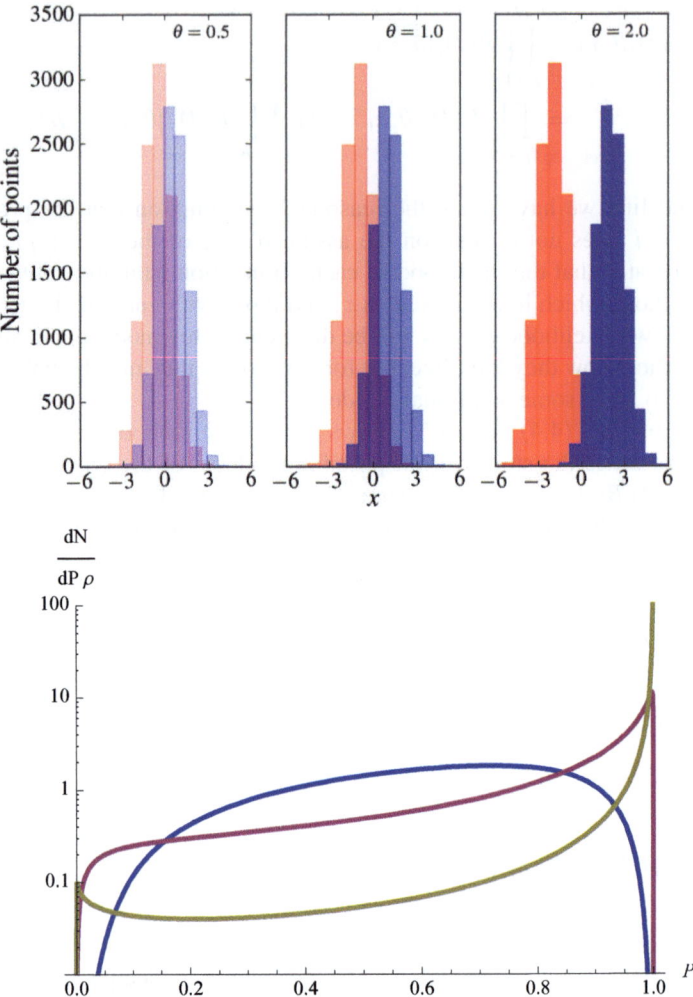

Fig. 4.2 Posterior probabilities: the top panel provides an illustration of the two toy distributions, in the case of $\theta = 0.5, 1.0, 2.0$ (left to right). The bottom panel shows the probability histogram density plots, or number of red points with a given probability, where $dN^{(r)}(P)$ is given in Eq. (4.14) for $\theta = 0.5$ (blue), 1 (red) and 2 (yellow).

and we will in the actual applications always use this probability and allow for a free A that we will marginalize over.

The equality of the variances of the two populations means that we are measuring the distance $\Delta = 2\theta$ between the two mean values in units of the standard deviation. We also allow for different numbers of points drawn from the red and blue Gaussians through a 'rate parameter' $\rho \in [0, 1]$ that gives the probability to draw a red point. If we draw N points in total, we will then have on average ρN red points and $(1-\rho)N$ blue points. The likelihood for a set of points $\{x_j\}$, with j running from 1 to N, is then

$$P(\{x_j\}|\theta) = \prod_{j=1}^{N} \frac{1}{\sqrt{2\pi}} \left(Pe^{-\frac{1}{2}(\theta-x_j)^2} + (1-P)e^{-\frac{1}{2}(\theta+x_j)^2} \right) \quad (4.10)$$

for $P = \rho$.

To simplify the analysis we assume that we are dealing with large samples so that θ is determined to high precision, with an error much smaller than σ. In this case (and since this is a toy model) we can take the parameter θ fixed. We also note that if we are running this in BEAMS with a true prior probability $P = \rho$ then we would find a normalization parameter $A = 1$, while for $P = 1/2$ we would obtain $A = \rho/(1-\rho)$, and we again assume that this parameter can be fixed to its true value. Then it is easy to see that if we leave the probability for point i, P_i, free, we find a Bayes factor

$$B = \frac{P(\{x_j\}|\theta, P_i = 1)}{P(\{x_j\}|\theta, P_i = 0)} = \frac{e^{-\frac{1}{2}(\theta-x_i)^2}}{e^{-\frac{1}{2}(\theta+x_i)^2}} = e^{2\theta x_i}. \quad (4.11)$$

In other words, $\ln(B) = x_i \Delta$, just the value of the data point times the separation of the means. If the point is exactly in between the two distributions, $x_i = 0$, then $B = 1$, i.e. its BEAMS posterior probability to be red or blue is equal. This means that if we want to think of the BEAMS posterior probability as the probability to be red or blue, we should update the Bayes factor with A, i.e. use $\tilde{B} = BA$, with an associated probability $P = \tilde{B}/(1 + \tilde{B})$. We also see that the probability to be red increases exponentially as x_i increases. As we will see below, this reflects the fact that the number of red points relative to the blue points increases in the same way. The rapidity of this increase is governed by the separation, Δ, of the two distributions.

What is the distribution of the posterior probabilities, i.e. the histogram of probability values, and what determines how well BEAMS does as a typer in this example? The number of red points in an interval $[x, x + dx]$ is just given by the 'red' probability distribution function at this value, times dx. To plot this function in terms of P we also need

$$x(P) = \frac{\ln(B)}{\Delta} = \frac{\ln(P/(1-P))}{\Delta} \quad (4.12)$$

$$\frac{dP}{dx} = \Delta P(1-P). \quad (4.13)$$

The probability histograms for the red (r) and blue (b) points, normalized to ρ and $1 - \rho$ respectively, then are:

$$dN^{(r)}(P) = \frac{\rho}{\sqrt{2\pi}\Delta} \frac{dP}{P(1-P)} \exp\left\{-\frac{1}{2}\left(\frac{\ln[P/(1-P)]}{\Delta} - \theta\right)^2\right\}, \quad (4.14)$$

$$dN^{(b)}(P) = \frac{1-\rho}{\sqrt{2\pi}\Delta} \frac{dP}{P(1-P)} \exp\left\{-\frac{1}{2}\left(\frac{\ln[P/(1-P)]}{\Delta} + \theta\right)^2\right\}. \quad (4.15)$$

We plot $dN^{(r)}/dP/\rho$ for $\theta = 0.5$, 1 and 2 in the lower panel of Figure 4.2. We see how the values become more concentrated around $P = 1$ for larger separation of the distributions, i.e. BEAMS becomes a "better" typer. But for very large separations there are also suddenly more supernovae at low P (yellow curve). The reason is that BEAMS does not try to be the best possible typer, instead it respects the condition that the probabilities have to be unbiased, in the sense that

$$\frac{dN^{(r)}}{dN^{(b)}} = \left(\frac{P}{1-P}\right)\left(\frac{\rho}{1-\rho}\right) = BA = \tilde{B}. \quad (4.16)$$

Since BEAMS only uses the information coming from the distribution of the values, its power, as reflected in the distribution of probability values $dN(P)$, is given by how strongly the distributions are separated. If they are identical ($\theta = 0$) then BEAMS can only return $P = 1/2$ while for larger θ there is a stronger preference for one type over another. But given the two populations, we can in principle derive the probability histogram by just looking at the ratio of data points of either type at each point in data space, there is nothing else BEAMS can do. Also, in order for the probabilities to be unbiased (up to the rates which are taken into account by A) if there are, say, 200 red points in the $P = 0.9$ bin and only 10 in the $P = 0.8$ bin, then we need to find about two blue points in the $P = 0.8$ bin, but 20 in the $P = 0.9$ bin. Although this looks like a significant misclassification problem, it is just a reflection of Eq. (4.16) and is actually the desired behavior: BEAMS is not a classification algorithm (see e.g. [13–15] for efforts in that direction) but instead a way to compute the posterior pdf of the parameters θ. The property of unbiased probabilities is required to get unbiased parameter constraints, and indeed for that purpose we never classify any data points. Instead we leave them in a superposition of different types, weighted by the associated probabilities as encoded in the marginalization over τ in Eq. (4.1).

4.3 Application of BEAMS to Supernova Observations

In this section we will complete first our discussion of the posterior (4.7) by providing explicit expressions for the two likelihood functions. We will say a few words on the numerical strategy used to explore the posterior parameter distribution and check the performance of the algorithm with a range of tests. This section and the next is based on the results obtained in [10].

Before entering the likelihood discussion, we would like to remind the reader that supernova data is given in the form of a distance modulus μ as a function of redshift z. In addition, the distance modulus depends on a set of cosmological and nuisance parameters θ. The cosmological parameters $\{H_0, \Omega_m, \Omega_\Lambda\}$ are the true quantities of interest for us. Here H_0 is the expansion rate of the universe today (the Hubble constant), and the Ω_j are the relative energy densities in matter m and a cosmological constant Λ. See for example the book by Scott Dodelson [16] for a good introduction to cosmology.

The distance modulus is related to the cosmological model via:

$$\mu(z,\theta) = 5\log d_L(z,\theta) + 25, \tag{4.17}$$

where

$$d_L(z,\theta) = \frac{c(1+z)}{\sqrt{\Omega_k}H_0}\sinh\left(\sqrt{\Omega_k}\int_0^z \frac{dz}{E(z)}\right) \tag{4.18}$$

is the luminosity distance measured in Megaparsec (Mpc), and the normalized expansion rate is given by

$$E(z) \equiv \frac{H(z)}{H_0} = \sqrt{\Omega_m(1+z)^3 + \Omega_k(1+z)^2 + \Omega_\Lambda}. \tag{4.19}$$

The relative energy densities of matter (Ω_m), curvature (Ω_k) and the cosmological constant (Ω_Λ) obey the relation $\Omega_m + \Omega_k + \Omega_\Lambda = 1$, which we use to express Ω_k in terms of the other Ω's. Notice that $\Omega_k < 0$ is possible, in which case $\sqrt{\Omega_k}$ in Eq. (4.18) becomes imaginary and the hyperbolic sine becomes a normal sine function instead – the limit $\Omega_k = 0$ is also well defined. The distance modulus is defined as the difference between the absolute and apparent magnitudes of the supernova, $\mu = m - M$, with additional corrections made to the apparent magnitude for the correlations between brightness, color and stretch and a K-correction term related to the difference between the observer and rest-frame filters, for example. The corrections are typically made within the model employed in a light-curve fitter, such as that for MLCS2k2.

In this application of BEAMS we have assumed that the distance modulus μ is obtained directly from the light-curve fitter (such as is the case for fitters which use the MLCS2k2 light-curve model), however this is not an implicit assumption. In the case of the SALT light-curve fitter, the distance modulus would be reconstructed using a framework such as that outlined in [17] before including in the BEAMS algorithm. We will also always assume that the distance modulus has been obtained *under the assumption that all supernovae are of type Ia*. This means that it is straightforward to write down the likelihood for type-Ia supernovae, but that we need to do extra work for the non-Ia supernovae. It would of course be preferable to have distance moduli for all possible supernova types, but this is still an active research topic in astronomy.

4.3.1 Choice of Likelihood

4.3.1.1 Likelihood of Type-Ia Supernovae

Following the standard astronomical literature, the Ia likelihood is modeled as a Gaussian probability distribution function (pdf) for the observed distance modulus μ_i centered around the theoretical value $\mu(z, \theta)$ with a variance $\sigma_{\text{tot},i}^2$:

$$P(\mu_i|\theta, \tau_i = 1) = \frac{1}{\sqrt{2\pi}\sigma_{\text{tot},i}} \exp\left(-\frac{(\mu_i - \mu(z_i, \theta))^2}{2\sigma_{\text{tot},i}^2}\right). \quad (4.20)$$

Again following standard practice, we model the error on the distance modulus of each supernova as a sum in quadrature of several independent contributions,

$$\sigma_{\text{tot},i}^2 = \sigma_{\mu,i}^2 + \sigma_\tau^2 + \sigma_{\mu,z}^2, \quad (4.21)$$

where $\sigma_{\mu,i}$ is the error obtained from fits to the SN light-curve, σ_τ is the characteristic intrinsic dispersion of the supernova population, which we add as an additional global parameter to the vector θ with Jeffreys' prior. The constraints do not depend strongly on the prior used for the intrinsic dispersion. The error term $\sigma_{\mu,z}$ converts the uncertainty in redshift due to measurement errors and peculiar velocities into an error in the distance of the supernova as:

$$\sigma_{\mu,z} = \frac{5}{\ln(10)} \frac{1+z}{z(1+z/2)} \sqrt{\sigma_z^2 + (v_{pec}/c)^2}, \quad (4.22)$$

with σ_z as redshift error, and v_{pec} as the typical amplitude of the peculiar velocity of the supernova, which we take as 300 kms^{-1} [6, 18].

4.3.1.2 Likelihood of All Other Supernovae

The general form of the non-SNIa likelihood will be complicated since there are several sub-populations. Given the limited number of non-SNIa in the SDSS-II SN data set however, (see Figure 4.8) we will model it with a single mean and a dispersion. If one chooses to describe a population using only a mean and a variance, statistically the least-informative (maximum entropy) choice of pdf in this case is also a Gaussian [19],

$$P(\mu_i|\theta, \tau_i = 0) = \frac{1}{\sqrt{2\pi}s_{\text{tot},i}} \exp\left(-\frac{(\mu_i - \eta(z_i, \theta))^2}{2s_{\text{tot},i}^2}\right). \quad (4.23)$$

As we do not know the mean η and variance $s_{\text{tot},i}^2$ of the non-Ia population, we describe them with additional parameters. We will keep the parametrization of the mean very general (see below) but for the variance we restrict ourselves to the same form as for the Type Ia supernovae, Eq. (4.21), but with a potentially different intrinsic dispersion s_τ^2 described by an independent parameter (again with a Jeffreys' prior). We assume that the measurement errors and the contribution from the peculiar velocities enter in the same way for Type Ia and other supernovae and so keep these terms identical.

We do not know what to expect for the mean of the non-Ia pdf and so we allow for a range of possibilities. As the brightness is linked to the luminosity distance through Eq. (4.17), we describe the expected non-Ia distance modulus (as provided by the light-curve fitter which assumes actually a type-Ia supernova) as a deviation from the theoretical value, $\eta(z, \theta) = \mu(z, \theta) + \Upsilon(z)$, where we consider the following Taylor expansions of the difference as a function of redshift:

$$\Upsilon(z) = \eta(z, \theta) - \mu(z, \theta) \propto \sum_{i=0}^{3} (a_i z^i)/(1 + dz). \tag{4.24}$$

We consider the cases where we set different combinations of the parameters (a_i, d), to zero, and employ a criterion based on model probability to decide which of these functions to use. We note that the explicit link of $\eta(z, \theta)$ to $\mu(z, \theta)$ carries a risk that the non-Ia likelihood can influence the posterior estimation of the cosmological parameters. For this reason we verify that the contours do not shift when we set directly $\eta(z, \theta) = \Upsilon(z)$, although we will need a higher-order expansion in general (and of course the recovered parameters of the function $\Upsilon(z)$ will change). In general, as long as the basis assumed has enough freedom to fit the deviation in the distance modulus of the non-Ia population from the Ia model, the inferred cosmology will not be biased.

For a cosmological analysis we just marginalize over the values of the parameters in $\Upsilon(z)$, but these parameters contain information on the distribution of non-Ia type SN and thus their posterior is of interest as well, allowing us to gain insight into the distribution characteristics on the non-Ia population at no additional 'cost'.

The simple binomial case considered here, where the non-Ia population consists of all types of core-collapse SNe, is probably too simplistic to accurately describe the distribution of non-Ia supernovae. In general one could include multiple populations, one for each supernova type, which would yield a sum of Gaussian terms in the full posterior. In addition, the forms describing the distance modulus of the non-Ia population are chosen to minimize the cosmological information from the non-Ia's (we always test for a deviation from the cosmological distance modulus), however, the parameterization of the non-Ia distance modulus could be improved by investigating the distance modulus residuals from simulations, as the major contributions to the distance modulus residuals appear to be the core-collapse luminosity functions, along with the specific survey selection criteria and limiting magnitude, see [20]. While current SN samples do not include a large enough sample of non-Ia data to test for this, larger data sets (such as the data from the BOSS SN survey) will allow for a detailed analysis of the number (and form) of distributions describing the contaminant population.

4.3.2 Numerical Methodology

In this work, the BEAMS algorithm is implemented within a Markov Chain Monte Carlo (MCMC) framework, and the Metropolis-Hastings [21] acceptance criterion was used. We use the cosmological parameters $\{\Omega_m, \Omega_\Lambda, H_0\}$ in the case of the χ^2 approach on the *spectro* and *cut* samples described below, and add additional parameters $\{A, \sigma_\tau, s_\tau, \vec{a}\}$ in the case of the BEAMS application. The parameters $\vec{a} = \{a^0, a^1, a^2\}; d = a^3 = 0$ are for the quadratic model, in the other models for $\Upsilon(z)$ we adjust the parameters accordingly. The chains were in general run for around 100,000 steps per model; this was sufficient to ensure convergence. We test for convergence using the techniques described in [22]. We impose positivity priors on the energy densities of matter and dark energy, and impose a flat prior on the Hubble parameter between $20 < H_0 < 100$ kms^{-1}Mpc^{-1}. The Hubble parameter is marginalized over given that we do not know the intrinsic brightness of the supernovae, but through the distance modulus are only sensitive to the *relative* brightness of the supernovae. We impose broad Gaussian priors on the parameters of the non-Ia likelihood function, and step logarithmically in the probability normalization parameter A, as well as the intrinsic dispersion parameters of both the Ia and non-Ia distributions.

4.3.3 Comparison to Standard χ^2 Methods

The primary difference between BEAMS and current methods is that the latter either require that all data are spectroscopically confirmed, or apply a range of quality cuts based on selection criteria. Here we will compare the performance of BEAMS to these two approaches, by processing the data that pass the required selection criteria using the Ia likelihood, Eq. (4.20). We will hereafter refer to this as the χ^2 approach.

We use the following samples[1]:

- **spectro sample**:
  The sample containing only spectroscopically confirmed supernovae. In addition to spectroscopic confirmation we will also apply a cut on the goodness-of-fit probability from the light-curve templates within the MLCS2k2 model, $P_{\text{fit}} > 0.01$, and a cut on the light-curve fitter parameter $\Delta > -0.4$, where Δ is a parameter in the MLCS2k2 model describing the light-curve width-luminosity correlation. MLCS2k2 was trained using the range $-0.4 < \Delta < 1.7$ [23], hence we restrict the sample to $\Delta > -0.4$, which is a cut typical in current SN surveys, and so we introduce the cut to provide comparison between datasets. We process this *spectro* sample using the χ^2 approach.
- **cut sample**:
  This larger sample is selected both by removing 5σ outliers from a moving

---

[1] We apologize for the use of technical jargon in the description of the samples.

average fit to the Hubble diagram including both photometric and spectroscopically confirmed data and applying a cut to the sample, including only data with a high enough probability, $P_{\text{typer}} > 0.9$ (where the probability comes from a general supernova typing procedure, such as PSNID, described in [24, 25]). We choose to use the PSNID probabilities to make the probability cut on the sample ($P_{\text{typer}} > 0.9$); if the MLCS2k2 probabilities had themselves been used to make a *cut* sample, then objects would only be included if they had probabilities greater than, for example, $P_{\text{fit}} > 0.9$ In addition, we impose a cut on the goodness-of-fit of the light-curve data to the Type Ia typer, $\chi^2_{lc} < 1.8$, a cut on the goodness-of-fit probability from the light-curve templates within the MLCS2k2 model, $P_{\text{fit}} > 0.01$, and a cut on $\Delta > -0.4$. In this *cut* sample case we then use standard the χ^2 cosmological fitting procedure on the sample, and so set the Ia probability of all points to one.

- **photo sample**:
This sample is the one to which BEAMS will be applied, and will include all the photometric data with host galaxy redshifts. As in the previous two cases, we include only data which have $P_{\text{fit}} > 0.01, \Delta > -0.4$.

Note that the *spectro* sample will be included in all the three samples described above. While the *spectro* and *cut* samples have by definition $P_{\text{Ia}} = 1$ (as they are analyzed in the χ^2 approach), we do not set the probabilities to unity when applying BEAMS to the full sample – the *spectro* subsample within the larger *photo* sample will be treated 'blindly' by BEAMS. The *spectro* sample is the one most similar to current cosmological samples, and will be used to check for consistency in the derived parameters between BEAMS applied to the photo sample and the χ^2 approach on the *spectro* sample.

4.3.4 Tests on Simulated Data

To test the BEAMS algorithm explicitly we need a completely controlled sample, where all variables (such as the non-Ia model and the SN-Ia probabilities) are directly known and where we can verify that the algorithm is able to recover them correctly. In addition, we use this data set to check that we recover the correct shape of the non-Ia distance modulus $\eta(z)$ since the true $\eta(z, \vec{\theta})$ is known for this sample only. We simulate a population of 50,000 SNe, with redshifts drawn from a Gaussian distribution, $z \sim \mathcal{N}(0.3, 0.15)$, and distance moduli drawn from a flat ΛCDM universe with $(\Omega_m, \Omega_\Lambda, H_0) = (0.3, 0.7, 70)$. The non-Ia population includes a contribution to the distance modulus, $\eta(z, \vec{\theta}) = \mu(z, \vec{\theta}) + a^0 + a^1 z + a^2 z^2$, where we choose $(a^0, a^1, a^2) = (1.5, 1, -3)$. We assign P_{Ia} probabilities from a model $dN/dP_{Ia} = A_1 P_{Ia} + A_2 P_{Ia}^2$; with $A1 = -0.9$; $A2 = 1.9$. We then assign the *types* from the two samples (of Ia's and non-Ia's), i.e. we choose a random number t and if $t < P_{Ia}$ (i.e. the type also follows the same linear relationship as the probability) we take the data point to be a Ia, and if $t > P_{Ia}$ we assign it as a non-Ia, until we

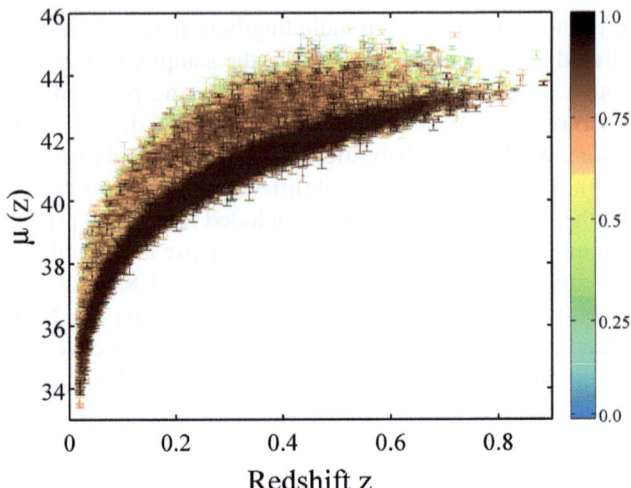

Fig. 4.3 Gaussian data: 37529 points simulated according to a Gaussian distributions around a distance modulus in a flat ΛCDM model for the Ia population (25000 points) and with extra terms up to quadratic order in redshift for the non-Ia population. The points are colored according to their simulated probabilities from blue (low probability) to dark brown (high probability).

run out of data points from either sample. This procedure reduces the sample size from 50,000 to 37,529, but guarantees unbiased probabilities.

We assign a 'measurement error' to each distance modulus of $\sigma_\mu = 0.1$; add an intrinsic error $\sigma_\tau = 0.16$ and a peculiar velocity error based on Eq. (4.21), with $v_{pec} = 300 \text{kms}^{-1}$. We then randomly scatter the data points based on the total errorbar. To mimic what happens in a light-curve fitter, only the measurement error is recorded, however. When performing parameter estimation on the points we either add this measurement error in quadrature to the other terms whose amplitudes are fixed (in the case of the χ^2 approach), or we estimate the magnitudes of the intrinsic dispersion when we apply the BEAMS algorithm. We randomly choose 10% of the Ia data and assign *spectro* status; this represents the data that are followed up by large telescopes on the ground. This *spectro* sample is drawn so that we can compare the BEAMS-estimated result to the χ^2 approach on a smaller sample. The data are shown in Figure 4.3. In the BEAMS analysis we checked on a small number of simulated samples that the results obtained were unbiased – a full Monte Carlo simulation of bias is beyond the scope of this work.

4.3.4.1 Performance on Cosmological Parameters

We show in Figure 4.4 the 2σ confidence contours in the Ω_m, Ω_Λ plane when analyzing the Gaussian simulation with BEAMS (filled contours), the 'spectroscopic' sample (dashed contours) and the 'cut' sample (solid contours). We see that both BEAMS and the spectroscopic sample are consistent with the input cosmology

4 BEAMS

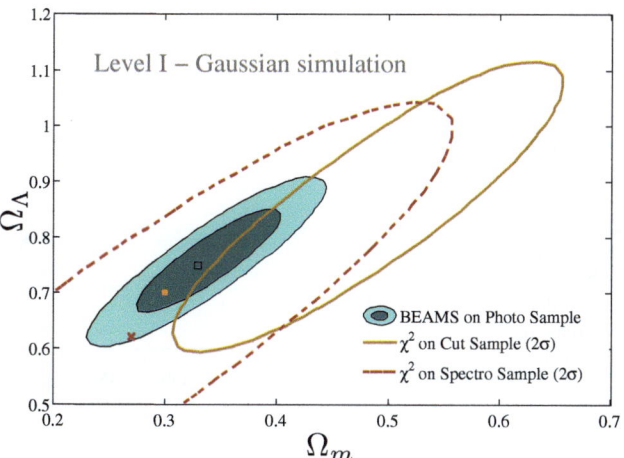

Fig. 4.4 Analysis of Gaussian simulation: We show the 2σ contours in the Ω_m, Ω_Λ plane for the simulated Gaussian data. The BEAMS constraints (filled contours with best fit indicated by the black open square) are consistent with the input cosmology (brown square), as is the 'spectroscopic' sample (dashed contours, best-fit indicated by brown cross). The 'cut' sample on the other hand is biased by over 2σ in spite of a relatively stringent cut on probability of $P_{\text{cut}} = 0.9$; stronger cuts will recover the true cosmology at the cost of sample size.

(filled brown square at $\Omega_m = 0.3$, $\Omega_\Lambda = 0.7$), but BEAMS can use the additional information in the data and is able to provide much tighter constraints. The cut sample has also smaller contours than the spectroscopic sample (although larger than BEAMS) but is biased with respect to the input cosmology.

As discussed in [8] for the one-dimensional case, the effective number of SNe that result when applying BEAMS scales as the number of spectroscopic SNe and the average probability of the dataset multiplied by the remainder of the photometric sample, $\sigma \to \sigma/\sqrt{N_{\text{spec}} + \langle P_{\text{Ia}} \rangle N_{\text{photo}}}$. In the two-dimensional case, the square root would be removed as the area of the ellipse scales with the increase in the effective number of supernovae. In our applications we have, however, not used the fact that we know that some points are confirmed as Type Ia. In other words, the probability of each data point was taken from the light-curve fitter and was not adjusted to one or zero depending on the known type. Hence we expect the size of the contours in the $i - j$ plane to scale as

$$C_{ij}^{1/2} \to \frac{C_{ij}^{1/2}}{\langle P_{\text{Ia}} \rangle N_{\text{photo}}} \quad (4.25)$$

We compute the size of the error ellipse for various Gaussian simulations as a function of the size of the simulation, shown in Figure 4.5, for one particular model of the probabilities, and hence one value of $\langle P \rangle$. We impose a prior on the densities, and hence the ellipses are not closed for smaller samples. For large enough sample sizes the ellipse is closed and we observe that the error ellipses scale in area as $\propto 1/\langle P_{\text{Ia}} \rangle N$, which is consistent with earlier results [8]. In general then, one

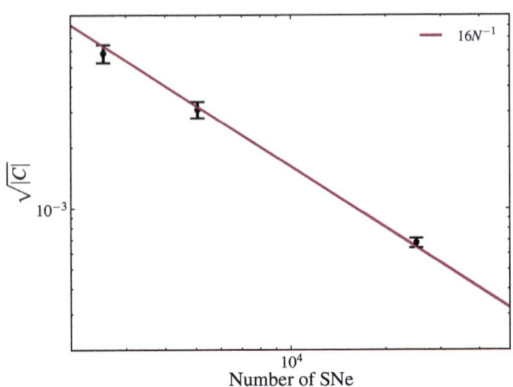

Fig. 4.5 Errors scale with number of SNe: the size of the error ellipse, approximated by the square root of the determinant of the two-dimensional chain of Ω_m, Ω_Λ shows the reduction in size with increasing the number of SNe in the simulation.

would obtain a different constant factor $\langle P \rangle$ in Figure 4.5 for different simulated probability distributions.

4.3.4.2 Constraining $\Upsilon(z)$ Forms for the Non-Ia Population

The Gaussian simulation described in this section uses a quadratic model for the differences between the standard ΛCDM $\mu(z)$ and the non-Ia distance modulus. We test here that assuming a different functional form while performing parameter estimation does not significantly bias the inferred cosmology. We define the effective χ^2 as $-2\ln\mathcal{L}$, where the posterior $\mathcal{L}$ is given by Equation (4.7), and we provide values relative to the simplest linear model for $\Upsilon(z)$. The goodness-of-fit of the distributions to the data is summarized in Table 4.1.

In Figure 4.6 we show that BEAMS is reasonably insensitive to the assumed form of the non-Ia likelihood, provided it is allowed enough freedom to capture the underlying model. A linear model fails to recover the correct cosmology, as it does not have enough freedom to recover the difference between the Ia and non-Ia distribution. It correspondingly has a very high χ^2 relative to the other approaches. The higher-order functions recover consistent cosmologies, and the χ^2 of these models improves by $\Delta\chi^2 < 0.5$, even though the models have increased the number of parameters by one.

4.3.4.3 Dependence on Probability

The BEAMS algorithm naturally uses some indication of the probability of a data point to belong to the Ia population, whether it is some measure of the goodness-of-fit of the data to a type Ia light-curve template, or something more robust such as the relative probability that the point is a Ia compared to the probability of being of a different type. By including a normalization factor, we can correct

4 BEAMS

Table 4.1 Comparison of non-Ia likelihood models for Gaussian simulation: χ^2 values for the fits using various forms of the non-Ia likelihood for the Gaussian simulations, where the true underlying model is quadratic. The constraints on Ω_m, Ω_Λ are shown in Figure 4.6. $\Delta\chi^2_{\text{eff}}$ is difference in the effective χ^2 between a given model and the linear case, which has $\chi^2_{\text{eff}} = 42526.2$.

Model	$\Delta\chi^2_{\text{eff}}$	Parameters
$\Upsilon(z) = az + c$	0	2
$\Upsilon(z) = az + bz^2 + c$	-192.9	3
$\Upsilon(z) = az + bz^2 + cz^3 + d$	-193.3	4
$\Upsilon(z) = (az + bz^2 + c)/(1 + dz)$	-193.4	4

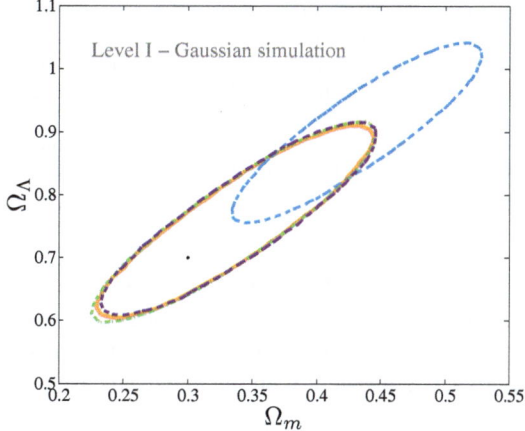

Fig. 4.6 Different $\Upsilon(z)$ for the non-Ia likelihoods: 2σ constraints in the Ω_m, Ω_Λ plane for different versions of the non-Ia distance modulus function, for the Gaussian simulation. We simulated a quadratic model, and ran BEAMS assuming a linear, quadratic, cubic and Padé form for $\Upsilon(z)$, as described in Section 4.3.1.2. As expected, the linear model does not have enough freedom to capture the non-Ia distribution.

for general biases in the probabilities of the Ia points. One might still question, however, how sensitive BEAMS is to the input probability of the objects.

For the Gaussian simulation, where we assign the probabilities, P_{Ia}, we can directly change the relationship between the true *underlying* distribution of the types (i.e. the ratio of Ias to non-Ias in the sample) and the *input probability value* (the number we input into the BEAMS algorithm as the P_{Ia}). If the probabilities are unbiased then the distribution of types should follow the probability distribution of the data, in other words 60% of the points with $P_{\text{Ia}} = 0.6$ should be Type Ia SNe. This is the standard case. We then modify the probabilities by assigning a probability of $P_{\text{Ia}} = 0.3$ to all points (which we know will be biased since the mean probability of the sample is 0.667).

We compare the constraints in the two cases in Figure 4.7. If we ignore all probability information and set it to a (biased) value of $P_{\text{Ia}} = 0.3$, the probability information is essentially controlled by the normalization parameter. A tends to a value of 4.7, which, when inserted into Equation (4.9) yields a 'normalized' probability

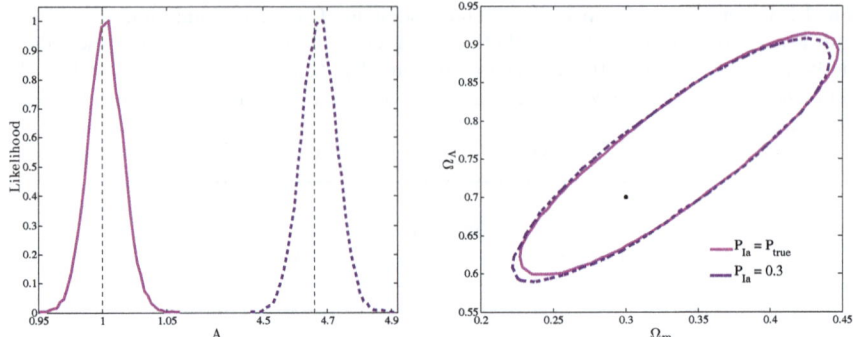

Fig. 4.7 BEAMS corrects for biased input probability: the marginalized one-dimensional likelihood for the normalization parameter A (top panel) and estimated contours (bottom panel) for Level I Gaussian simulation under two forms of the probability distribution. The pink curve and contours correspond to the nominal case, where the probabilities are generated in a linear model, and the types are assigned according to the probabilities. The purple dashed contours correspond to assigning a probability of $P_{\text{Ia}} = 0.3$ to all points. The dashed vertical lines show the expected value of the parameter A such that the true input mean probability of $P_{\text{Ia}} = 0.667$ is recovered. Note that the x-axis in the top panel has been shortened to allow for comparison of the two distributions.

of $P_{\text{Ia}} = 0.668$. Hence BEAMS uses the normalization parameter to remap the mean of the *given* probabilities to ones that have a mean that fits the true *unbiased* probabilities. In correcting for this effect, BEAMS manages to recover cosmological parameters consistent with the unbiased case.

4.4 Results from the SDSS-II SN data

The Sloan Digital Sky Survey Supernova Search operated for three, three-month long seasons during 2005 to 2007. We use the photometric supernovae from all three seasons of the SDSS-II SN survey which also had host galaxy redshifts from the SDSS survey. The analysis and cosmological interpretation of the first season of data (hereafter Fall 2005) are described in [6, 18, 26] and [27]. The SDSS CCD camera is located on a 2.5 m telescope at the Apache Point Observatory in New Mexico. The camera operated in the five Sloan optical bands *ugriz* [28]. The telescope made repeated drift scans of Stripe 82, a roughly 300 square degree region centered on the celestial equator in the Southern Galactic hemisphere, with a cadence of roughly four to five days, accounting for problems with weather and instrumentation.

The images were scanned and objects were flagged as candidate supernovae [24]. Candidate light-curves were compared to a set of supernova light-curve templates in the g, r, i bands (consisting of both core-collapse and Type Ia supernovae) as a function of redshift, intrinsic luminosity and extinction. Likely SNIa candidates were

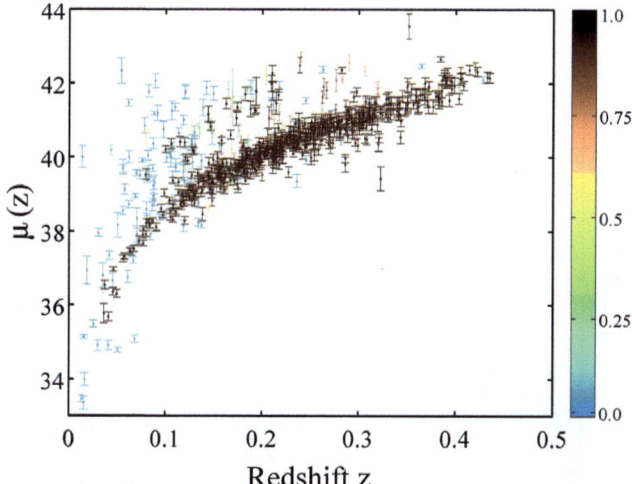

Fig. 4.8 SDSS-II SN data: the photometric sample of the full three seasons of SDSS-II SN survey. The 792 points are all those with host galaxy spectroscopic redshifts. The sample includes 297 spectroscopically confirmed SNe, and the data points are color coded using probabilities from the PSNID typer [24, 25] from low (blue) to high (dark brown).

preferentially followed up with spectroscopic observations of both the candidates and their host galaxies (where possible) on various larger telescopes (see [24]).

In addition to the spectroscopically confirmed SNeIa discovered in the SDSS-II SN, many high-quality candidates without spectroscopic confirmation (i.e. only photometric observations were made of the SNe) but which, by chance, have a host galaxy spectroscopic redshift, are present in the SDSS sample[2].

We include these SNe in both the *cut* sample and the full *photo* sample, but do not set the probabilities of the spectroscopically confirmed *spectro* sample points to unity in the latter. These supernovae are fit with the MLCS2k2 model [23] to obtain a distance modulus for each supernova, *assuming* the supernova is a type Ia.

As outlined in Section 4.3.3, we impose the standard selection cuts on the probability of the fit to the MLCS2k2 light-curve template $P_{\text{fit}} > 0.01$ and $\Delta > -0.4$ to all data, and require that the data used have spectroscopic host galaxy redshift information. Applying these cuts to the full three year data yields a photometric sample of 792 SNe, with a spectroscopic subsample of 297 SNe. The *spectro* sample consists of the objects which have been spectroscopically confirmed by other ground-based telescopes, while the *cut* sample consists of the data points which have a typer probability of $P_{\text{typer}} > 0.9$ and a goodness-of-fit to the light-curve templates within the PSNID typer [24, 25], $\chi^2_{lc} < 1.8$.

---

[2] The BOSS survey recently obtained host galaxy redshifts of all high-quality SN candidates from all three seasons of the SDSS-II Supernova Search. This work does not use the additional BOSS information and only uses the host galaxy redshifts obtained during the running of the SDSS-II survey.

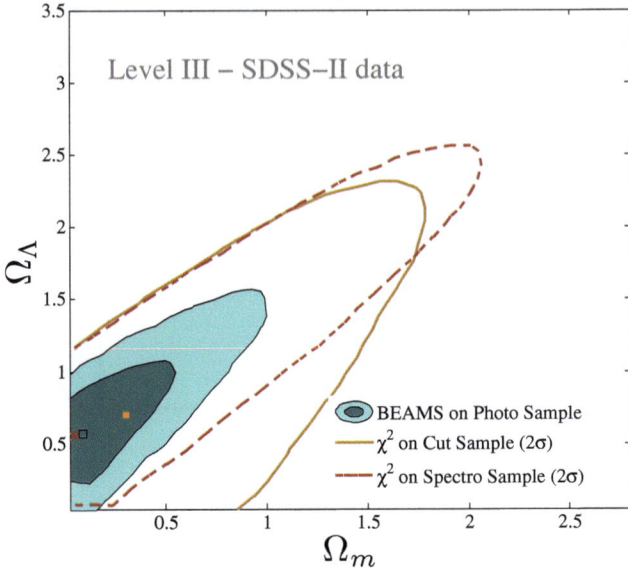

Fig. 4.9 Analysis of the SDSS-II SN data: We show the 2σ contours in the Ω_m, Ω_Λ plane for the SDSS-II SN data. All the constraints – BEAMS (filled contours with best fit indicated by black open square), the 'spectroscopic' sample (dashed contours, best-fit indicated by brown cross) and the 'cut' sample – are consistent with the concordance cosmology (filled brown square). The best-fit BEAMS point is given by the black square, while the best-fit cosmology from the spectroscopic data is indicated by the brown cross. BEAMS provides the smallest contours on the SDSS-II data set, while still being consistent with the constraints from the spectroscopic subsample.

As is shown in Figure 4.9, BEAMS estimates parameters consistent with the *spectro* sample as well as the concordance cosmology in the case of the SDSS-II SN data. Moreover, the BEAMS contours are three times smaller than when using the *spectro* sample alone. In the Gaussian simulations (see Figure 4.4), the BEAMS contours using all the points are $\simeq 16\%$ of the size of the *spectro* sample. This highlights the potential of photometric supernova cosmology to drastically reduce the size of error contours with larger samples while remaining unbiased relative to the 'known' spectroscopic case.

4.5 Conclusions and Outlook

Bayesian Estimation Applied to Multiple Species (BEAMS) is a statistically robust method for parameter estimation in the presence of contamination. The key power of BEAMS is in the fact that it makes use of all available data, hence reducing the statistical error of the measurement, whether or not the purity of the sample can be guaranteed. Rather than discarding data, the probability that the data are "pure" is used as a weight in the full Bayesian posterior, reducing potential bias from the interloper distribution.

Here we have presented the algorithm, and discussed in some detail the role of the probabilities. We have tested BEAMS on an ideal Gaussian simulation of two populations and demonstrated that it recovers the input parameters. We have also shown that the BEAMS errors scale as expected with sample size, and that it provides smaller errors than some of the traditional approaches. Using the Gaussian simulation we have further verified that we can detect the correct form of the non-Ia likelihood and correct for a bias in the probabilities.

We have then applied BEAMS to the SDSS-II SN data set of 792 SNe, using photometric data points with host galaxy spectroscopic redshifts, and showed that the BEAMS contours are three times smaller than those obtained when using only the spectroscopically confirmed sample of 297 SNe Ia.

We have restricted ourselves to the binomial case of a SN-Ia population and one general core-collapse, or non-Ia, population. While this assumption is valid for the SDSS-II SN data, we expect that for larger samples a more complicated model with at least two separate non-Ia Gaussians is more appropriate. On the other hand, large supernova surveys will not only increase the total number of type Ia SNe candidates, but will also allow to investigate systematics about the SNe populations directly. The BEAMS algorithm is designed to include and adapt to information about the non-Ia population easily. By adapting the form of the non-Ia population, and including more than one population group, one could use BEAMS to gain insight into the contaminant distribution.

As we move into the era of huge astronomical surveys that will provide data on thousands of supernovae, BEAMS provides a platform to learn more about the SN populations while at the same time tackling the fundamental questions about the constituents of the universe.

Acknowledgements We thank Michelle Knights and the SDSS-II SN team (especially Rick Kessler, John Marriner and Masao Sako) for helpful comments. RH thanks Jo Dunkley, Olaf Davis, David Marsh, Sarah Miller and Joe Zuntz for useful discussions, and thanks the Kavli Institute for Cosmological Physics, Chicago, the South African Astronomical Observatory, the University of Cape Town, and the University of Geneva for hospitality while this work was being completed. MK would like to thank AIMS for hospitality during part of the work. RH acknowledges funding from the Rhodes Trust and Christ Church. MK acknowledges funding by the Swiss NSF. BB acknowledges funding from the NRF and DST. Part of the numerical calculations for this paper were performed on the Andromeda cluster of the University of Geneva.

Funding for the SDSS and SDSS-II has been provided by the Alfred P. Sloan Foundation, the Participating Institutions, the National Science Foundation, the U.S. Department of Energy, the National Aeronautics and Space Administration, the Japanese Monbukagakusho, the Max Planck Society, and the Higher Education Funding Council for England. The SDSS Web Site is http://www.sdss.org/. The SDSS is managed by the Astrophysical Research Consortium for the Participating Institutions. The Participating Institutions are the American Museum of Natural History, Astrophysical Institute Potsdam, University of Basel, University of Cambridge, Case Western Reserve University, University of Chicago, Drexel University, Fermilab, the Institute for Advanced Study, the Japan Participation Group, Johns Hopkins University, the Joint Institute for Nuclear Astrophysics, the Kavli Institute for Particle Astrophysics and Cosmology, the Korean Scientist Group, the Chinese Academy of Sciences (LAMOST), Los Alamos National Laboratory, the Max-

Planck-Institute for Astronomy (MPIA), the Max-Planck-Institute for Astrophysics (MPA), New Mexico State University, Ohio State University, University of Pittsburgh, University of Portsmouth, Princeton University, the United States Naval Observatory, and the University of Washington.

References

1. Riess, A.G., Filippenko, A.V., Challis, P., Clocchiatti, A., Diercks, A., Garnavich, P.M., Gilliland, R.L., Hogan, C.J., Jha, S., Kirshner, R.P., Leibundgut, B., Phillips, M.M., Reiss, D., Schmidt, B.P., Schommer, R.A., Smith, R.C., Spyromilio, J., Stubbs, C., Suntzeff, N.B., Tonry, J.: Observational Evidence from Supernovae for an Accelerating Universe and a Cosmological Constant. Astron. J. **116**, 1009–1038 (1998). DOI 10.1086/300499. ArXiv:astro-ph/9805201
2. Perlmutter, S., Aldering, G., Goldhaber, G., Knop, R.A., Nugent, P., Castro, P.G., Deustua, S., Fabbro, S., Goobar, A., Groom, D.E., Hook, I.M., Kim, A.G., Kim, M.Y., Lee, J.C., Nunes, N.J., Pain, R., Pennypacker, C.R., Quimby, R., Lidman, C., Ellis, R.S., Irwin, M., McMahon, R.G., Ruiz-Lapuente, P., Walton, N., Schaefer, B., Boyle, B.J., Filippenko, A.V., Matheson, T., Fruchter, A.S., Panagia, N., Newberg, H.J.M., Couch, W.J., The Supernova Cosmology Project: Measurements of Omega and Lambda from 42 High-Redshift Supernovae. Astrophys. J. **517**, 565–586 (1999). DOI 10.1086/307221. ArXiv:astro-ph/9812133
3. Cooke, J., Sullivan, M., Barton, E., Bullock, J., Carlberg, R., et al.: Type IIn supernovae at z 2 from archival data. Nature **460**, 237–239 (2009). ArXiv:0907.1928
4. Komatsu, E., et al.: Seven-Year Wilkinson Microwave Anisotropy Probe (WMAP) Observations: Cosmological Interpretation. Astrophys. J. Suppl. **192**, 18 (2011). DOI 10.1088/0067-0049/192/2/18. ArXiv:1001.4538
5. Holtzman, J.A., Marriner, J., Kessler, R., Sako, M., Dilday, B., Frieman, J.A., Schneider, D.P., Bassett, B., Becker, A., Cinabro, D., DeJongh, F., Depoy, D.L., Doi, M., Garnavich, P.M., Hogan, C.J., Jha, S., Konishi, K., Lampeitl, H., Marshall, J.L., McGinnis, D., Miknaitis, G., Nichol, R.C., Prieto, J.L., Riess, A.G., Richmond, M.W., Romani, R., Smith, M., Takanashi, N., Tokita, K., van der Heyden, K., Yasuda, N., Zheng, C.: The Sloan Digital Sky Survey-II: Photometry and Supernova IA Light Curves from the 2005 Data. Astron. J. **136**, 2306–2320 (2008). DOI 10.1088/0004-6256/136/6/2306. ArXiv:0908.4277
6. Kessler, R., Becker, A.C., Cinabro, D., Vanderplas, J., Frieman, J.A., Marriner, J., Davis, T.M., Dilday, B., Holtzman, J., Jha, S.W., Lampeitl, H., Sako, M., Smith, M., Zheng, C., Nichol, R.C., Bassett, B., Bender, R., Depoy, D.L., Doi, M., Elson, E., Filippenko, A.V., Foley, R.J., Garnavich, P.M., Hopp, U., Ihara, Y., Ketzeback, W., Kollatschny, W., Konishi, K., Marshall, J.L., Mc Millan, R.J., Miknaitis, G., Morokuma, T., Mörtsell, E., Pan, K., Prieto, J.L., Richmond, M.W., Riess, A.G., Romani, R., Schneider, D.P., Sollerman, J., Takanashi, N., Tokita, K., van der Heyden, K., Wheeler, J.C., Yasuda, N., York, D.: First-Year Sloan Digital Sky Survey-II Supernova Results: Hubble Diagram and Cosmological Parameters. Astrophys. J. Suppl. **185**, 32–84 (2009). DOI 10.1088/0067-0049/185/1/32. ArXiv:0908.4274
7. Abell, P.A., et al.: LSST Science Book, Version 2.0. ArXiv e-prints (2009). ArXiv:0912.0201
8. Kunz, M., Bassett, B.A., Hlozek, R.A.: Bayesian estimation applied to multiple species. Phys. Rev. D **75**(10), 103,508 (2007). DOI 10.1103/PhysRevD.75.103508. ArXiv:astro-ph/0611004
9. Newling, J., Bassett, B., Hlozek, R., Kunz, M., Smith, M., et al.: Parameter Estimation with BEAMS in the presence of biases and correlations. ArXiv e-prints (2011). ArXiv:1110.6178
10. Hlozek, R., Kunz, M., Bassett, B., Smith, M., Newling, J., et al.: Photometric Supernova Cosmology with BEAMS and SDSS-II. ArXiv e-prints (2011). ArXiv:1111.5328
11. Press, W.H.: Understanding data better with Bayesian and global statistical methods. ArXiv e-prints (1996). Astro-ph/9604126
12. Gong, Y., Cooray, A., Chen, X.: Cosmology with Photometric Surveys of Type Ia Supernovae. Astrophys. J. **709**, 1420–1428 (2010). DOI 10.1088/0004-637X/709/2/1420. ArXiv:0909.2692

13. Kessler, R., Conley, A., Jha, S., Kuhlmann, S.: Supernova Photometric Classification Challenge. ArXiv e-prints (2010). ArXiv:1001.5210
14. Newling, J., Varughese, M., Bassett, B., Campbell, H., Hlozek, R., et al.: Statistical Classification Techniques for Photometric Supernova Typing. Mon. Not. Roy. Astron. Soc. **414**, 1987–2004 (2011). DOI 10.1111/j.1365-2966.2011.18514.x. ArXiv:1010.1005
15. Kessler, R., Bassett, B., Belov, P., Bhatnagar, V., Campbell, H., et al.: Results from the Supernova Photometric Classification Challenge. Publ. Astron. Soc. Pac. **122**, 1415–1431 (2010). DOI 10.1086/657607. ArXiv:1008.1024
16. Dodelson, S.: Modern cosmology. Academic Press (2003). http://home.fnal.gov/~dodelson/book.html
17. Marriner, J., Bernstein, J.P., Kessler, R., Lampeitl, H., Miquel, R., Mosher, J., Nichol, R.C., Sako, M., Schneider, D.P., Smith, M.: A More General Model for the Intrinsic Scatter in Type Ia Supernova Distance Moduli. Astrophys. J. **740**, 72 (2011). DOI 10.1088/0004-637X/740/2/72. ArXiv:1107.4631
18. Lampeitl, H., Nichol, R.C., Seo, H., Giannantonio, T., Shapiro, C., Bassett, B., Percival, W.J., Davis, T.M., Dilday, B., Frieman, J., Garnavich, P., Sako, M., Smith, M., Sollerman, J., Becker, A.C., Cinabro, D., Filippenko, A.V., Foley, R.J., Hogan, C.J., Holtzman, J.A., Jha, S.W., Konishi, K., Marriner, J., Richmond, M.W., Riess, A.G., Schneider, D.P., Stritzinger, M., van der Heyden, K.J., VanderPlas, J.T., Wheeler, J.C., Zheng, C.: First-year Sloan Digital Sky Survey-II (SDSS-II) supernova results: consistency and constraints with other intermediate-redshift datasets. ArXiv e-prints (2009). ArXiv:0910.2193
19. Jaynes, E.T., Bretthorst, G.L.: Probability Theory, by E. T. Jaynes and Edited by G. Larry Bretthorst. Cambridge University Press (2003)
20. Falck, B.L., Riess, A.G., Hlozek, R.: Characterizing the Contaminating Distance Distribution for Bayesian Supernova Cosmology. Astrophys. J. **723**, 398–408 (2010). DOI 10.1088/0004-637X/723/1/398. ArXiv:1009.1903
21. Metropolis, N., Rosenbluth, A., Rosenbluth, M., Teller, A., Teller, E.: Equation of state calculations by fast computing machines. J. Chem. Phys. **21**, 1087–1092 (1953). DOI 10.1063/1.1699114
22. Dunkley, J., Bucher, M., Ferreira, P.G., Moodley, K., Skordis, C.: Fast and reliable Markov chain Monte Carlo technique for cosmological parameter estimation. Mon. Not. R. Astron. Soc. **356**, 925–936 (2005). DOI 10.1111/j.1365-2966.2004.08464.x. ArXiv:astro-ph/0405462
23. Jha, S., Riess, A.G., Kirshner, R.P.: Improved Distances to Type Ia Supernovae with Multicolor Light-Curve Shapes: MLCS2k2. Astrophys. J. **659**, 122–148 (2007). DOI 10.1086/512054. ArXiv:astro-ph/0612666
24. Sako, M., Bassett, B., Becker, A., Cinabro, D., DeJongh, F., Depoy, D.L., Dilday, B., Doi, M., Frieman, J.A., Garnavich, P.M., Hogan, C.J., Holtzman, J., Jha, S., Kessler, R., Konishi, K., Lampeitl, H., Marriner, J., Miknaitis, G., Nichol, R.C., Prieto, J.L., Riess, A.G., Richmond, M.W., Romani, R., Schneider, D.P., Smith, M., Subba Rao, M., Takanashi, N., Tokita, K., van der Heyden, K., Yasuda, N., Zheng, C., Barentine, J., Brewington, H., Choi, C., Dembicky, J., Harnavek, M., Ihara, Y., Im, M., Ketzeback, W., Kleinman, S.J., Krzesiński, J., Long, D.C., Malanushenko, E., Malanushenko, V., McMillan, R.J., Morokuma, T., Nitta, A., Pan, K., Saurage, G., Snedden, S.A.: The Sloan Digital Sky Survey-II Supernova Survey: Search Algorithm and Follow-Up Observations. Astron. J. **135**, 348–373 (2008). DOI 10.1088/0004-6256/135/1/348. ArXiv:0708.2750
25. Sako, M., Bassett, B., Connolly, B., Dilday, B., Campbell, H., Frieman, J., Gladney, L., Kessler, R., Lampeitl, H., Marriner, J., Miquel, R., Nichol, R., Schneider, D., Smith, M., Sollerman, J.: Photometric SN Ia Candidates from the Three-Year SDSS-II SN Survey Data. ArXiv e-prints (2011). ArXiv:1107.5106
26. Frieman, J.A., Bassett, B., Becker, A., Choi, C., Cinabro, D., DeJongh, F., Depoy, D.L., Dilday, B., Doi, M., Garnavich, P.M., Hogan, C.J., Holtzman, J., Im, M., Jha, S., Kessler, R., Konishi, K., Lampeitl, H., Marriner, J., Marshall, J.L., McGinnis, D., Miknaitis, G., Nichol, R.C., Prieto, J.L., Riess, A.G., Richmond, M.W., Romani, R., Sako, M., Schneider, D.P., Smith, M., Takanashi, N., Tokita, K., van der Heyden, K., Yasuda, N., Zheng, C., Adelman-

McCarthy, J., Annis, J., Assef, R.J., Barentine, J., Bender, R., Blandford, R.D., Boroski, W.N., Bremer, M., Brewington, H., Collins, C.A., Crotts, A., Dembicky, J., Eastman, J., Edge, A., Edmondson, E., Elson, E., Eyler, M.E., Filippenko, A.V., Foley, R.J., Frank, S., Goobar, A., Gueth, T., Gunn, J.E., Harvanek, M., Hopp, U., Ihara, Y., Ivezić, Ž., Kahn, S., Kaplan, J., Kent, S., Ketzeback, W., Kleinman, S.J., Kollatschny, W., Kron, R.G., Krzesiński, J., Lamenti, D., Leloudas, G., Lin, H., Long, D.C., Lucey, J., Lupton, R.H., Malanushenko, E., Malanushenko, V., McMillan, R.J., Mendez, J., Morgan, C.W., Morokuma, T., Nitta, A., Ostman, L., Pan, K., Rockosi, C.M., Romer, A.K., Ruiz-Lapuente, P., Saurage, G., Schlesinger, K., Snedden, S.A., Sollerman, J., Stoughton, C., Stritzinger, M., Subba Rao, M., Tucker, D., Vaisanen, P., Watson, L.C., Watters, S., Wheeler, J.C., Yanny, B., York, D.: The Sloan Digital Sky Survey-II Supernova Survey: Technical Summary. Astron. J. **135**, 338–347 (2008). DOI 10.1088/0004-6256/135/1/338. ArXiv:0708.2749
27. Sollerman, J., Mörtsell, E., Davis, T.M., Blomqvist, M., Bassett, B., Becker, A.C., Cinabro, D., Filippenko, A.V., Foley, R.J., Frieman, J., Garnavich, P., Lampeitl, H., Marriner, J., Miquel, R., Nichol, R.C., Richmond, M.W., Sako, M., Schneider, D.P., Smith, M., Vanderplas, J.T., Wheeler, J.C.: First-Year Sloan Digital Sky Survey-II (SDSS-II) Supernova Results: Constraints on Nonstandard Cosmological Models. Astrophys. J. **703**, 1374–1385 (2009). DOI 10.1088/0004-637X/703/2/1374. ArXiv:0908.4276
28. Fukugita, M., Ichikawa, T., Gunn, J.E., Doi, M., Shimasaku, K., Schneider, D.P.: The Sloan Digital Sky Survey Photometric System. Astron. J. **111**, 1748 (1996). DOI 10.1086/117915

Chapter 5
Gaussian Random Fields in Cosmostatistics*

Benjamin D. Wandelt

Abstract A gentle and pedagogical introduction to Gaussian random fields and their applications in cosmological statistics.

5.1 Motivation

Gaussian random fields (GRFs) are ubiquitous in cosmological statistics. There are good physical reasons for this. Calculations of quantum fluctuations produced during the epoch of inflation predict very nearly Gaussian primordial perturbations. Even for models which are said to produce "large" non-Gaussianity, deviations from Gaussianity are limited by observational tests to less than fractions of a percent. To very high accuracy the cosmic microwave background is a linear map of the primordial perturbations. We will see below that linear maps preserve Gaussianity—so to the extent that the primordial perturbations are Gaussian, the Cosmic Microwave Background (CMB) is a Gaussian field on the sphere. Even the large scale distribution of galaxies ("large scale structure" or LSS) can be approximately modeled as a Gaussian random field, at least on very large scales, where gravitational evolution is still well-described in the linear approximation. The goal of comostatistics is to extract cosmological information from observations of these stochastic fluctuations across all scales, illustrated in Figure 5.1.

Benjamin D. Wandelt
Institut d'Astrophysique de Paris, Université Pierre et Marie Curie, 98 Boulevard Arago 75014 Paris, France,
e-mail: wandelt@iap.fr

* These notes originated as a supplement to the first four lectures on "Cosmostatistics and Large Surveys" given at the Institut d'Astrophysique de Paris at the Université Pierre et Marie Curie in February 2011.

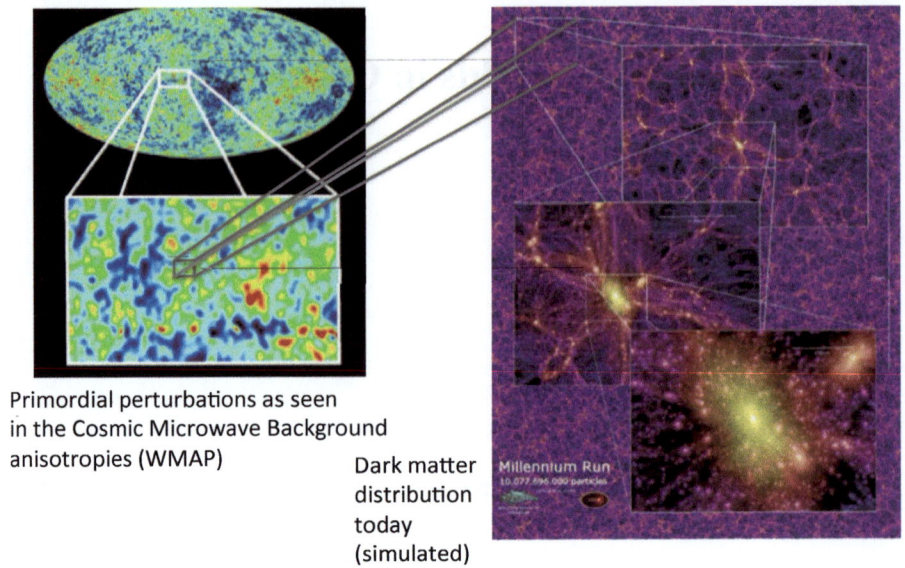

Fig. 5.1 The goal of cosmostatistics is to infer cosmological information from observables of stochastic fields such as the cosmic microwave background or the large scale distribution of matter in the Universe.

5.2 What is a Gaussian Random Field?

Defining GRFs is the easy part, at least if you allow me to talk about finite-dimensional GRFs. If you are faced with a continuous GRF (which is infinite-dimensional), simply discretize it. If you choose your discretization scale to be small enough you will not lose any information. In practice, any observation you might want to model as a GRF will already be discretized.

An n-dimensional vector x is a GRF (I will often say "is Gaussian") with mean μ and covariance C if it has the following probability density function (pdf)

$$p(x|\mu, C) = \frac{e^{\frac{1}{2}(x-\mu)^T C^{-1}(x-\mu)}}{\sqrt{|2\pi C|}}. \tag{5.1}$$

It is easy to check that the mean is really μ and the covariance is really C: just do the Gaussian integrals. You just show that

$$\langle x \rangle = \int_x x p(x|\mu, C) dx = \mu \tag{5.2}$$

$$\langle (x-\mu)(x-\mu)^T \rangle = \int_x (x-\mu)(x-\mu)^T p(x|\mu, C) dx = C \tag{5.3}$$

by doing the integrals (dx is the n-dimensional volume element). In fact I will not do this here, though you are welcome to train your mathematical muscle doing this

5 Gaussian Random Fields in Cosmostatistics

(Exercise!), because there is a much more powerful way to do Gaussian integrals like this. I will show you two tricks that make working with GRFs much easier than with other random fields, namely what I like to call *integration by differentiation*.

5.2.1 Trick for Doing Gaussian Integrals: Integration by Differentiation

First, consider Eq. (5.2). Since the Gaussian pdf has only one global maximum, and it is symmetric about the mean, we can find the mean by finding the maximum (often called the *mode* of the distribution). Since the exponential is a monotonic function, and the denominator of the pdf does not depend on x, we can just find the maximum of the exponent.

$$\partial_x \left(\frac{1}{2} (x_{\max} - \mu)^T C^{-1} (x_{\max} - \mu) \right) = 0 \qquad (5.4)$$

$$C^{-1}(x_{\max} - \mu)) = 0 \qquad (5.5)$$

So as long as C^{-1} does not have any zero eigenvalues (what would that mean? *(Exercise!)*) we find $x_{\max} = \langle x \rangle = \mu$. No integration required!

Once you have proven to yourself the long way round that Eq. (5.3) is correct you can use an even easier short cut to find the covariance of any complicated Gaussian pdf you might come across without doing any integration. Looking at the pdf what would you do to extract C from it? It is easy to verify that

$$\partial_x \partial_{x^T} (\ln p(x|\mu, C)) = -C^{-1} \qquad (5.6)$$

So the recipe for finding the covariance of any quantity is to look at the exponent, select the coefficient matrix of all quadratic terms in that quantity, invert and multiply by -1.

5.3 Generating GRFs with a Given Mean and Covariance

If you would actually like to simulate a model universe you need some way to generate a *realization* of the initial perturbations. In other applications, e.g., when actually working with data from large surveys of the CMB or LSS, generating fake data sets with known inputs gives a way to test whether these inputs are recovered by the data analysis, or to calibrate the analysis.

There are any number of software packages that allow generating single Gaussian random variates with mean 0 and variance 1, i.e., *normal variates*, e.g., using the well-known Box-Müller method. Using an n-vector ξ of such normal variates we can generate random realizations of a GRF with covariance C and mean μ by simply taking any matrix $\sqrt{C}$ that satisfies $\sqrt{C}\sqrt{C}^T = C$ and computing

$$x = \sqrt{C}\xi + \mu. \tag{5.7}$$

One general way to generate $\sqrt{C}$ under the condition that C has only positive definite eigenvalues is to use the so-called Cholesky decomposition, implemented in many numerical packages. It is easy to verify that x has the right mean and covariance using that $\langle \xi \xi^T \rangle = \mathbf{1}$.

5.3.1 Higher Order Moments of Gaussian Random Fields

So we can calculate $\langle x \rangle$ and $\langle xx^T \rangle$. What about higher order moments? Let us focus on central moments, e.g., $\langle (x-\mu)(x-\mu)^T \rangle$, since it is always easy to put back the mean back in. Equivalently, we look at the moments for $\mu = 0$, which we will assume until further notice:

$$\mu = 0 \quad \text{from now on.} \tag{5.8}$$

I will also put back the explicit indices on the vectors.

Any odd (central) moments, e.g., the third ($\langle x_i x_j x_k \rangle$), fifth ($\langle x_i x_j x_k x_l x_m \rangle$), etc., are obviously zero by symmetry.

The higher even ones (e.g., the fourth, sixth, etc.) can be evaluated through brute force calculation or through the application of Wick's theorem. Simply connect up all pairs of xs and write down the covariance matrix for each pair. Example:

$$\begin{aligned}\langle x_i x_j x_k x_l \rangle &= \langle x_i x_j \rangle \langle x_k x_l \rangle + \langle x_i x_k \rangle \langle x_j x_l \rangle + \langle x_i x_l \rangle \langle x_j x_k \rangle \\ &= C_{ij}C_{kl} + C_{ik}C_{jl} + C_{il}C_{jk}\end{aligned} \tag{5.9}$$

5.4 Marginals and Conditionals for GRFs

Let us say you model some physical system as a Gaussian random field and your model predicts its mean and covariance. Possible examples are many: inflation as a model of the large scale perturbations in the Universe, with zero mean and homogeneous and isotropic covariance (one of the main subjects of this chapter); the gas velocities in a molecular cloud (quite plausible); hourly prices of your favorite stock this week (I would like to see your model!).

In any probabilistic model of this type one can ask and answer questions regarding the *marginals* and *conditionals* of the random field. Let us take a simple example and model the current temperature in Paris T_P and in London T_L as random field with $n = 2$ at any given time with pdf $p(T_P, T_L)$.

5 Gaussian Random Fields in Cosmostatistics

You already know from probability theory that

$$p(T_P, T_L) = p(T_P|T_L)p(T_L) = p(T_L|T_P)p(T_P). \tag{5.10}$$

This identity is nothing other than the decomposition of the joint pdf into products of conditional and marginal densities, which we will define now.

Let us say I am interested in the temperature in Paris but I know nothing of the temperature in London. Then the model predicts the *marginal* pdf $p(T_P) = \int dT_L\, p(T_P, T_L)$.

Now I measure T_P very precisely and would like to guess T_L. What does my model tell me? I am interested in the pdf of T_L "given" T_P, so the conditional density $p(T_L|T_P)$ which we can easily calculate from the identity above. The Mathematica notebook in Figure 5.2 gives examples of all these cases.

```
mu = {21, 24}
covar = N[{{4, 1.8}, {1.8, 1}}]
{21, 24}
{{4., 1.8}, {1.8, 1.}}
pdf[x_, y_] :=
  Exp[-1/2 ({x, y} - mu).Inverse[covar].({x, y} - mu)] / Sqrt[Det[2 Pi covar]]
Plot3D[pdf[x, y], {x, 14, 28}, {y, 20, 28},
  PlotRange -> All, PlotPoints -> {50, 50}, AxesLabel -> {x, y}]
```

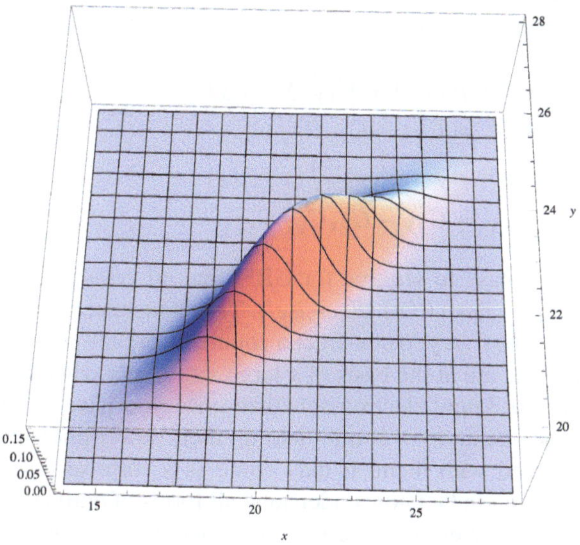

```
MarginalX[x_] := NIntegrate[pdf[x, y], {y, -Infinity, Infinity}]
MarginalY[y_] := NIntegrate[pdf[x, y], {x, -Infinity, Infinity}]
ConditionalXgivenY[x_, y_] := pdf[x, y] / MarginalY[y]
ConditionalYgivenX[y_, x_] := pdf[x, y] / MarginalX[x]
```

Fig. 5.2 A short Mathematica notebook illustrating the notions of 1-D marginal and conditional densities of a 2-D joint density.

Easy computation of marginal and conditional pdfs is a very convenient property of GRFs. First of all, all marginal and conditional densities of GRFs are Gaussian. So all we need to calculate are their means and covariances. Let us the split the GRF up into two parts x and y, so that

$$\mu = \begin{pmatrix} \mu_x \\ \mu_y \end{pmatrix} \tag{5.11}$$

and

$$C = \begin{pmatrix} C_{xx} & C_{xy} \\ C_{yx} & C_{yy} \end{pmatrix} \tag{5.12}$$

$C_{xy} = C_{yx}$ by symmetry.

First for the marginal pdfs,

$$\mu_x = \mu_x \tag{5.13}$$
$$C_{xx} = C_{xx} \tag{5.14}$$
$$\mu_y = \mu_y \tag{5.15}$$
$$C_{yy} = C_{yy}. \tag{5.16}$$

I know these expressions are entirely tautological, but the point is obvious: the marginal mean and marginal covariances are just the corresponding parts of the joint mean and covariance.

Less trivially, for the conditional means

$$\mu_{x|y} = \mu_x + C_{xy} C_{yy}^{-1} (y - \mu_y) \tag{5.17}$$
$$C_{x|y} = C_{xx} - C_{xy} C_{yy}^{-1} C_{yx} \tag{5.18}$$
$$\mu_{y|x} = \mu_y + C_{yx} C_{xx}^{-1} (x - \mu_x) \tag{5.19}$$
$$C_{y|x} = C_{yy} - C_{yx} C_{xx}^{-1} C_{xy}. \tag{5.20}$$

From these expressions it is easy to see that for GRFs lack of covariance implies *independence*, i.e., $P(x, y) = P(x)P(y)$ *(Exercise!)*. This is most certainly not the case for general random fields. Can you construct a counterexample? *(Exercise!)*

These innocuous equations are at the basis of a large set of very powerful methods. But just like you do not know electromagnetism once you have memorized Maxwell's equations, these equations will not yield up their power until you have become intimately acquainted with them. It may come as a surprise that having understood these equations you in fact now understand all forms of optimal filtering (in the least square sense) and the entire Bayesian linear model. After some meditation over these formulas and with just a little additional reading you can figure out the ideas behind 90% of scientific data compression, interpolation, extrapolation, and many surprisingly powerful data analysis tools.

5.5 Transformations of Random Processes

In case you are interested in exploring beyond Gaussian random fields, or to understand better the range of applicability of GRFs it is useful to be able to compute the pdfs for transformations of Gaussian random fields. For me, the easiest way to think about transformations of a random field in general is to write down a delta function that deterministically links the transformed field to the original field

$$\delta^n(z - f(x)) \tag{5.21}$$

and then to explicitly compute the integral

$$p(z) = \int dx \, \delta^n(z - f(x)) p(x). \tag{5.22}$$

You can read integrals like this in a very intuitive fashion as follows. The $p(x)$-weighted integral describes the Monte Carlo simulation that you would have to perform to generate a histogram of z from $p(z)$. You draw stochastically from $p(x)$ and the delta function tells you to transform deterministically from x to z.

The bad news is that this integral can only be done analytically in a very limited number of special cases (though it is always worth trying, just in case).

For the special case of a linear transformation $z = Lx$ this integral *can* be done. Use this fact to show that any linear combination of the elements of a GRF is Gaussian, by showing that the transformed density is still of right form to define a GRF *(Exercise!)*.

5.6 GRFs with Special Symmetries

Standard inflation produces a homogeneous and isotropic space on all scales accessible to observation, so we would like to be able to model GRFs that respect these symmetries. It follows that the parameters μ and C of the Gaussian random field modeling inflationary perturbations have to respect these symmetries.

What does homogeneity mean? Homogeneity is invariance under translations. It is easy to see that this implies $\mu = const.$ Since we are modeling fluctuations, $\mu = 0$.

For the covariance between the fluctuations at pairs of space points r_i and r_j, the translational symmetry means that $C(r_i, r_j) = C(r_i + s, r_j + s)$ for any displacement s (here we think of C as a function of two variables). The symmetry allows reducing the number of variables from two to one, since $C_{ij} = C(r_i, r_j) = C(r_i - r_j) = C(r_j - r_i) = C(\Delta r_{ij})$ satisfies this requirement.

Matrices with this property are quite special and so they have a special name: Toeplitz matrices. Since our covariance matrix is also symmetric they are fully specified by specifying the elements of the first row. (Write down the Toeplitz matrix for a 1-dimensional space with first row $(3, 2, 1)$ *(Exercise!)*).

If we impose periodic boundary conditions on Δ_{ij} (think of space points on a ring, or d-torus in d dimensions) then the elements are further constrained. Such matrices are called *circulant*. (Write down the circulant matrix for a periodic 1-dimensional space, also known as a circle (!), with first row $(3,2,1)$ *(Exercise!)*.[1])

It turns out that circulant matrices are intimately connected with the *Fourier* basis, which should not be surprising since you have probably encountered Fourier bases in the context of periodic functions.

Explicitly, for a 1-D circle with n elements, where $0 \leq \Delta < n$ is an integer measuring the angle between two elements in units of $2\pi/n$ we can expand

$$C(\Delta) = \sum_{k=0}^{n-1} P_k e^{2\pi i k \Delta / n}. \tag{5.23}$$

Note that the reality of C implies $P_{n-k} = P_k^*$ *(Exercise!)*. Symmetry of the covariance together with the periodic boundary conditions implies that $C(n - \Delta) = C(\Delta)$ and hence $P_k = P_{n-k}$ *(Exercise!)*. These two conditions together imply that $P(k)$ is real and that only $\lfloor n/2 \rfloor + 1$ of the P_k are independent (where $\lfloor x \rfloor$ denotes the largest integer less than x).

The P_k have an important physical interpretation. Define the forward Fourier transformation as

$$\tilde{f}_k = \frac{1}{n} \sum_{j=0}^{n-1} f_j e^{-2\pi i k j / n} \tag{5.24}$$

and the inverse transformation as in Eq. (5.23),

$$f_j = \sum_{k=0}^{n-1} \tilde{f}_k e^{2\pi i k j / n}. \tag{5.25}$$

Orthonormality of the Fourier modes gives $\tilde{\tilde{f}}_i = f_i$.

Then it easy to prove (by transforming the two indices of C_{ij}) that the covariance matrix for the Fourier coefficients of a homogeneous (periodic) Gaussian random field is *diagonal* with diagonal elements P_k *(Exercise!)*. This fact is often referred to by the fancy name of Wiener-Khinchin theorem. If you find the analytic proofs tedious you can verify this empirically, along the lines shown in the Mathematica notebook in Figure 5.3 which does this transform for an example of a 4-element circle.

In other words, the k modes are not correlated, because there is no *co*-variance, only variance. In that sense it is easy to see that the variance, or fluctuation power, of

---

[1] If you get stuck doing this, then good! You should run into trouble since there is no solution. You are only free to chose 2 different values, not three due to the additional periodic boundary conditions.

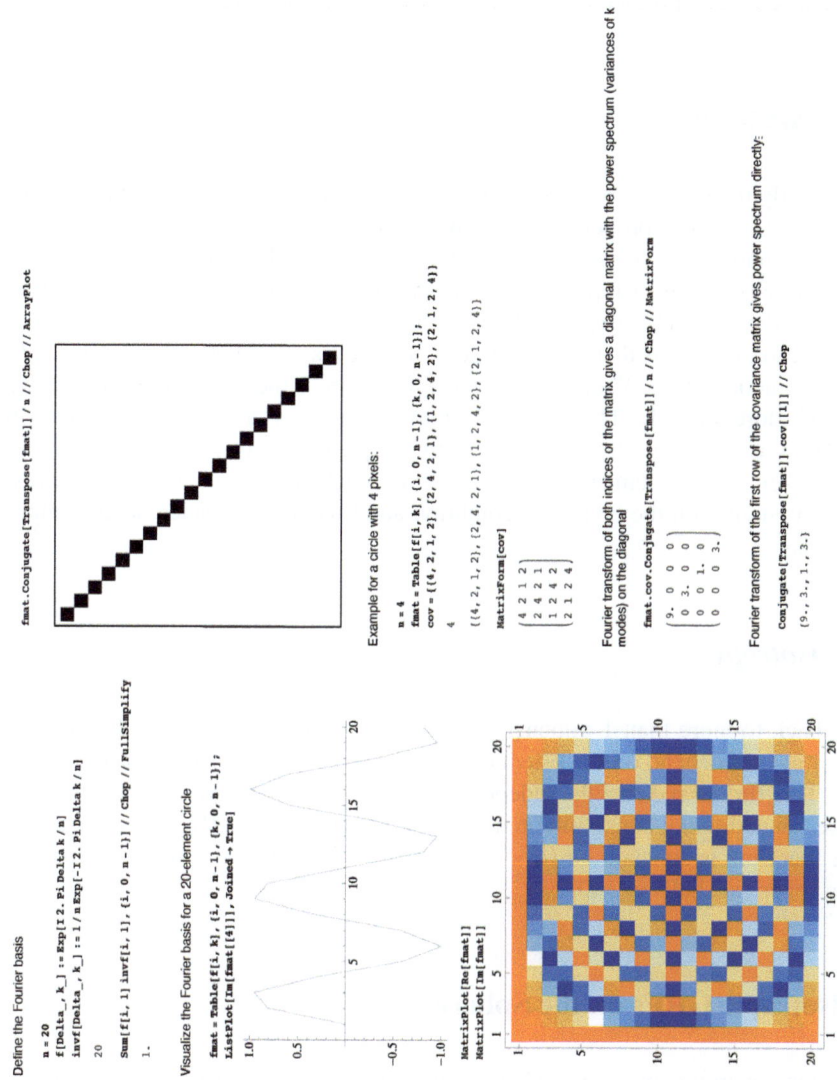

Fig. 5.3 A numerical verification of the Wiener-Khinchin theorem.

the entire GRF can be thought of as being partitioned amongst the different Fourier modes in a mutually exclusive and collectively exhaustive way. For this reason P_k is referred to as the *power spectrum*. Since there are no off-diagonal elements in the Fourier space covariance matrix we can see that the field is very easy to describe in Fourier space. Each Fourier k-mode fluctuates completely independently.

5.6.1 Summary

Let us briefly review what we have done. We started with imposing translation symmetry on the statistical properties of the fluctuations and obtained the conditions the covariance matrix have to satisfy. Then it turned out that all covariance matrices satisfying these properties simplify dramatically when expressed in the Fourier basis. All the coefficients in the expansion decouple.

The reason for this is that Fourier modes provide a systematic and unique way to expand a function on $(\mathcal{S}^1)^n$ or $\mathcal{R}^n$ into terms which break the translational symmetry more and more strongly. Think about it: that is exactly what sines and cosines do as function of increasing frequency. For zero frequency the translational symmetry is not broken. The fundamental mode ($n = 1$) breaks the symmetry at the lowest order consistent with the boundary condition and higher frequencies break it more and more.

5.6.2 Isotropy

Going back to more than 1-dimensional space, if we impose *isotropy* this implies that Δ_{ij} cannot depend on the orientation of the separation between space points i and j. This further reduces the independent elements of the covariance matrix and enforces an additional constraint on the P_k, namely that $P_k = P_{|k|}$. This means the power spectrum is constant on $(d-1)$–spheres in k-space, i.e., on circles for 2-D and on spherical shells in 3-D.

5.7 Isotropic GRFs on the Sphere

An analogous argument on the sphere singles out spherical harmonics as a special basis for the covariance of the CMB anisotropies on the sphere. Again, isotropy requires $\mu = const$ which we can always absorb into the background and use $\mu = 0$.

Let us look at the covariance. For isotropy to hold, the covariance has to be invariant (symmetric) under *rotations*, $C_{ij} \equiv C(n_i, n_j) = C(R(\Psi) \cdot n_i, R(\Psi) \cdot n_j)$, where n_i is a unit vector pointing to the 2-sphere in 3D and $R(\Psi)$ is the rotation specified by the Euler angles Ψ.

5 Gaussian Random Fields in Cosmostatistics

It is easy to verify using the orthogonality of rotations that $C_{ij} = C(n_i \cdot n_j) = C(\cos \sphericalangle(i,j))$ is the form of covariance matrix that fits the bill *(Exercise!)*.

Again, symmetry has reduced the covariance matrix from 2-dimensional to and effectively 1-dimensional object. As a function of $z = \cos\theta$ we can expand the covariance in terms of Legendre polynomials

$$C(z) = \sum_{l=0}^{l_{max}} \frac{2l+1}{4\pi} C_l P_l(z). \tag{5.26}$$

Owing to the orthogonality of Legendre polynomials

$$\int_{z=-1}^{z=1} P_l(z) P_{l'}(z) = \frac{2}{2l+1} \delta_{ll'} \tag{5.27}$$

we can invert this relationship to express the expansion coefficients in terms of the covariance function for two points an angle θ apart $C(z(\theta))$ and obtain

$$C_l = 2\pi \int dz\, C(z) P_l(z). \tag{5.28}$$

On the sphere (S^2), what are the analogous quantities to Fourier modes? Can we similarly find modes that break isotropy systematically, order by order?

It turns out the Spherical Harmonics $Y_{lm}(n)$ are exactly that set of functions. We can expand the CMB temperature anisotropy $T(n)$ in

$$T(n) = \sum_{lm} a_{lm} Y_{lm}(n). \tag{5.29}$$

A convenient notation for the sums over l and m which I have used in the preceding equation and will use in all that follow is to sum over all terms which are non-zero. Since the Y_{lm} are zero for $|m| > l$ there is no need explicitly to write the limits on the sums. In principle, all sums run over all l s.t. $0 \leq l < \infty$, but in all practical cases we will consider the field to be smooth below some scale θ_{smooth} which implies that at any desired level of accuracy there is some c such that we can neglect the a_{lm} for all $l > l_{max} = c/\theta_{smooth}$.

The spherical harmonics are orthonormal on the sphere

$$\int_{S^2} d^2n\, Y_{lm}(n) Y^*_{l'm'}(n) = \delta_{ll'} \delta_{mm'} \tag{5.30}$$

and so we can invert Eq. (5.29) to get

$$a_{lm} = \int d^2n\, T(n) Y^*_{lm}(n). \tag{5.31}$$

This operation is implemented numerically in the HEALPIX [1] code `anafast`, while synthesizing a map $T(n)$ from a set of a_{lm} can be done easily using `synfast`.

They are linked to the Legendre Polynomials through the addition theorem for spherical harmonics

$$\sum_m Y_{lm}(n_1) Y_{lm}^*(n_2) = \frac{(2l+1)}{4\pi} P_l(n_1.n_2). \tag{5.32}$$

Only the $m = 0$ spherical harmonics have support at the pole (it does not make sense for a (scalar) function to have a variation as a function of angle on a single point!). So, if $n_1 = n_2 = \hat{z}$, we get

$$|Y_{l0}(\hat{z})|^2 = \frac{2l+1}{4\pi} \tag{5.33}$$

since the Legendre polynomials are normalized $P_l(1) = 1$. Choosing one direction not to be $\hat{z}$, we get

$$Y_{l0}(\theta) = \sqrt{\frac{(2l+1)}{4\pi}} P_l(\cos\theta). \tag{5.34}$$

With just these formulas we can already do some interesting calculations.

Let us try the following *(Exercise!)*: verify that the C_l are the diagonal elements of the covariance matrix of the a_{lm}. There are a couple of ways of doing this, either starting with the covariance matrix in pixel space (Eq. (5.26)) or by directly calculating the covariance between two a_{lm}.

In other words, let us calculate

$$\langle a_{lm} a_{l'm'}^* \rangle = ? \tag{5.35}$$

The C_l are therefore the analogue quantities for isotropic random fields on the sphere to the power spectrum coefficients P_k we found for homogeneous random fields with periodic boundary conditions. Knowing what we now know about GRFs, we understand the claim, often heard in talks on this subject, or seen in the relevant literature, that for isotropic GRFs on the sphere, the power spectrum contains *all* of the information.

This is why inferring the power spectrum from CMB data forms such a central part of the analysis of CMB data sets. A quantitative description of what we know about the power spectrum (a few thousand numbers) of data from the Planck [2] mission (Terabytes of data) contains all of the information about cosmological parameters from Planck, if the primordial perturbation is Gaussian and isotropic.

5.8 The Cosmic Microwave Background and Primordial Perturbations

Linearizing gravity to first order in the Newtonian potential, which is accurate to about 1 part in 10^5 up to the time of recombination, the CMB is a linear map of the primordial perturbation. Let us write the perturbation in terms of its Fourier mode amplitudes, ϕ_k because they have the simple property of having diagonal covariance, as we have shown,

$$a_{lm} = \int d^3k \, \phi_k g_{lmk}. \tag{5.36}$$

Owing to homogeneity and isotropy, the linear transfer may only depend on l and on the modulus (the length) of $g_{lmk} = g_l(|k|) i^l Y_{lm}^*(\hat{k})$. These transfer functions (also sometimes referred to as $\Delta(l, k)$ in the literature) encode all of the complicated physics, such as the baryon acoustic oscillations of the photon-baryon plasma, the growth of dark matter perturbations, the impact of the geometry of the Universe on the CMB, the impact of anisotropies of the cosmic neutrino background on the CMB, etc. These transfer functions are the objects that cosmological Boltzmann codes computer. Examples of such codes are CAMB [3] or CMBFAST [4] and you can extract these quantities from these codes. Equation 5.36 separates out nicely the *dynamics* and the *initial conditions*. The observed CMB constrains both of these together.

So, to predict the CMB power spectrum we need to work out the C_l from Eq. (5.36). It is an easy but instructive exercise to show that this results in *(Exercise!)*

$$C_l = \int k^2 dk \, g_l(k)^2 P(k). \tag{5.37}$$

At this point it becomes very useful to be able to expand exponentials in terms of spherical harmonics—luckily there is a well-known formula for that:

$$e^{ik \cdot x} = 4\pi \sum_{lm} i^l j_l(|k||x|) Y_{lm}(\hat{x}) Y_{lm}^*(\hat{k}), \tag{5.38}$$

where the $j_l(x)$ are the spherical Bessel functions. This is the crucial link between the 3-D perturbations and the anisotropies on the sphere.

An example application of this formula is to derive the useful alternative to Eq. (5.36) in terms of an explicit radial integral over the light cone

$$a_{lm} = \int dr \, r^2 \phi_{lm}(r) \alpha_l(r) \tag{5.39}$$

where

$$\alpha_l(r) = \frac{2}{\pi} \int dk \, k^2 g_l(k) j_l(|k||r|). \tag{5.40}$$

This is a slightly more challenging calculation, which I will leave as an *(Exercise!)*.

To a rough approximation, which is quite good on large scales, the CMB shows you a spherical slice of the primordial perturbations at the time of last scattering t_{CMB}. This is known as the Sachs–Wolfe effect (derived in a pedagogical way in [5])

$$T(n) = -\frac{1}{3}\phi(n, r_{CMB}) \tag{5.41}$$

Starting from this result we can predict the a_{lm} and hence the C_l of the CMB temperature anisotropies if it were purely due to the Sachs–Wolfe effect. The result is

$$C_l \propto \int k^2 dk P(k) j_l(|k||r_{CMB}|)^2. \tag{5.42}$$

Comparing this with the general equation, we can see that $j_l(|k||r|)$ takes the role of a transfer function. We set up the problem in a purely geometric way. So we learn that the set of spherical Bessel functions play the role of a *geometric* transfer function, which corresponds to just taking a spherical slice through a 3-D field at a distance r.

As a bonus, it turns out that this integral can actually be done analytically for $P(k) \propto 1/k^3$ and the answer is that

$$C_l \propto \frac{1}{l(l+1)}, \tag{5.43}$$

to a good approximation for l larger than a few.

5.8.1 Example Inference Problem: Reconstructing the Primordial Perturbations from the CMB

All this may appear elementary to you, but let me show you that you now have all the tools to solve the problem we solved in [6], published only a few years ago (see [7, 8] for the extension to polarization). In this paper we derived the optimal reconstruction for the primordial perturbations ϕ on any spherical slice from the cosmic microwave background anisotropy in the limit of small non-Gaussianity. You know how to build optimal filters because we understand the conditional densities of Gaussian random fields. Remember that the optimal estimate for y given x is the conditional mean of $p(y|x)$, Eq. (5.17), with the width (or covariance) of this distribution describing the uncertainty of this estimate. From the perspective of GRFs introduced in these lectures, the entire problem can be solved immediately by considering the joint GRF for $z = (T, \phi(r))$ including the CMB and a slices of ϕ at the radius r^* where we would like to reconstruct it *(Exercise!)*.

5.9 Large Scale Structure and the Primordial Perturbations

Now let us see how the large scale density perturbations we observe in galaxy surveys are related to the primordial perturbations. I will present a simplified treatment at first, which considers only the perturbations in the cold dark matter. This approximation ignores the effects of baryons and neutrinos on the dark matter perturbations. Both of these give important effects, which I will describe qualitatively but not derive. For further details, a good and accessible textbook reference is Lyth and Liddle [9].

I have discussed the primordial perturbations in terms of ϕ, a variable which has a well-defined meaning in relativistic perturbation theory and has the virtue of reducing to the gravitational potential on small scales.

Standard inflation produces primordial perturbations with a nearly scale-invariant power spectrum of $P_\phi(k) = \frac{A}{k^{3+(1-n)}}$ where n is the *scalar spectral index*. You can check *(Exercise!)* that for $n = 1$, the variance of ϕ at a space-point can be divided up in to equal amounts from each decade (or constant logarithmic interval) in k-space.

In particular, it is related to the density perturbation $\delta = \frac{\delta\rho}{\rho}$ by the relativistic generalization of Poisson's equation at linear order

$$\frac{k^2}{a^2}\phi_k = 4\pi G\rho\delta_k. \tag{5.44}$$

We can read off the fact that inflation predicts $P_\delta(k) \propto k^n$ for the density perturbations.

The density perturbation grows differently in the radiation dominated epoch and in the matter dominated epoch, so we have to distinguish between these two regimes.

Remember small scale modes enter the Hubble scale first. For small enough modes, this occurs during radiation domination. During that time the growth of perturbations was severely damped, which caused the cold dark matter perturbations merely to grow logarithmically in time.

Modes with $k < k_{eq}$, the wave number corresponding to the size of the Hubble scale matter-radiation equality, entered the Hubble scale during the epoch of matter domination, with growth proportional to the scale factor $a \propto t^{\frac{2}{3}}$.

So the smallest scales modes had to wait the longest to start growing more than logarithmically. Large scale modes entering now start growing without delay. This means that the largest scale density perturbations track the inflationary prediction but scales smaller than k_{eq} are suppressed by a factor

$$T(k) = \frac{k_{eq}^2}{k^2}\ln\frac{k}{k_{eq}}, \qquad k \gg k_{eq} \tag{5.45}$$

and $T(k) = 1$ for $k \ll k_{eq}$.

Since all of this is linear theory we can again write the observed perturbations as a transfer function in terms of the primordial fluctuations

$$\delta_{lin,k}(z) = D(z)T(k)k^2\phi_k \qquad (5.46)$$

where $D(z)$ is the growth function.

From this it is easy to derive the large and small scale behaviors of the power spectrum of density perturbations today *(Exercise!)* which come out as $P(k) \propto k$ on larger scales (small k) and $P(k) \propto (\ln k)/k^3$.

This treatment neglected the fact that even though baryons and cold dark matter only interact weakly, i.e., negligibly, baryons do have mass. Since the universe is dark energy dominated, baryons actually make up a less negligible fraction of the matter density than it would in a Universe without dark energy. This means that when the baryons can finally cluster after recombination, since photon pressure has suddenly disappeared, the dark matter does react to the baryon perturbations and the baryon oscillation features imprint on the dark matter perturbations. So while posing to us the puzzle of "what is dark energy?" Nature kindly provided us with a means to probe it, namely the baryon acoustic oscillation (BAO) features in $P_\delta(k)$, which measure Hubble scale at recombination and thus represent a fixed angular scale which we measure at different redshifts in galaxy redshift surveys. In fact, BAO are just one of many observational probe of dark energy but that discussion would take us outside the scope of these notes.

5.9.1 CMB versus Density Perturbations

What are the relative merits of probing the CMB and density perturbations in terms of learning about the primordial perturbations? We have seen that the CMB is a fairly direct probe of the *potential* whereas the large scale structure probes *density*. These two quantities are related by a factor k^2 which suppresses the density perturbations on large scales. This suggests that the CMB has important information on very large scales (very small k), which is true.

On small scales the CMB sourced by the primary anisotropy becomes exceedingly smooth because effect called "Silk damping" smooths out the CMB beyond $l \sim 2000$, which corresponds to perturbations on angular scales of ~ 4 arc minutes, or about 10 Mpc co-moving. This roughly sets the minimum scale we can usefully probe in the CMB[2].

Density perturbations on the other hand have been processed by non-linear gravitational effects below the non-linearity scale l_{nl}. We can estimate this by calculating the expected rms overdensity fluctuation of the linear density field when averaged

---

[2] Note that measurements of the CMB on even smaller scales are ongoing in several experiments.

5 Gaussian Random Fields in Cosmostatistics   103

over spherical regions of radius l *(Exercise!)*. We then set l_{nl} to be the scale on which we get fluctuations of order 1. Information on scales smaller than that will have been modified in non-linear ways and partially erased. So let us check whether the large scale structure provides us with useful information (modes that are still in the linear regime) on smaller scales than the CMB.

Here is one way to do this calculation: due to homogeneity, the average fluctuation will be the same wherever we center our sphere, so we can just calculate the average fluctuation within a sphere of radius l centered at the origin

$$\delta(l, r = 0) = \frac{1}{C} \int_{|r'|<l} \delta(r') d^3 r'. \quad (5.47)$$

C normalizes the average. We can extend the range of integration if we introduce a radial *window function* $W(|r|/l)$ which is 1 if $r/l < 1$ and 0 otherwise. Squaring and taking the expectation value gives a double integral over the covariance. We can Fourier expand this in terms of the the power spectrum

$$\sigma_l^2 = C^2 \iint W(r'l) W(r''l) \int P(|k|) e^{ik \cdot (r'-r'')} d^3k d^3r' d^3r''. \quad (5.48)$$

We notice that the r-integrals factorize and we obtain the modulus square of the Fourier space window function,

$$\sigma_l^2 = C^2 4\pi k^2 \int_{k=0}^{k=\infty} P(|k|) |W(kl)|^2 dk. \quad (5.49)$$

Nothing in the derivation used the fact that we chose the sharp spherical top hat window so $W(r/l)$ can be anything as long as we set $C = 4\pi \int r^2 W(r/l) = W(kl = 0)$. The scale where this quantity becomes 1 is ~ 15 Mpc today which is larger than the 10 Mpc we found as the cutoff for the CMB. So at first glance it looks as if the CMB would win across the scale spectrum.

But we should not forgot the effect of the relative number of independent modes we can measure. Since the CMB is a two-dimensional object on the sphere, there are about l^2 modes to be measured in total *(Exercise!)*.

The large scale structure is a 3-D object so in principle we have access to a total of $\sim (k_{max}/k_{min})^3 = (l_{max}/l_{min})^3$ modes, where l_{max} is the largest scale observable and l_{min} is the smallest scale. Therefore, in principle, large scale structure is much more informative on small scales than the CMB. The largest scale is limited by causality and the smallest scale the non-linearity scale, the scale where our linear transfer function analysis breaks down.

But remember that we actually observe on the light-cone. So the largest scale accessible to observation and the non-linearity scale vary as a function of the cosmological redshift that we look back to. The co-moving scale of the onset of non-linearity is larger today than it was at any point in the past and scales that are non-linear today were less non-linear in the past.

So, in summary we expect the CMB to have much more signal on very large scales, whereas probes of density should win on intermediate scales just due to the much larger number of modes accessible in a 3-D field. In addition, it is possible in principle to access smaller scales at higher redshift, where non-linearities play less of a role.

5.10 Conclusion

Gaussian random fields are physically motivated tools to model the primordial perturbations, probed through observations of the CMB and large scale galaxy distribution. GRFs in cosmology respect certain symmetries, such as those expected of the galaxy distribution and the temperature anisotropies of the CMB and I explained how to construct and model GRFs with these symmetries. Starting from these basic concepts I showed examples of how to compute correlations and how to set up inference problems.

I have added a short appendix containing a brief discussion of three of the most common and confusing misstatements about (or unnecessarily restrictive definitions of) GRFs in the literature, which often arise from conflating homogeneity and isotropy with Gaussianity.

These notes provide a point of departure for further investigations into, for example

- signal reconstruction and the inference of the correlation properties from incomplete and noisy observations of a GRF;
- the computation of cross-correlations between different tracers of the (processed) primordial perturbations;
- extracting 3-D information from the combination of measurements of a set of 2-D shells at different redshifts as would result from a photometric redshift survey;

and many other possible applications.

Appendix: Common misconceptions about GRFs

- *Gaussian fields have independent Fourier (or momentum) modes*
  No—the authors confuse homogeneity and Gaussianity.
- *Gaussian random fields are those with "random phases"*
  No. Besides being wrong, the statement is imprecise because it does not define what exactly is meant by "random" without specifying the pdf. The phases that are being referred to are the angles of the complex Fourier amplitudes. If the real and imaginary parts are independent, which is true for a *homogeneous* Gaussian field, then it is true that the phases are also independently drawn from a uniform distribution between 0 and 2π. This is indeed something that *follows* only

for a Gaussian field. However, the argument does not work the other way. It is trivial to construct examples with independent (between different k), uniformly distributed phases where the real and imaginary parts are not independent and drawn to give a non-Gaussian field.
- *Histograms of Gaussian Random Fields are Gaussian*
This is not true in general. The histogram of a Gaussian random field with different variances in different pixels will be a sample from a density which is a *mixture* of Gaussians with the marginal means and variances,

$$hist \leftarrow \frac{1}{n} \sum_{i=1}^{n} g(x_i|\mu_i, C_{ii}). \qquad (5.50)$$

So depending on the GRF one can obtain any pdf that can be represented in this form. Note that this is only Gaussian if all the marginal means and variances are the same, but not in any other case.

Acknowledgements I am grateful for comments from students in my 2011 class on "Cosmostatistics and large surveys" which helped improve these notes. BDW holds a Chaire Internationale at the Université Pierre et Marie Curie (Paris 6) and is supported by NSF grants AST 0908902, AST 09-08693 ARRA, and AST 07-08849, and an ANR Chaire d'Excellence.

References

1. Górski, K.M., Hivon, E., Banday, A.J., Wandelt, B.D., Hansen, F.K., Reinecke, M., Bartelmann, M.: HEALPix: A Framework for High-Resolution Discretization and Fast Analysis of Data Distributed on the Sphere. Astrophys. J. **622**, 759–771 (2005). DOI 10.1086/427976
2. Planck Collaboration, Ade, P.A.R., Aghanim, N., Arnaud, M., Ashdown, M., Aumont, J., Baccigalupi, C., Baker, M., Balbi, A., Banday, A.J., et al.: Planck early results. I. The Planck mission. Astron. Astrophys. **536**, A1 (2011). DOI 10.1051/0004-6361/201116464
3. Lewis, A., Challinor, A., Lasenby, A.: Efficient Computation of Cosmic Microwave Background Anisotropies in Closed Friedmann-Robertson-Walker Models. Astrophys. J. **538**, 473–476 (2000). DOI 10.1086/309179
4. Seljak, U., Zaldarriaga, M.: A Line-of-Sight Integration Approach to Cosmic Microwave Background Anisotropies. Astrophys. J. **469**, 437 (1996). DOI 10.1086/177793
5. White, M., Hu, W.: The Sachs-Wolfe effect. Astron. Astrophys. **321**, 8–9 (1997)
6. Komatsu, E., Spergel, D.N., Wandelt, B.D.: Measuring Primordial Non-Gaussianity in the Cosmic Microwave Background. Astrophys. J. **634**, 14–19 (2005). DOI 10.1086/491724
7. Yadav, A.P., Wandelt, B.D.: CMB tomography: Reconstruction of adiabatic primordial scalar potential using temperature and polarization maps. Phys. Rev. D **71**(12), 123004 (2005). DOI 10.1103/PhysRevD.71.123004
8. Yadav, A.P., Komatsu, E., Wandelt, B.D.: Fast Estimator of Primordial Non-Gaussianity from Temperature and Polarization Anisotropies in the Cosmic Microwave Background. Astrophys. J. **664**, 680–686 (2007). DOI 10.1086/519071
9. Lyth, D.H., Liddle, A.R.: The Primordial Density Perturbation: Cosmology, Inflation and the Origin of Structure. Cambridge University Press (2009)

Chapter 6
Recent Advances in Bayesian Inference in Cosmology and Astroparticle Physics Thanks to the MultiNest Algorithm

Roberto Trotta, Farhan Feroz, Mike Hobson, and Roberto Ruiz de Austri

Abstract We present a new algorithm, called MultiNest, which is a highly efficient alternative to traditional Markov Chain Monte Carlo (MCMC) sampling of posterior distributions. MultiNest is more efficient than MCMC, can deal with highly multi-modal likelihoods and returns the Bayesian evidence (or model likelihood, the prime quantity for Bayesian model comparison) together with posterior samples. It can thus be used as an all-around Bayesian inference engine. When appropriately tuned, it also provides an exploration of the profile likelihood that is competitive with what can be obtained with dedicated algorithms.

We demonstrate the power and flexibility of MultiNest for Bayesian inference for multi-dimensional, multimodal-likelihoods, for Bayesian model selection and for profile likelihood evaluation for multi-modal, multi-scale likelihoods. Applications in cosmology and astroparticle physics are presented, including gravitational waves astronomy, inflationary Bayesian model comparison and supersymmetric parameter spaces exploration.

6.1 Introduction

Cosmology (the study of the large-scale properties of the Universe) and astroparticle physics (the intersection of particle physics and astrophysics, studying high-energy particles emitted by astrophysical bodies) present many interesting inference challenges for statisticians and physicists alike. Both disciplines have seen a great surge in the quality and quantity of data in the last decade, and this has in turn spurred the development of more sophisticated statistical modelling and analysis methods. The launch of the ISI Astrostatistics Section, which is present for the first time at

---

Roberto Trotta
Imperial College London, Blackett Laboratory, Prince Consort Road, London, SW7 2AZ, UK
e-mail: r.trotta@imperial.ac.uk

this ISI World Congress, is a testament to the growing importance of astrostatistics, the application of advanced statistical techniques to astrophysical and cosmological problems.

Among the many different statistical challenges that large and complex astrophysical data sets pose, three classes of problems arise often in very disparate contexts. They can be broadly described as follows:

1. *Bayesian inference for multi-dimensional, multi-modal likelihoods.* Problems in as disparate fields as Supersymmetry phenomenology and gravitational wave detection have likelihood functions that are multi-modal, with modes presenting widely different characteristic scales. Numerical exploration of the full posterior distribution is a challenging problem for traditional Markov Chain Monte Carlo (MCMC) methods in this context, which risk getting stuck in local modes.
2. *Bayesian model selection aimed at model building or object detection.* Bayesian model selection is used to decide whether one or several extra parameters are needed in a model to explain the data. This has applications to questions such as deciding whether the Universe is flat, whether dark energy is a cosmological constant, and astronomical object detection (is there a galaxy cluster in a weak lensing image? Does this star have one or several exo-planets orbiting it?). This requires a reliable and efficient algorithm to compute the Bayesian evidence (or model likelihood).
3. *Profile likelihood evaluation for multi-modal likelihoods.* An alternative to posterior-based inference is represented by the (frequentist) construction of confidence regions based on the profile likelihood ratio. For highly non-Gaussian, multi-modal problems, Bayesian credible regions can differ very considerably from confidence regions (especially in problems in high dimensions and/or when the prior is informative). Evaluation of the profile likelihood is in general a much more challenging task than exploring the posterior distribution, and the question of the accuracy of the recovered credible intervals is a relevant one in domains such as supersymmetry phenomenology (i.e., the reconstruction of properties of particles beyond the Standard Model from data from accelerators, cosmology and other astrophysical probes).

In this paper, we present the MULTINEST algorithm (section 6.2), an implementation of nested sampling, and we give examples of its successfull application to problems in cosmology and astroparticle physics in the above three classes (section 6.3).

6.2 Nested Sampling and the MULTINEST Algorithm

Bayesian inference is often the statistical framework of choice in cosmology (see e.g. [1, 2]), and, increasingly so, in astroparticle physics. The posterior pdf $p(\Theta|d, \mathcal{M})$ for the n-dimensional parameters vector Θ of a model $\mathcal{M}$ is given by

$$p(\Theta|d, \mathcal{M}) = \frac{p(d|\Theta, \mathcal{M}) p(\Theta|\mathcal{M})}{p(d|\mathcal{M})}. \tag{6.1}$$

Here, $p(\Theta|\mathcal{M})$ is the prior, $p(d|\Theta, \mathcal{M})$ the likelihood and $p(d|\mathcal{M})$ the model likelihood, or marginal likelihood (usually called "Bayesian evidence" by physicists). The Bayesian evaluation of a model's performance in the light of the data is based on the Bayesian evidence, the normalization integral on the right-hand-side of Bayes' theorem, Eq. (6.1):

$$p(d|\mathcal{M}) \equiv \int p(d|\Theta, \mathcal{M}) p(\Theta|\mathcal{M}) d^n \Theta. \tag{6.2}$$

Thus the Bayesian evidence is the average of the likelihood under the prior for a specific model choice. From the evidence, the model posterior probability given the data is obtained by using Bayes' Theorem to invert the order of conditioning:

$$p(\mathcal{M}|d) \propto p(\mathcal{M}) p(d|\mathcal{M}), \tag{6.3}$$

where $p(\mathcal{M})$ is the prior probability assigned to the model itself. Usually this is taken to be non-committal and equal to $1/N_m$ if one considers N_m different models. When comparing two models, $\mathcal{M}_0$ versus $\mathcal{M}_1$, one is interested in the ratio of the posterior probabilities, or *posterior odds*, given by

$$\frac{p(\mathcal{M}_0|d)}{p(\mathcal{M}_1|d)} = B_{01} \frac{p(\mathcal{M}_0)}{p(\mathcal{M}_1)} \tag{6.4}$$

and the *Bayes factor* B_{01} (see e.g. [3] for details) is the ratio of the models' evidences:

$$B_{01} \equiv \frac{p(d|\mathcal{M}_0)}{p(d|\mathcal{M}_1)} \quad \text{(Bayes factor).} \tag{6.5}$$

A value $B_{01} > (<) 1$ represents an increase (decrease) of the support in favour of model 0 versus model 1 given the observed data. From Eq. (6.4) it follows that the Bayes factor gives the factor by which the relative odds between the two models have changed after the arrival of the data, regardless of what we thought of the relative plausibility of the models before the data, given by the ratio of the prior models' probabilities. Bayes factors are usually interpreted against the Jeffreys' scale [4] for the strength of evidence. This is an empirically calibrated scale, with thresholds at values of the odds of about 3:1, 12:1 and 150:1, representing weak, moderate and strong evidence, respectively.

Nested sampling [5, 6] is a Monte Carlo technique aimed at an efficient evaluation of the Bayesian evidence, which also produces posterior inferences as a byproduct. It calculates the evidence by transforming the multi-dimensional evidence integral of Eq. (6.2) into a one-dimensional integral that is easy to evaluate numerically. This is accomplished by defining the prior volume X as $dX = p(\Theta) d^n \Theta$, so that

$$X(\lambda) = \int_{\mathcal{L}(\Theta) > \lambda} p(\Theta) d^n \Theta, \tag{6.6}$$

where $\mathcal{L}(\Theta) \equiv p(d|\Theta)$ is the likelihood function and the integral extends over the region(s) of parameter space contained within the iso-likelihood contour $\mathcal{L}(\Theta) = \lambda$ (in this section we drop the explicit conditioning on model $\mathcal{M}$, as this is understood). Assuming that $\mathcal{L}(X)$, i.e. the inverse of (6.6), is a monotonically decreasing function of X (which is trivially satisfied for most posteriors), the evidence integral (6.2) can then be written as

$$\mathcal{Z} \equiv p(d) = \int_0^1 \mathcal{L}(X) \mathrm{d}X, \qquad (6.7)$$

Thus, if one can evaluate the likelihoods $\mathcal{L}_j = \mathcal{L}(X_j)$, where X_j is a sequence of decreasing values,

$$0 < X_M < \cdots < X_2 < X_1 < X_0 = 1, \qquad (6.8)$$

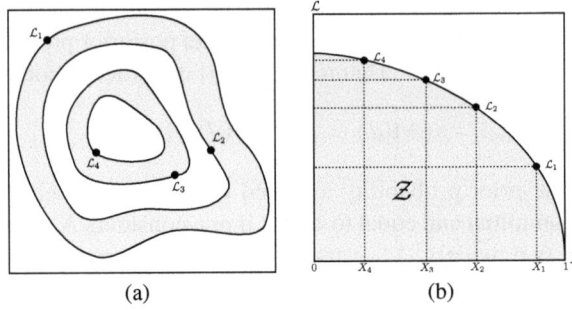

Fig. 6.1 Cartoon illustrating (a) the likelihood of a two-dimensional problem; and (b) the transformed $\mathcal{L}(X)$ function where the prior volumes X_i are associated with each likelihood $\mathcal{L}_i$.

as shown schematically in Fig. 6.1, the evidence can be approximated numerically using standard quadrature methods as a weighted sum

$$\mathcal{Z} = \sum_{i=1}^{M} \mathcal{L}_i w_i. \qquad (6.9)$$

If one uses a simple trapezium rule, the weights are given by $w_i = \frac{1}{2}(X_{i-1} - X_{i+1})$. An example of a posterior in two dimensions and its associated function $\mathcal{L}(X)$ is shown in Fig. 6.1.

6.2.1 Evaluation of the Bayesian Evidence

The nested sampling algorithm performs the summation (6.9) as follows. To begin, the iteration counter is set to $i = 0$ and N "live" (or "active") samples are drawn from the full prior $p(\Theta)$ (which is often simply the uniform distribution over the prior range), so the initial prior volume is $X_0 = 1$. The samples are then sorted in order of their likelihood and the smallest (with likelihood $\mathcal{L}_0$) is removed from the live set and replaced by a point drawn from the prior subject to the constraint that the point has a likelihood $\mathcal{L} > \mathcal{L}_0$. The corresponding prior volume contained within this iso-likelihood contour will be a random variable given by $X_1 = t_1 X_0$, where t_1 follows the distribution $p(t) = Nt^{N-1}$ (i.e. the probability distribution for the largest of N samples drawn uniformly from the interval $[0, 1]$). At each subsequent iteration i, the discarding of the lowest likelihood point $\mathcal{L}_i$ in the live set, the drawing of a replacement with $\mathcal{L} > \mathcal{L}_i$ and the reduction of the corresponding prior volume $X_i = t_i X_{i-1}$ are repeated, until the entire prior volume has been traversed. The algorithm thus travels through nested shells of likelihood as the prior volume is reduced.

The expectation value and standard deviation of $\log t$, which dominates the geometrical exploration, are:

$$E[\log t] = -\frac{1}{N}, \qquad \sigma[\log t] = \frac{1}{N}. \tag{6.10}$$

Since each value of $\log t$ is independent, after i iterations the prior volume will shrink down such that $\log X_i \approx -(i \pm \sqrt{i})/N$. Thus, one takes $X_i = \exp(-i/N)$.

The nested sampling algorithm is terminated on determining the evidence to some specified precision. An adequate and robust condition [5] is given by the upper limit on the evidence that can be determined from the remaining set of current active points. By selecting the maximum-likelihood value $\mathcal{L}_{\max}$ in the set of active points, one can safely assume that an upper bound to the evidence contribution from the remaining portion of the posterior is $\Delta \mathcal{Z}_i = \mathcal{L}_{\max} X_i$, i.e. the product of the remaining prior volume and maximum likelihood value. We choose to stop when this quantity would no longer change the final evidence estimate by some user-defined value, described by a tolerance parameter, tol. As described in section 6.3.3 below, it is important to adjust this tolerance value appropriately if one wants to use MULTI-NEST for profile likelihood evaluation.

6.2.2 Posterior Inferences

Once the evidence $\mathcal{Z}$ is found, posterior inferences are obtained from the full sequence of discarded samples from the nested sampling process, as well as the live set at termination. Each such samples is assigned the probability weight

$$p_i = \frac{\mathcal{L}_i w_i}{\mathcal{Z}}. \tag{6.11}$$

These samples can then be used to calculate inferences of posterior parameters such as means, standard deviations, covariances and so on, or to construct marginalised posterior distributions. The use of the posterior sample to approximate the profile likelihood is further discussed below.

6.2.3 The MULTINEST Algorithm

The most challenging task in implementing the nested sampling algorithm is drawing samples from the prior within the hard constraint $\mathcal{L} > \mathcal{L}_i$ at each iteration i. Employing a naive approach that draws blindly from the prior would result in a steady decrease in the acceptance rate of new samples with decreasing prior volume (and increasing likelihood). Ellipsoidal nested sampling [7] tries to overcome this problem by approximating the iso-likelihood contour of the point to be replaced by an n-dimensional ellipsoid determined from the covariance matrix of the current set of live points. New points are then selected from the prior within this (enlarged) ellipsoidal bound until one is obtained that has a likelihood exceeding that of the discarded lowest-likelihood point. In the limit that the ellipsoid coincides with the true iso-likelihood contour, the acceptance rate tends to unity.

Ellipsoidal nested sampling is efficient for simple uni-modal posterior distributions without pronounced degeneracies, but is not well suited to multi-modal distributions. As advocated by [8–10] and shown in Fig. 6.2, the sampling efficiency can be substantially improved by identifying distinct *clusters* of live points that are well separated and constructing an individual ellipsoid for each cluster. In some problems, however, some modes of the posterior might possess very pronounced curving degeneracies, which are problematic to explore efficiently. To sample with maximum efficiency from such distributions, the MULTINEST algorithm of [10] divides the live point set into sub-clusters which are then enclosed in ellipsoids and a new point is then drawn uniformly from the region enclosed by these 'overlapping' ellipsoids, see Fig. 6.3. The number of points in an individual sub-cluster and the total number of sub-clusters is decided by a an 'expectation-maximization' algorithm so that the total sampling volume, which is equal to the sum of volumes of the ellipsoids enclosing the sub-clusters, is minimized. This allows maximum flexibility and efficiency by breaking up a mode resembling a Gaussian into relatively fewer number of sub-clusters, while using a larger number of small overlapping ellipsoid in the presence of severe, curving degeneracies.

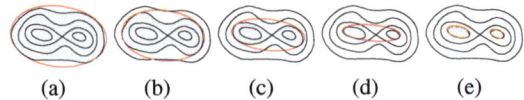

Fig. 6.2 Cartoon of ellipsoidal nested sampling from a simple bimodal distribution. In (a), the red ellipsoid represents a good bound to the active region. From (a) to (d), as we nest inward we can see that the acceptance rate will rapidly decrease as the bound steadily worsens. Figure (e) illustrates the increase in efficiency obtained by sampling from each clustered region separately. From [10].

Fig. 6.3 Illustration of the ellipsoidal decompositions used by MULTINEST to deal with degeneracies in the posterior distribution: the points given as input are overlaid on the resulting ellipsoids. 1000 points were sampled uniformly from: (a) two non-intersecting ellipsoids; and (b) a torus. From [10].

6.3 Applications of MULTINEST

The MULTINEST algorithm sketched above has been successfully applied to a broad range of challenging inference problems, including cosmological model selection [12], exo-planets detection and characterization [13], astronomical object detection [14], inference on Supersymmetric parameters [11, 15, 16], gravitational waves analysis [17], cosmic ray propagation models [18], and others. As an illustration, we briefly present one example for each of the three classes of problems mentioned in the Introduction.

6.3.1 Sampling of Multi-modal Posteriors in Gravitational Waves Astronomy

The proposed space-based gravitational wave detector, the Laser Interferometer Space Antenna (LISA) is expected to observe thousands of gravitational wave signals from many different types of sources, including galactic compact binaries, inspiral and mergers of supermassive black holes (SMBH) and extreme mass ratio inspirals. Detecting and characterizing this many sources presents a significant data analysis challenge. In order to encourage the research in this area, a program of Mock LISA Data Challenges (MLDC) (e.g., [19]) has been taking place, with each round consisting of one or more datasets containing simulated instrumental noise and gravitational waves from sources with undisclosed parameters. Several of the LISA sources exhibit degeneracies in their parameter space resulting in multiple modes in the posterior pdf, which proved to be a very challenging problem for traditional MCMC methods. With its ability to explore highly multi-modal distributions efficiently, MULTINEST is well poised to tackle data analysis problems in gravitational wave astronomy, as demonstrated in [17].

MULTINEST was used for the detection and characterisation of cosmic string burst sources in mock LISA data of increasing realism (including for example instrumental noise and partially subtracted galactic background). As a search tool, the algorithm was successful in finding the three cosmic string bursts that were present in the MLDC challenge data set. These sources, and the five sources in the MLDC training data, were correctly identified in the sense that the full signal-to-noise ratio of the injected source was recovered, and a posterior distribution for the parameters obtained. The maximum likelihood and maximum a-posteriori parameters were not particularly close to the true parameters of the injected signals, but this was a consequence of the intrinsic degeneracies in the cosmic string model parameter space (shown in Fig. 6.4) and in all cases the true parameters were consistent with the recovered posterior distributions.

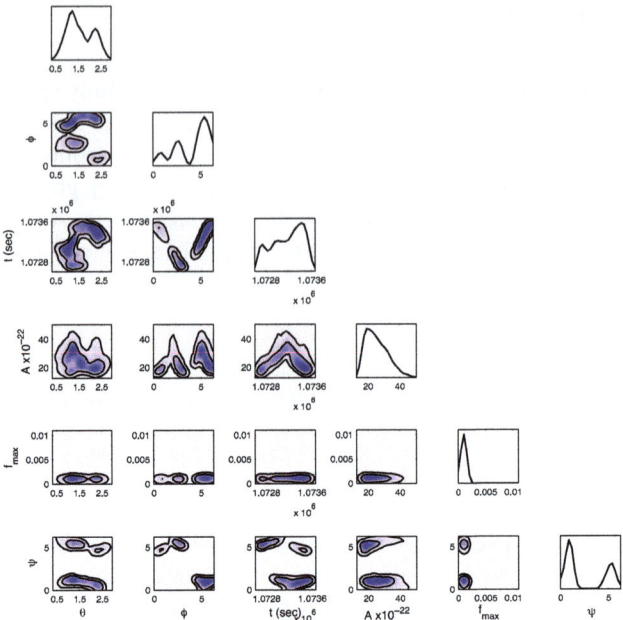

Fig. 6.4 2D and 1D marginalised posteriors as recovered by MULTINEST in the search for the second of the cosmic string bursts in the LISA gravitational wave detector training data set. The parameters, from top-to-bottom and left-to-right, are colatitude, longitude, burst time, burst amplitude, burst break frequency and waveform polarization. From [17].

6.3.2 Inflationary Bayesian Model Comparison

A second example is the inflationary model comparison carried out in Ref. [12]. Although the technical details are fairly involved, the underlying idea can be sketched as follows.

The term "inflation" describes a period of exponential expansion of the Universe in the very first instants of its life, some 10^{-32} seconds after the Big Bang, during which the size of the Universe increased by at least 25 orders of magnitude. This huge and extremely fast expansion is required to explain the observed isotropy of the cosmic microwave background on large scales. It is believed that inflation was powered by one or more "scalar fields". The behaviour of the scalar field during inflation is determined by the shape of its potential, which is a real-valued function $V(\phi)$ (where ϕ denotes the value of the scalar field). The detailed shape of $V(\phi)$ controls the duration of inflation, but also the spatial distribution of inhomogeneities (perturbations) in the distribution of matter and radiation which emerge from inflation. It is from those perturbations that galaxies and cluster form out of gravitational collapse. Hence the shape of the scalar field can be constrained by observations of the large scale structures of the Universe and of cosmic microwave background anisotropies.

Theories of physics beyond the Standard Model motivate certain functional forms of $V(\phi)$, which however typically have a number of free parameters, Θ. The fundamental model selection question is to use cosmological observations to discriminate between alternative models for $V(\phi)$ (and hence alternative fundamental theories). The major obstacle to this programme is that very little if anything at all is known *a priori* about the free parameters Θ describing the inflationary potential. What is worse, such parameters can assume values across several orders of magnitude, according to the theory. Hence the Occam's razor effect of Bayesian model comparison can vary in a very significant way depending on the prior choices for Θ. Furthermore, a non-linear reparameterization of the problem (which leaves the physics invariant) does in general change the Occam's razor factor, and hence the model comparison result.

In Ref. [12] a first attempt was made to tackle inflationary model selection from a principled point of view. The main result of the analysis is shown in Fig. 6.5, which presents the Bayes factors between models, obtained using MULTINEST (suitably normalized w.r.t. a reference model, here the so-called LFI_2 model). Two classes of models for $V(\phi)$ have been considered, namely so-called Small Field Inflation (SFI) models and Large Field Inflation (LFI) models. The two classes of model differ in the parameterized form of $V(\phi)$, and have different sets of parameters, differing in dimensionality, as well. Within each class of models, sub-classes are defined (denoted by subscripts in Fig. 6.5) based on theoretical considerations, e.g. by fixing some of the parameters to certain values. The priors on the models' parameters have been chosen based on theoretical considerations of possible values achievable under each class of models. Typical priors are uniform on the log of the parameter

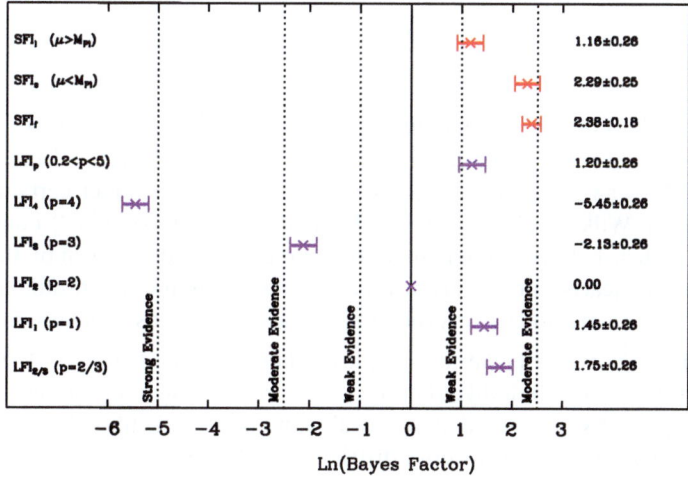

Fig. 6.5 Results of Bayesian model comparison between 9 inflationary models (vertical axis), subdivided in two categories (SFI models and LFI models), from Ref. [12]. Errorbars reflect the 68% uncertainty on the value of the Bayes factor from the numerical evaluation performed with MULTINEST.

(to reflect indifference w.r.t. the characteristic scale of the quantity), within a range chosen as a reflection of physical model building. The models' priors are chosen in such a way to lead to non-committal priors for the two classes as a whole, i.e. $p(\text{SFI}) = p(\text{LFI}) = 1/2$.

Fig. 6.5 shows that some models in the LFI class are fairly strongly disfavoured by the data (e.g., LFI_3 and LFI_4), while the model comparison is inconclusive in most other cases. One finds that the posterior probability for the SFI model class evaluates to $p(\text{SFI}|d) \approx 0.77$. Therefore, the probability of the SFI class has increased from 50% in the prior to about 77% in the posterior, signalling a weak preference for this type of models in the light of the data.

6.3.3 Challenges of Profile Likelihood Evaluation

For highly non-Gaussian problems like supersymmetric (SUSY) parameter determination, inference can depend strongly on whether one chooses to work with the posterior distribution (Bayesian) or profile likelihood (frequentist) [11, 20, 21]. There is a growing consensus that both the posterior and the profile likelihood ought to be explored in order to obtain a fuller picture of the statistical constraints from present-day and future data. This begs the question of the algorithmic solutions available to reliably explore both the posterior and the profile likelihood in the context of SUSY phenomenology.

When Θ is composed of parameters of interest, θ, and nuisance parameters, ψ, the profile likelihood ratio is defined as

$$\lambda(\theta) \equiv \frac{\mathcal{L}(\theta, \hat{\hat{\psi}})}{\mathcal{L}(\hat{\theta}, \hat{\psi})}. \qquad (6.12)$$

where $\hat{\hat{\psi}}$ is the conditional maximum likelihood estimate (MLE) of ψ with θ fixed and $\hat{\theta}, \hat{\psi}$ are the unconditional MLEs. The profile likelihood ratio defined in Eq. (6.12) is an attractive choice as a test statistics, for under certain regularity conditions, Wilks [22] showed that the distribution of $-2 \ln \lambda(\theta)$ converges to a chi-square distribution with a number of degrees of freedom given by the dimensionality of θ. Clearly, for any given value of θ, evaluation of the profile likelihood requires solving a maximization problem in many dimensions to determine the conditional MLE $\hat{\hat{\psi}}$. While posterior samples obtained with MULTINEST have been used to estimate the profile likelihood, the accuracy of such an estimate has been questioned [23]. As mentioned above, evaluating profile likelihoods is much more challenging than evaluating posterior distributions. Therefore, one should not expect that a vanilla setup for MULTINEST (which is adequate for an accurate exploration of the posterior distribution) will automatically be optimal for profile likelihoods evaluation. In Ref. [24] the question of the accuracy of profile likelihood evaluation from MULTINEST was investigated in detail. We report below the main results.

The two most important parameters that control the parameter space exploration in MULTINEST are the number of live points N – which determines the resolution at which the parameter space is explored – and a tolerance parameter tol, which defines the termination criterion based on the accuracy of the evidence. Generally, a larger number of live points is necessary to explore profile likelihoods accurately. Moreover, setting tol to a smaller value results in MULTINEST gathering a larger number of samples in the high likelihood regions (as termination is delayed). This is usually not necessary for the posterior distributions, as the prior volume occupied by high likelihood regions is usually very small and therefore these regions have relatively small probability mass. For profile likelihoods, however, getting as close as possible to the true global maximum is crucial and therefore one should set tol to a relatively smaller value. In Ref. [24] it was found that $N = 20{,}000$ and tol $= 1 \times 10^{-4}$ produce a sufficiently accurate exploration of the profile likelihood in toy models that reproduce the most important features of the parameter space of a popular supersymmetric scenario.

In principle, the profile likelihood does not depend on the choice of priors. However, in order to explore the parameter space using any Monte Carlo technique, a set of priors needs to be defined. Different choices of priors will generally lead to different regions of the parameter space to be explored in greater or lesser detail, according to their posterior density. As a consequence, the resulting profile likelihoods might be slightly different, purely on numerical grounds. We can obtain more robust profile likelihoods by simply merging samples obtained from scans with different choices of Bayesian priors. This does not come at a greater computational cost, given that a responsible Bayesian analysis would estimate sensitivity to the choice of prior as well. The results of such a scan are shown in Fig. 6.6, which was obtained by tuning MULTINEST with the above configuration, appropriate for an

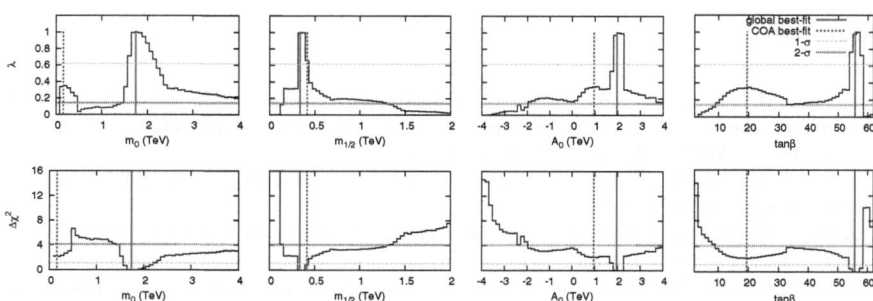

Fig. 6.6 1-D profile likelihoods from present-day data for the parameters of interest of a Supersymmetric model (so-called CMSSM), normalized to the global best-fit point. The red solid and blue dotted vertical lines represent the global best-fit point ($\chi^2 = 9.26$, located in the focus point region) and the best-fit point found in the stau co-annihilation region ($\chi^2 = 11.38$) respectively. The upper and lower panel show the profile likelihood and $\Delta\chi^2$ values, respectively. Green (magenta) horizontal lines represent the 1σ (2σ) approximate confidence intervals. MULTINEST was run with 20,000 live points and tol $= 1\times 10^{-4}$ (a configuration deemed appropriate for profile likelihood estimation), requiring approximately 11 million likelihood evaluations. From [24].

accurate profile likelihood exploration, and by merging the posterior samples from two different choices of priors (see [24] for details). This high-resolution profile likelihood scan using MULTINEST compares favourably with the results obtained by adopting a dedicated Genetic Algorithm technique [23], although at a slightly higher computational cost (a factor of ~ 4). In general, an accurate profile likelihood evaluation was about an order of magnitude more computationally expensive than mapping out the Bayesian posterior.

6.4 Conclusions

We have given a short overview of the capabilities of the MULTINEST algorithm, which has been used very successfully in a variety of problems to (a) reconstruct multi-modal posterior distributions; (b) evaluate the model likelihood for Bayesian model comparison and (c) approximate the profile likelihood in the challenging case of multi-modal likelihoods in many dimensions. Current statistical challenges in cosmology and astroparticle physics still present many open questions, and no doubt this will provide fertile ground for an increased collaboration between physicists and statisticians in the burgeoning field of astrostatistics.

Acknowledgements We would like to thank Joseph Hilbe for organizing the first Astrostatistics session at an ISI World Congress and for his excellent efforts at getting the Astrostatistics section organized within ISI. We would also like to thank our colleagues for many stimulating discussions over the years: J. Berger., R. Cousin, K. Cranmer, A. Jaffe, M. Kunz, O. Lahav, A. Liddle, T. Loredo, L. Lyons and D. van Dyk.

References

1. Hobson, M. et al. (eds.): Bayesian Methods in Cosmology. Cambridge University Press (2010)
2. Trotta, R.: Bayes in the sky: Bayesian inference and model selection in cosmology. Contemp. Phys. **49**, 71 (2008). [arXiv:0803.4089 [astro-ph]]
3. Sellke, T., Bayarri, M.J., Berger, J.: Calibration of P-values for testing precise null hypotheses. The American Statistician **55**, 62–71 (2001)
4. Jeffreys, H.: Theory of probability, 3rd edn, Oxford Classics series (reprinted 1998). Oxford University Press, Oxford, UK (1961)
5. Skilling, J.: Nested sampling. In: Fischer, R., Preuss, R., von Toussaint, U. (eds.) Bayesian Inference and Maximum Entropy Methods in Science and Engineering, 735, pp. 395–405. Amer. Inst. Phys. Conf. Proc. (2004)
6. Skilling, J.: Bayesian Analysis **1**, 833–861 (2006)
7. Mukherjee, P., Parkinson, D. Liddle, A. R.: A Nested Sampling Algorithm for Cosmological Model Selection. Astrophys. J. **638**, L51–L54 (2006)
8. Shaw, R., Bridges, M., Hobson, M. P.: Efficient Bayesian inference for multimodal problems in cosmology. Mon. Not. Roy. Astron. Soc. **378**, 1365–1370 (2007)

9. Feroz, F., Hobson, M. P.: Multimodal nested sampling: an efficient and robust alternative to MCMC methods for astronomical data analysis. Mon. Not. Roy. Astron. Soc. **384**, 449–463 (2008)
10. Feroz, F., Hobson, M. P., Bridges, M.: MultiNest: an efficient and robust Bayesian inference tool for cosmology and particle physics. Mon. Not. Roy. Astron. Soc. **398**, 1601–1614 (2009). [arXiv:0809.3437 [astro-ph]]
11. Trotta, R., Feroz, F., Hobson, M. P., Roszkowski, L., Ruiz de Austri, R.: The Impact of priors and observables on parameter inferences in the Constrained MSSM. JHEP **0812**, 024 (2008). [arXiv:0809.3792 [hep-ph]]
12. Martin, J., Ringeval, C., Trotta, R.: Hunting Down the Best Model of Inflation with Bayesian Evidence. Phys. Rev. D **83**, 063524 (2011)
13. Feroz, F., Balan, S. T., Hobson, M. P.: Detecting extrasolar planets from stellar radial velocities using Bayesian evidence. Mon. Not. Roy. Astron. Soc. **415**, 3462–3472 (2011)
14. Feroz, F., Hobson, M. P., Zwart, J. T. L., Saunders. R. D. E., Grainge, K. J. B.: Bayesian modelling of clusters of galaxies from multi-frequency pointed Sunyaev–Zel'dovich observations. Mon. Not. Roy. Astron. Soc. **398**, 2049–2060 (2009). [arXiv:0811.1199 [astro-ph]]
15. Feroz, F., Allanach, B. C., Hobson, M., AbdusSalam, S. S., Trotta, R., Weber, A. M.: Bayesian Selection of sign(mu) within mSUGRA in Global Fits Including WMAP5 Results. JHEP **0810**, 064 (2008)
16. Bertone, G., Cerdeno, D. G., Fornasa, M., Ruiz de Austri, R., Trotta, R.: Identification of Dark Matter particles with LHC and direct detection data. Phys. Rev. D **82**, 055008 (2010)
17. Feroz, F., Gair, J. R., Graff, P., Hobson, M. P., Lasenby, A.: Classifying LISA gravitational wave burst signals using Bayesian evidence. Classical and Quantum Gravity **27**, 075010 (2010)
18. Trotta, R., Johannesson, G., Moskalenko, I. V., Porter, T. A., Ruiz de Austri, R., Strong, A. W.: Constraints on Cosmic-ray Propagation Models from a Global Bayesian Analysis. Astrophys. J. **729**, 106 (2011)
19. Babak S. et al.: The Mock LISA Data Challenges: from challenge 3 to challenge 4. Classical and Quantum Gravity **27**(8), 084009 (2010)
20. Allanach, B. C., Cranmer, K., Lester, C. G., Weber, A. M.: Natural priors, CMSSM fits and LHC weather forecasts. JHEP **0708**, 023 (2007)
21. Scott. P., Conrad, J., Edsjö, J., Bergström, L., Farnier, C., Akrami, Y.: Direct constraints on minimal supersymmetry from Fermi-LAT observations of the dwarf galaxy Segue 1. JCAP **1001**, 031 (2010)
22. Wilks, S.: The large-sample distribution of the likelihood ratio for testing composite hypotheses. Ann. Math. Statist. **9**, 60–62 (1938)
23. Akrami, Y., Scott, P., Edsjö, J., Conrad, J., Bergström, L.: A Profile Likelihood Analysis of the Constrained MSSM with Genetic Algorithms. JHEP **04**, 057 (2010)
24. Feroz, F., Cranmer, K., Hobson, M., Ruiz de Austri, R., Trotta, R.: Challenges of profile likelihood evaluation in multi-dimensional SUSY scans. JHEP **06**, 42 (2011)

Chapter 7
Extra-solar Planets via Bayesian Fusion MCMC

Philip C. Gregory

Abstract A Bayesian multi-planet Kepler periodogram has been developed based on a fusion Markov chain Monte Carlo algorithm (FMCMC). FMCMC is a new general purpose tool for nonlinear model fitting. It incorporates parallel tempering, simulated annealing and genetic crossover operations. Each of these features facilitate the detection of a global minimum in chi-squared in a highly multi-modal environment. By combining all three, the algorithm greatly increases the probability of realizing this goal.

The FMCMC is controlled by a unique adaptive control system that automates the tuning of the proposal distributions for efficient exploration of the model parameter space even when the parameters are highly correlated. This controlled statistical fusion approach has the potential to integrate other relevant statistical tools as desired. The FMCMC algorithm is implemented in *Mathematica* using parallelized code and runs on an 8 core PC. The performance of the algorithm is illustrated with some recent successes in the exoplanet field where it has facilitated the detection of a number of new planets. Bayesian model selection is accomplished with a new method for computing marginal likelihoods called nested restricted Monte Carlo (NRMC).

7.1 Introduction

A remarkable array of new ground based and space based astronomical tools are providing astronomers access to other solar systems. Over 700 planets have been discovered to date, starting from the pioneering work of [1–4]. One example of the fruits of this work is the detection of a super earth in the habitable zone surrounding Gliese 581 [5]. Fig. 7.1 illustrates the pace of discovery up to Dec. 2011.

Because a typical star is approximately a billion times brighter than a planet, only a small fraction of the planets have been detected by direct imaging. The majority

---

P.C. Gregory
Department of Physics and Astronomy, University of British Columbia, Vancouver, BC V6T 1Z1, Canada, e-mail: gregory@phas.ubc.ca

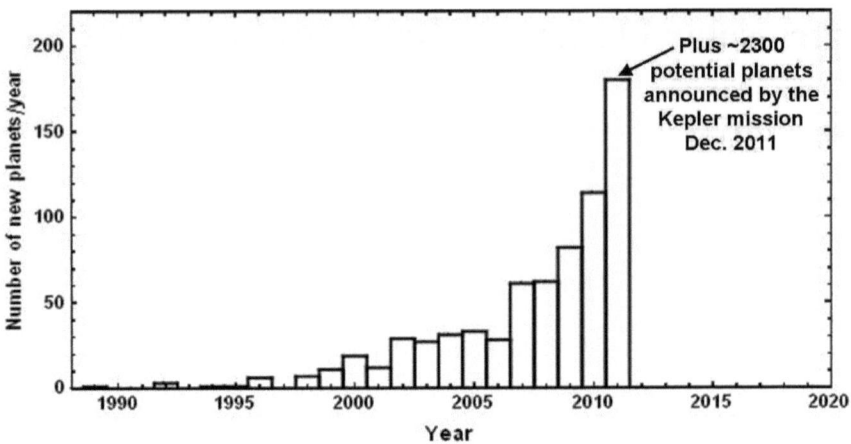

Fig. 7.1 The pace of exoplanet discoveries.

of planets have been detected by studying the reflex motion of the star caused by the gravitational tug of unseen planets, using precision radial velocity (RV) measurements. There are currently 84 known multiple planet systems, the largest of which has seven planets [6]. Recently the Kepler space mission has detected 2326 planetary candidates using the transit detection method. These are awaiting confirmation using the RV method or transit timing variation analysis. More than thirty of these candidates have a radius smaller than or equal to the radius of the earth.

These successes on the part of the observers has spurred a significant effort to improve the statistical tools for analyzing data in this field (e.g., [7–16]). Much of this work has highlighted a Bayesian MCMC approach as a way to better understand parameter uncertainties and degeneracies and to compute model probabilities. MCMC algorithms provide a powerful means for efficiently computing the required Bayesian integrals in many dimensions (e.g., an 8 planet model has 41 unknown parameters). The output at each iteration of the MCMC is a vector of the model parameters. After an initial burn-in period, the MCMC produces an equilibrium distribution of samples in model parameter space such that the density of samples is proportional to the joint posterior probability distribution of the parameters. The marginal posterior probability density function (PDF) for any single parameter is given by a histogram of that component of the vector for all post burn-in iterations.

Frequently, MCMC algorithms have been augmented with an additional tool such as parallel tempering, simulated annealing or differential evolution depending on the complexity of the problem. My approach [17] has been to fuse together the advantages of all of the above tools together with a genetic crossover operation in a single MCMC algorithm to facilitate the detection of a global minimum in χ^2 (maximum posterior probability in the Bayesian context). The FMCMC is controlled by a unique multi-stage adaptive control system that automates the tuning of the proposal distributions for efficient exploration of the model parameter space even when the parameters are highly correlated. The FMCMC algorithm is implemented in *Mathematica* using parallelized code and run on an 8 core PC. It is designed to be a very general tool for nonlinear model fitting. When implemented with a multi-planet Kepler model[1], it is able to identify any significant periodic signal component in the data that satisfies Kepler's laws and is able to function as a multi-planet Kepler periodogram[2].

The different components of Fusion MCMC are described in Section 7.2. In Section 7.3, the performance of the algorithm is illustrated with some recent successes in the exoplanet field where it has facilitated the detection of a number of new planets. Section 7.4 deals with the challenges of Bayesian model selection in this arena.

7.2 Adaptive Fusion MCMC

The adaptive fusion MCMC (FMCMC) is a very general Bayesian nonlinear model fitting program. After specifying the model, M_i, the data, D, and priors, I, Bayes' theorem dictates the target joint probability distribution for the model parameters which is given by

$$p(\vec{X}|D, M_i, I) = C\ p(\vec{X}|M_i, I) \times p(D|M_i, \vec{X}, I). \tag{7.1}$$

where C is the normalization constant which is not required for parameter estimation purposes and $\vec{X}$ represent the vector of model parameters. The term, $p(\vec{X}|M_i, I)$, is the prior probability distribution of $\vec{X}$, prior to the consideration of the current data D. The term, $p(D|\vec{X}, M_i, I)$, is called the likelihood and it is the probability that we would have obtained the measured data D for this particular choice of parameter vector $\vec{X}$, model M_i, and prior information I. At the very least, the prior information, I, must specify the class of alternative models being considered (hypothesis space of interest) and the relationship between the models and the data (how to compute the likelihood). In some simple cases the log of the likelihood is simply proportional to the familiar χ^2 statistic. For further details of the likelihood function for this type of problem see Gregory [11].

An important feature that prevents the fusion MCMC from becoming stuck in a local probability maximum is *parallel tempering* [20] (and re-invented under the name *exchange Monte Carlo* [21]). Multiple MCMC chains are run in parallel. The joint distribution for the parameters of model M_i, for a particular chain, is given by

$$\pi(\vec{X}|D, M_i, I, \beta) \propto p(\vec{X}|M_i, I) \times p(D|\vec{X}, M_i, I)^\beta. \tag{7.2}$$

---

[1] For multiple planet models, there is no analytic expression for the exact radial velocity perturbation. In many cases, the radial velocity perturbation can be well modeled as the sum of multiple independent Keplerian orbits which is what has been assumed in this paper.

[2] Following on from the pioneering work on Bayesian periodograms by [18, 19]

Each MCMC chain corresponding to a different β, with the value of β ranging from zero to 1. When the exponent $\beta = 1$, the term on the LHS of the equation is the target joint probability distribution for the model parameters, $p(\vec{X}|D, M_i, I)$. For $\beta \ll 1$, the distribution is much flatter.

In equation 7.2, an exponent $\beta = 0$ yields a joint distribution equal to the prior. The reciprocal of β is analogous to a temperature, the higher the temperature the broader the distribution. For parameter estimation purposes 8 chains were employed. A representative set of β values is shown in Fig. 7.2. At an interval of 10 to 40 iterations, a pair of adjacent chains on the tempering ladder are chosen at random and a proposal made to swap their parameter states. A Monte Carlo acceptance rule determines the probability for the proposed swap to occur (e.g., Gregory [10], equation 12.12). This swap allows for an exchange of information across the population of parallel simulations. In low β (higher temperature) simulations, radically different configurations can arise, whereas in higher β (lower temperature) states, a configuration is given the chance to refine itself. The lower β chains can be likened to a series of scouts that explore the parameter terrain on different scales. The final samples are drawn from the $\beta = 1$ chain, which corresponds to the desired target probability distribution. The choice of β values can be checked by computing the swap acceptance rate. When they are too far apart the swap rate drops to very low values. In this work a typical swap acceptance rate of $\approx 30\%$ was employed but rates in a broad range from 0.15 to 0.5 were deemed acceptable as they did not exhibit any clear differences in performance. For a swap acceptance rate of 30%, jumps to adjacent chains will occur at an interval of ~ 230 to 920 iterations while information from more distant chains will diffuse much more slowly. Recently, Atchade et al. [22] have shown that under certain conditions, the optimal swap acceptance rate is

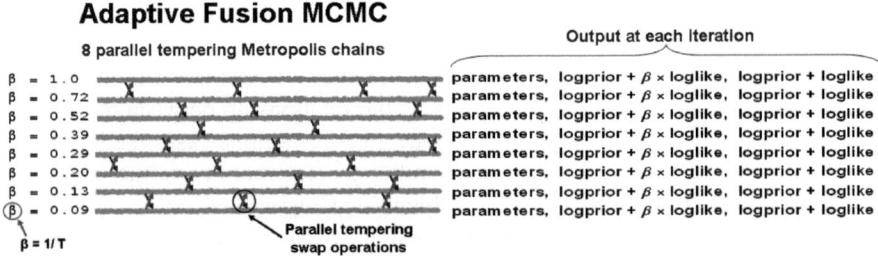

Fig. 7.2 Parallel tempering schematic.

0.234. A future goal for fusion MCMC is to extend the control system to automate the selection of an optimal set of β values as well.

At each iteration, a single joint proposal to jump to a new location in the parameter space is generated from independent Gaussian proposal distributions (centered on the current parameter location), one for each parameter. In general, the values of σ for these Gaussian proposal distributions are different because the parameters can be very different entities. If the values of σ are chosen too small, successive samples will be highly correlated and will require many iterations to obtain an equilibrium set of samples. If the values of σ are too large, then proposed samples will very rarely be accepted. The process of choosing a set of useful proposal values of σ when dealing with a large number of different parameters can be very time consuming. In parallel tempering MCMC, this problem is compounded because of the need for a separate set of Gaussian proposal distributions for each tempering chain. This process is automated by an innovative statistical control system [23, 24] in which the error signal is proportional to the difference between the current joint parameter acceptance rate and a target acceptance rate [25], λ (typically $\lambda \sim 0.25$). A schematic of the first two stages of the adaptive control system (CS) is shown[3] in Fig. 7.3. Further details on the operation of the control system can be found in Gregory [17] and references therein. A third stage that handles highly correlated parameters is described in Section 7.2.2.

The adaptive capability of the control system can be appreciated from an examination of Fig. 7.4. The upper left portion of the figure depicts the FMCMC iterations from the 8 parallel chains, each corresponding to a different tempering level β as

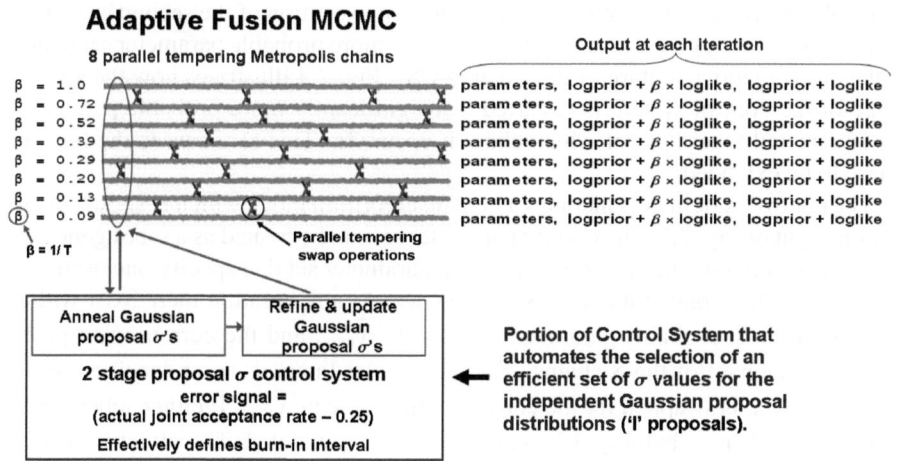

Fig. 7.3 First two stages of the adaptive control system.

---

[3] The interval between tempering swap operations is typically much smaller than is suggested by this schematic.

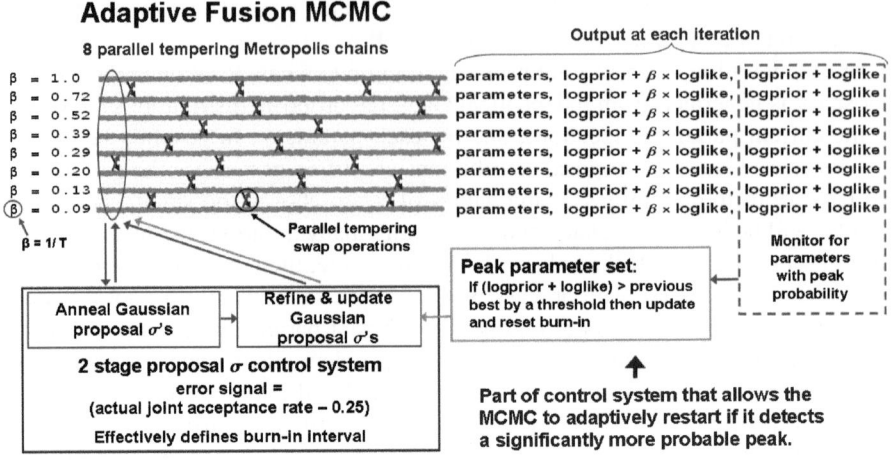

Fig. 7.4 Schematic illustrating how the second stage of the control system is restarted if a significantly more probable parameter set is detected.

indicated on the extreme left. One of the outputs obtained from each chain at every iteration (shown at the far right) is the log prior + log likelihood. This information is continuously fed to the CS which constantly updates the most probable parameter combination regardless of which chain the parameter set occurred in. This is passed to the 'Peak parameter set' block of the CS. Its job is to decide if a significantly more probable parameter set has emerged since the last execution of the second stage CS. If so, the second stage CS is re-run using the new more probable parameter set which is the basic adaptive feature of the existing CS[4]. Fig. 7.4 illustrates how the second stage of the control system is restarted if a significantly more probable parameter set is detected regardless of which chain it occurs in. This also causes the burn-in phase to be extended.

The control system also includes a genetic algorithm block which is shown in the bottom right of Fig. 7.5. The current parameter set can be treated as a set of genes. In the present version, one gene consists of the parameter set that specify one orbit. On this basis, a three planet model has three genes. At any iteration there exist within the CS the most probable parameter set to date $\vec{X}_{\max}$, and the current most probable parameter set of the 8 chains, $\vec{X}_{\text{cur}}$. At regular intervals (user specified) each gene from $\vec{X}_{\text{cur}}$ is swapped for the corresponding gene in $\vec{X}_{\max}$. If either substitution leads to a higher probability it is retained and $\vec{X}_{\max}$ updated. The effectiveness of

---

[4] *Mathematica* code that implements the version of fusion MCMC shown in Fig. 7.4 is available on the Cambridge University Press web site for my textbook [10], 'Bayesian Logical data Analysis for the Physical Sciences'. See the 'Additional book examples with *Mathematica* 8 tutorial' in the resource material. There you will find an example entitled, 'Markov chain Monte Carlo powered Kepler periodogram'. Non *Mathematica* users can download a free Wolfram CDF Player to view the resource material.

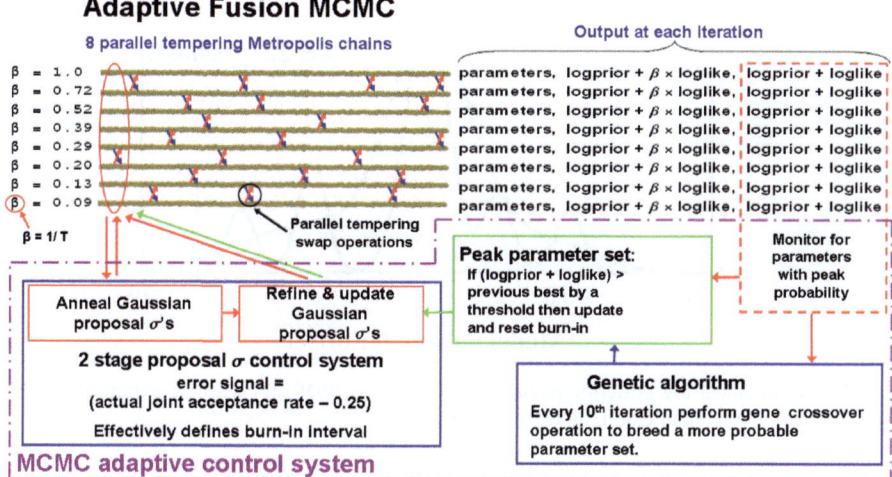

Fig. 7.5 This schematic shows how the genetic crossover operation is integrated into the adaptive control system.

this operation can be tested by comparing the number of times the gene crossover operation gives rise to a new value of $\vec{X}_{\max}$ compared to the number of new $\vec{X}_{\max}$ arising from the normal parallel tempering MCMC iterations. The gene crossover operations prove to be very effective, and give rise to new $\vec{X}_{\max}$ values ≈ 3 times more often than MCMC operations. Of course, most of these swaps lead to very minor changes in probability but occasionally big jumps are created. It turns out that individual gene swaps from $\vec{X}_{\text{cur}}$ to $\vec{X}_{\max}$ are much more effective (in one test by a factor of 17) than the other way around (reverse swaps). Since it costs just as much time to compute the probability for a swap either way we no longer carry out the reverse swaps. Instead, we have extended this operation to swaps from $\vec{X}_{\text{cur2}}$, the parameters of the second most probable current chain, to $\vec{X}_{\max}$. This gives rise to new values of $\vec{X}_{\max}$ at a rate approximately half that of swaps from $\vec{X}_{\text{cur}}$ to $\vec{X}_{\max}$. Crossover operations at a random point in the entire parameter set did not prove as effective except in the single planet case where there is only one gene.

7.2.1 Automatic Simulated Annealing and Noise Model

The annealing of the proposal sigma values occurs while the MCMC is homing in on any significant peaks in the target probability distribution. Concurrent with this, another aspect of the annealing operation takes place whenever the Markov chain is started from a location in parameter space that is far from the best fit values. This automatically arises because all the models considered incorporate an extra additive

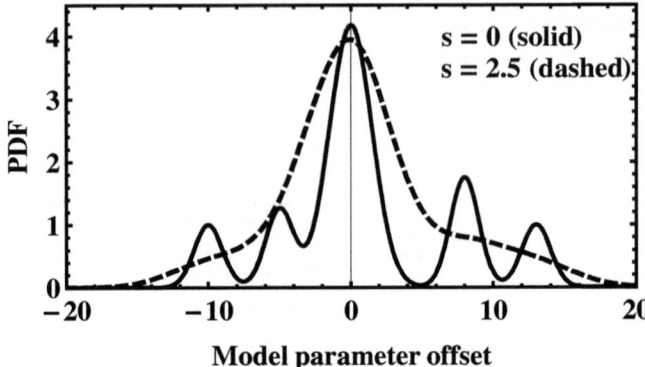

Fig. 7.6 A simulated toy posterior probability distribution (PDF) for a single parameter model with (dashed) and without (solid) an extra noise term s.

noise [11], whose probability distribution is Gaussian with zero mean and with an unknown standard deviation s. When the χ^2 of the fit is very large, the Bayesian Markov chain automatically inflates s to include anything in the data that cannot be accounted for by the model with the current set of parameters and the known measurement errors. This results in a smoothing out of the detailed structure in the χ^2 surface and, as pointed out by [13], allows the Markov chain to explore the large scale structure in parameter space more quickly. This is illustrated in Figure 7.6 which shows a simulated toy posterior probability distribution (PDF) for a single parameter model with (dashed) and without (solid) an extra noise term s. Figure 7.7 shows the behavior of $\text{Log}_{10}[\text{Prior} \times \text{Likelihood}]$ and s versus MCMC iteration for a some real data. In the early stages s is inflated to around 38 m s^{-1} and then decays to a value of ≈ 4 m s^{-1} over the first 9,000 iterations as $\text{Log}_{10}[\text{Prior} \times \text{Likelihood}]$ reaches a maximum. This is similar to simulated annealing, but does not require choosing a cooling scheme.

7.2.2 Highly Correlated Parameters

For some models the data is such that the resulting estimates of the model parameters are highly correlated and the MCMC exploration of the parameter space can be very inefficient. Fig. 7.8 shows an example of two highly correlated parameters and possible ways of dealing with this issue which includes a transformation to more orthogonal parameter set. It would be highly desirable to employ a method that automatically samples correlated parameters efficiently. One potential solution in the literature is Differential Evolution Markov Chain (DE-MC) [26]. DE-MC is a population MCMC algorithm, in which multiple chains are run in parallel, typically from 15 to 40. DE-MC solves an important problem in MCMC, namely that of choosing an appropriate scale and orientation for the jumping distribution.

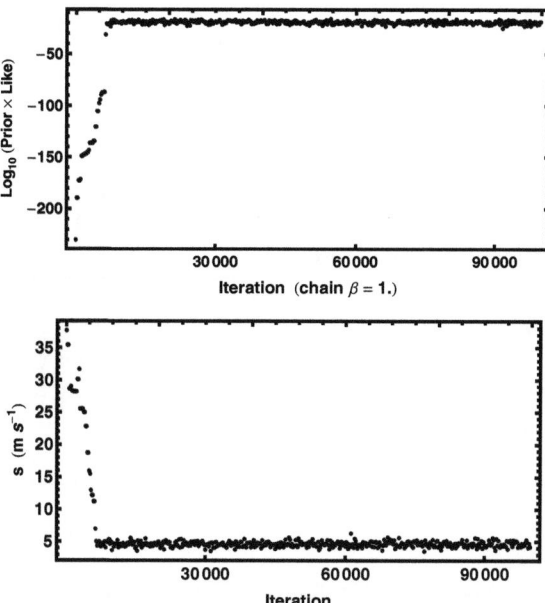

Fig. 7.7 The upper panel is a plot of the $\text{Log}_{10}[\text{Prior} \times \text{Likelihood}]$ versus MCMC iteration. The lower panel is a similar plot for the extra noise term s. Initially s is inflated and then rapidly decays to a much lower level as the best fit parameter values are approached.

For the fusion MCMC algorithm, I developed and tested a new method [27], in the spirit of DE, that automatically achieves efficient MCMC sampling in highly correlated parameter spaces without the need for additional chains. The block in the lower left panel of Fig. 7.9 automates the selection of efficient proposal distributions when working with model parameters that are independent or transformed to new independent parameters. New parameter values are jointly proposed based on independent Gaussian proposal distributions ('I' scheme), one for each parameter. Initially, only this 'I' proposal system is used and it is clear that if there are strong correlations between any parameters the σ values of the independent Gaussian proposals will need to be very small for any proposal to be accepted and consequently convergence will be very slow. However, the accepted 'I' proposals will generally cluster along the correlation path. In the optional third stage of the control system every second[5] accepted 'I' proposal is appended to a correlated sample buffer. There is a separate buffer for each parallel tempering level. Only the 300 most recent additions to the buffer are retained. A 'C' proposal is generated from the difference between a pair of randomly selected samples drawn from the correlated sample buffer for that tempering level, after multiplication by a constant. The value of this constant (for each tempering level) is computed automatically [27] by another control system module which ensures that the 'C' proposal acceptance rate is close to

---

[5] Thinning by a factor of 10 has already occurred meaning only every tenth iteration is recorded.

Highly correlated parameters

Top figure shows an exoplanet Example. For low eccentricity orbits the parameters ω and χ are not separately well determined. This shows up as a strong correlation between ω and χ.

One option re-parameterization

The combination $2\pi\chi+\omega$ is well determined for all eccentricities. Although $2\pi\chi-\omega$ is not well determined for low eccentricities, it is at least orthogonal to $2\pi\chi+\omega$ as shown.

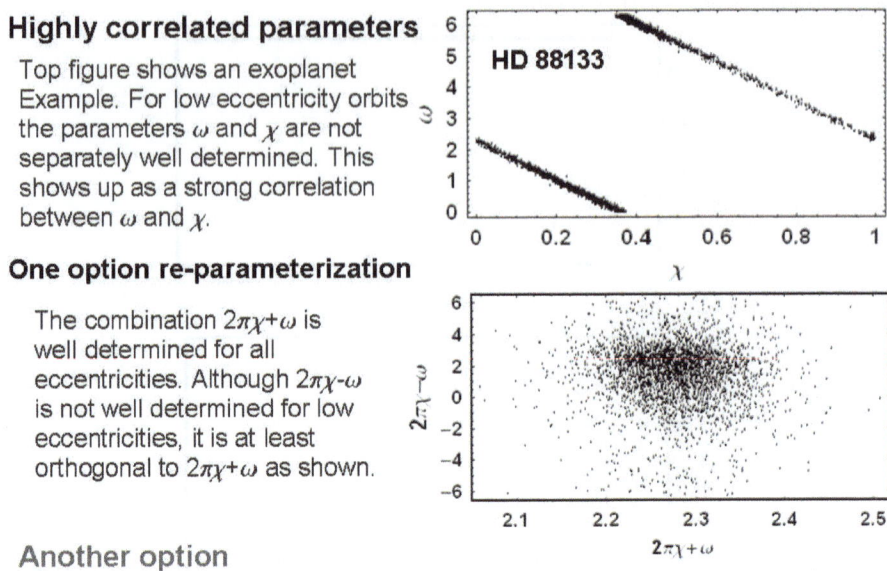

Another option

Algorithm learns about the parameter correlations during the burn-in and generates proposals with these statistical correlations.

Fig. 7.8 An example of two highly correlated parameters and possible ways of dealing with this issue which includes a transformation to more orthogonal parameter set.

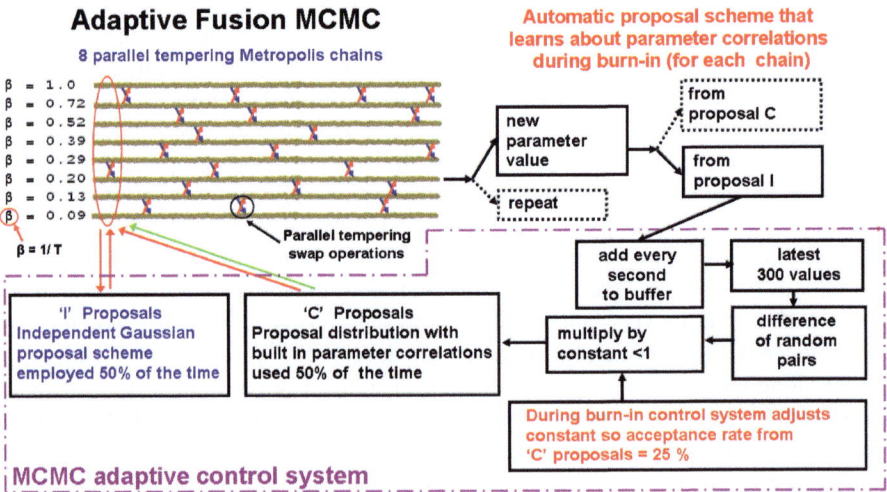

Fig. 7.9 This schematic illustrates the automatic proposal scheme for handling correlated ('C') parameters.

25 %. With very little computational overhead, the 'C' proposals provide the scale and direction for efficient jumps in a correlated parameter space.

The final proposal distribution is a random selection of 'I' and 'C' proposals such that each is employed 50% of the time. The combination ensures that the whole parameter space can be reached and that the FMCMC chain is aperiodic. The parallel tempering feature operates as before to avoid becoming trapped in a local probability maximum.

Because the 'C' proposals reflect the parameter correlations, large jumps are possible allowing for much more efficient movement in parameter space than can be achieved by the 'I' proposals alone. Once the first two stages of the control system have been turned off, the third stage continues until a minimum of an additional 300 accepted 'I' proposals have been added to the buffer and the 'C' proposal acceptance rate is within the range ≥ 0.22 and ≤ 0.28. At this point further additions to the buffer are terminated and this sets a lower bound on the burn-in period.

7.2.2.1 Tests of the 'C' Proposal Scheme

Gregory [27] carried out two tests of the 'C' proposal scheme using (a) simulated exoplanet astrometry data, and (b) a sample of real radial velocity data. In the latter test we analyzed a sample of seventeen HD 88133 precision radial velocity measurements [28] using a single planet model in three different ways. Fig. 7.10 shows a comparison of the resulting post burn-in marginal distributions for two correlated parameters χ and ω together with a comparison of the autocorrelation functions. The black trace corresponds to a search in χ and ω using only 'I' proposals. The red trace corresponds to a search in χ and ω with 'C' proposals turned on. The green trace corresponds to a search in the transformed orthogonal coordinates $\psi = 2\pi\chi + \omega$ and $\phi = 2\pi\chi - \omega$ using only 'I' proposals. It is clear that a search in χ and ω with 'C' proposals turned on achieves the same excellent results as a search in the transformed orthogonal coordinates ψ and ϕ using only 'I' proposals.

7.3 Exoplanet Applications

As previously mentioned the FMCMC algorithm is designed to be a very general tool for nonlinear model fitting. When implemented with a multi-planet Kepler model it is able to identify any significant periodic signal component in the data that satisfies Kepler's laws and is able to function as a multi-planet Kepler periodogram.

In this section we describe the model fitting equations and the selection of priors for the model parameters. For a one planet model the predicted radial velocity is given by

$$v(t_i) = V + K[\cos\{\theta(t_i + \chi P) + \omega\} + e\cos\omega], \qquad (7.3)$$

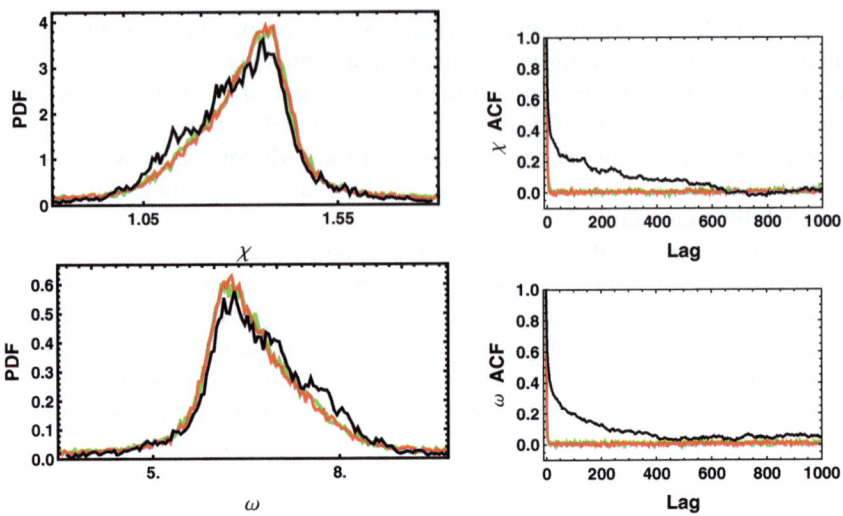

Fig. 7.10 The two panels on the left show a comparison of the post burn-in marginal distributions for χ and ω. The two panels on the right show a comparison of their MCMC autocorrelation functions. The black trace corresponds to a search in χ and ω using only 'I' proposals. The red trace corresponds to a search in χ and ω with 'C' proposals turned on. The green trace corresponds to a search in the transformed orthogonal coordinates $\psi = 2\pi\chi + \omega$ and $\phi = 2\pi\chi - \omega$ using only 'I' proposals.

and involves the 6 unknown parameters

V = a constant velocity.
K = velocity semi-amplitude.
P = the orbital period.
e = the orbital eccentricity.
ω = the longitude of periastron.
χ = the fraction of an orbit, prior to the start of data taking, that periastron occurred at. Thus, χP = the number of days prior to $t_i = 0$ that the star was at periastron, for an orbital period of P days.
$\theta(t_i + \chi P)$ = the true anomaly, the angle of the star in its orbit relative to periastron at time t_i.

We utilize this form of the equation because we obtain the dependence of θ on t_i by solving the conservation of angular momentum equation

$$\frac{d\theta}{dt} - \frac{2\pi[1 + e\cos\theta(t_i + \chi\ P)]^2}{P(1-e^2)^{3/2}} = 0. \tag{7.4}$$

Our algorithm is implemented in *Mathematica* and it proves faster for *Mathematica* to solve this differential equation than solve the equations relating the true anomaly to the mean anomaly via the eccentric anomaly. *Mathematica* generates an accurate interpolating function between t and θ so the differential equation does not need to be solved separately for each t_i. Evaluating the interpolating function for each t_i is very fast compared to solving the differential equation. Details on how equation 7.4 is implemented are given in the Appendix of [17].

We employed a re-parameterization of χ and ω to improve the MCMC convergence speed motivated by the work of Ford [13]. The two new parameters are $\psi = 2\pi\chi + \omega$ and $\phi = 2\pi\chi - \omega$. Parameter ψ is well determined for all eccentricities. Although ϕ is not well determined for low eccentricities, it is at least orthogonal to the ψ parameter. We use a uniform prior for ψ in the interval 0 to 4π and uniform prior for ϕ in the interval -2π to $+2\pi$. This insures that a prior that is wraparound continuous in (χ, ω) maps into a wraparound continuous distribution in (ψ, ϕ). To account for the Jacobian of this re-parameterization it is necessary to multiply the Bayesian integrals by a factor of $(4\pi)^{-nplan}$, where $nplan$ is the number of planets in the model. Also, by utilizing the orthogonal combination (ψ, ϕ) it was not necessary to make use of the 'C' proposal scheme outlined in Section 7.2.2 which typically saves about 25% in execution time.

In a Bayesian analysis we need to specify a suitable prior for each parameter. These are tabulated in Table 7.1. For the current problem, the prior given in equation 7.2 is the product of the individual parameter priors. Detailed arguments for the choice of each prior were given in [24, 29].

Table 7.1 Prior parameter probability distributions.

Parameter	Prior		Lower bound	Upper bound
Orbital frequency	$p(\ln f_1, \ln f_2, \cdots \ln f_n \lvert M_n, I) = \frac{n!}{[\ln(f_H/f_L)]^n}$ (n = number of planets)		1/1.1 d	1/1000 yr
Velocity K_i (m s^{-1})	Modified Jeffreys[i] $\frac{(K+K_0)^{-1}}{\ln\left[1 + \frac{K_{max}}{K_0}\left(\frac{P_{min}}{P_i}\right)^{1/3}\frac{1}{\sqrt{1-e_i^2}}\right]}$		0 ($K_0 = 1$)	$K_{max}\left(\frac{P_{min}}{P_i}\right)^{1/3}\frac{1}{\sqrt{1-e_i^2}}$ $K_{max} = 2129$
V (m s^{-1})	Uniform		$-K_{max}$	K_{max}
e_i Eccentricity	a) Uniform		0	1
	b) Ecc. noise bias correction filter		0	0.99
χ orbit fraction	Uniform		0	1
ω_i Longitude of periastron	Uniform		0	2π
s Extra noise (m s^{-1})	$\frac{(s+s_0)^{-1}}{\ln\left(1+\frac{s_{max}}{s_0}\right)}$		0 ($s_0 = 1$)	K_{max}

[i]Since the prior lower limits for K and s include zero, we used a modified scale invariant prior of the form

$$p(X|M, I) = \frac{1}{X + X_0} \frac{1}{\ln\left(1 + \frac{X_{max}}{X_0}\right)} \quad (7.5)$$

For $X \ll X_0$, $p(X|M, I)$ behaves like a uniform prior and for $X \gg X_0$ it behaves like a scale invariant prior.

The $\ln\left(1 + \frac{X_{max}}{X_0}\right)$ term in the denominator ensures that the prior is normalized in the interval 0 to X_{max}.

As mentioned in Section 7.2.1, all of the models considered in this paper incorporate an extra noise parameter, s, that can allow for any additional noise beyond the known measurement uncertainties[6]. We assume the noise variance is finite and adopt a Gaussian distribution with a variance s^2. Thus, the combination of the known errors and extra noise has a Gaussian distribution with variance $\sigma_i^2 + s^2$, where σ_i is the standard deviation of the known noise for ith data point. For example, suppose that the star actually has two planets, and the model assumes only one is present. In regard to the single planet model, the velocity variations induced by the unknown second planet acts like an additional unknown noise term. Other factors like star spots and chromospheric activity can also contribute to this extra velocity noise term which is often referred to as stellar jitter. In general, nature is more complicated than our model and known noise terms. Marginalizing s has the desirable effect of treating anything in the data that can't be explained by the model and known measurement errors as noise, leading to conservative estimates of orbital parameters. See Sections 9.2.3 and 9.2.4 of [10] for a tutorial demonstration of this point. If there is no extra noise then the posterior probability distribution for s will peak at $s = 0$. The upper limit on s was set equal to $K_{\max}$. We employed a modified Jeffrey's (scale invariant) prior for s with a knee, $s_0 = 1\mathrm{m\ s}^{-1}$.

Fig. 7.11 shows sample MCMC traces for a two planet fit [29] to a set of HD 208487 radial velocity data [31].

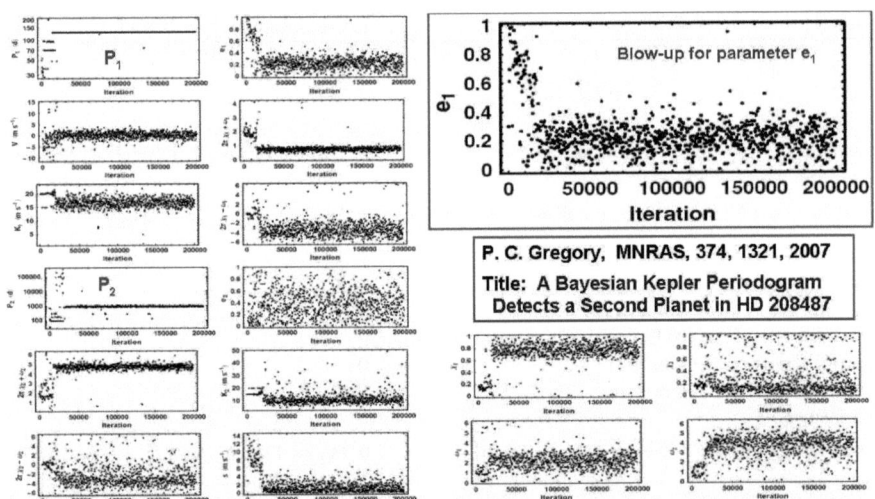

Fig. 7.11 Sample FMCMC traces for a two planet fit to HD 208487 radial velocity data.

---

[6] In the absence of detailed knowledge of the sampling distribution for the extra noise, we pick a Gaussian because for any given finite noise variance it is the distribution with the largest uncertainty as measured by the entropy, i.e., the maximum entropy distribution [30] and [10] (section 8.7.4.)

7.3.1 47 Ursae Majoris

47 Ursae Majoris (47 UMa) is a solar twin at a distance of 46 light years. A brief history of the analysis of the radial velocity data together with the recent 3 planet FMCMC fit of the Lick Observatory data [24] is shown in Fig. 7.12.

Five different models assuming 0, 1, 2, 3, & 4 planets were explored. Our Bayesian model selection analysis indicates that the 3 planet model is significantly [24] more probable than the others. Fig. 7.13 shows some of the FMCMC results for the preferred 3 planet model. The upper left graph shows the trace of the Log_{10}[Prior × Likelihood] versus iteration for the Lick telescope data. The lower left graph shows the corresponding trace of the three period parameters. The starting parameter values are indicated by the three arrows. The top right graph shows a plot of eccentricity versus period for the same run. There is clear evidence for three signals including one with a period of ∼2400d. The longest period of ∼10000d is not as well defined mainly because it corresponds to periods longer than the data duration of 7907d. Previous experience with the FMCMC periodogram indicates that it is capable of finding a global peak in a blind search of parameter space for a three planet model. The lower right graph shows the trace of the three period parameters for a second run with very different starting parameter periods of 5, 20, 100d. The algorithm readily finds the same set of final periods in both cases.

To test the evidence for three planets the analysis was repeated with the Lick data combined with the data [32] from the 9.2 m Hobby-Eberly Telescope (HET) and 2.7 m Harlam J. Smith (HJS) telescopes of the McDonald Observatory. Figure 7.14 shows a plot of eccentricity versus period for our 3 planet FMCMC fit to the

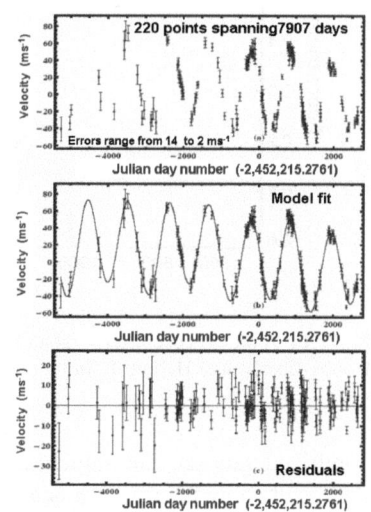

1) 1996, report of a P =1090 day companion
   Butler, R. P. & Marcy, G. W. 1996, ApJ, 464, L153

2) 2002, report of a 2nd companion, P = 2594±90 days
   Fischer, D. A., Marcy, G. W., Butler, R. P., Laughlin, G. L., and Vogt, S. S., 2002, ApJ, 564, 1028

3) 2004-2009, several papers report either no 2nd planet or a 2nd planet with P = 9660 days when eccentricity of 2nd planet set = 0.005, value found by Fischer et al., 2002.
   eg., Wittenmyer, R. A., Endl, M., Cochran, W. D., Levison, H. F., Henry, G. W., 2009, ApJS, 182, 97

4) 2010, Gregory, P. C., & Fischer, D. A., MNRAS, 403, 731, 2010
   FMCMC confirms 2400 d planet and finds evidence for a 3rd planet with P ~10000 days

Fig. 7.12 Brief history of 47 UMa radial velocity fits.

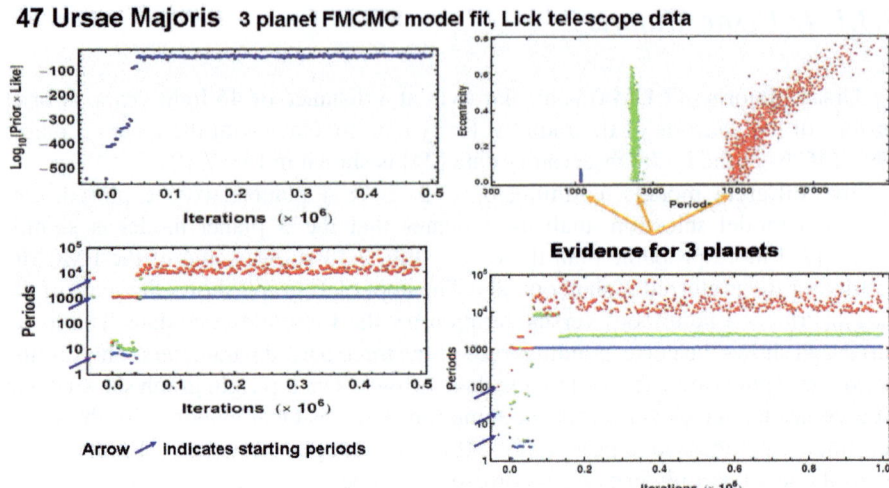

Fig. 7.13 The upper left graph shows the trace of the $\text{Log}_{10}[\text{Prior} \times \text{Likelihood}]$ versus iteration for the 3 planet FMCMC Kepler periodogram of the Lick telescope data for 47 UMa. The lower left graph shows the corresponding trace of the three period parameters. The starting parameter values are indicated by the three arrows. The top right graph shows a plot of eccentricity versus period for the same run. The lower right graph shows the trace of the three period parameters for a second run with very different starting parameter values.

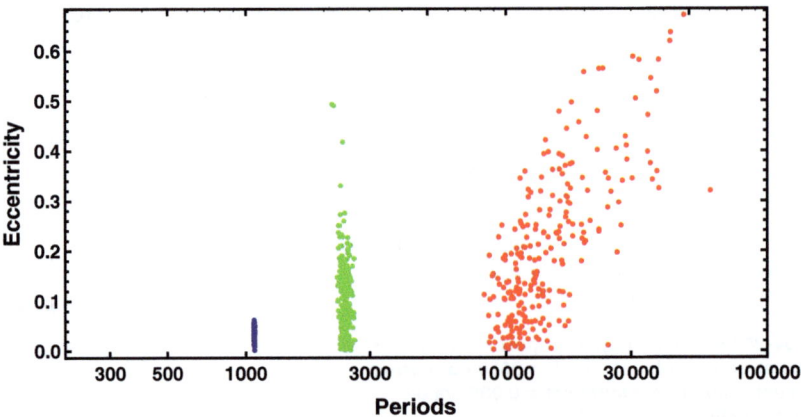

Fig. 7.14 A plot of eccentricity versus period for a 3 planet FMCMC fit [24] of the combined Lick, Hobby-Eberly, and Harlam J. Smith telescope data set for 47 UMa.

combined data set. The same three periods appear as before but with the extra data the results now favor low eccentricity orbits for all three periods. This is a particularly pleasing result as low eccentricity orbits are more likely to exhibit long term stability than high eccentricity orbits.

Figure 7.15 shows the final marginal distributions for the parameters for the three telescope combined analysis. One of the advantages of the Bayesian approach is the ability to deal with nuisance parameters such as unknown systematic residual velocity offsets when combining data from different observatories, and different detector dewars on the same telescope. The marginal distributions for these nuisance

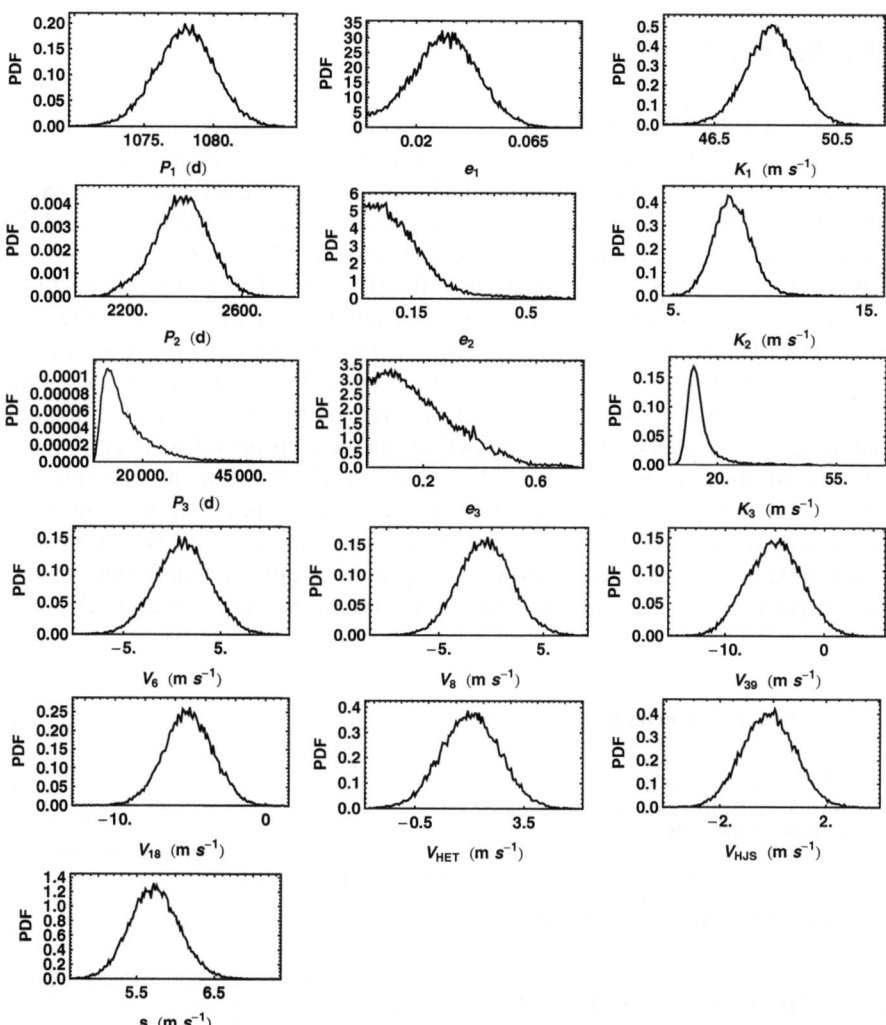

Fig. 7.15 A plot of parameter marginal distributions for a 3 planet FMCMC of the combined Lick, HET, and HJS telescope data set for 47 UMa. In addition to the extra noise parameter s distributions are also shown for 6 additional nuisance parameters. They are systematic residual offset velocity parameters relative to the Lick dewar 24. They are designated V_j, where $j = 6, 8, 39, 18$ correspond to the other Lick dewars and subscripts HET and HJS refer to the Hobby-Eberly Telescope and Harlam J. Smith telescopes.

parameters are included in Figure 7.15 together with the distribution of the extra noise parameter s. The systematic residual offset velocity parameters, relative to the Lick dewar number 24, are designated V_j, where $j = 6, 8, 39, 18$ correspond to the other Lick dewars and subscripts HET and HJS refer to the Hobby-Eberly Telescope and Harlam J. Smith telescopes.

7.3.2 Gliese 581

Gliese 581 (Gl 581) is an M dwarf with a mass of 0.31 times the mass of the sun at a distance of 20 light years which has received a lot of attention because of the possibility of two super-earths in the habitable zone where liquid water could exist. A brief history of the analysis of the radial velocity data together with our recent [17] 5 planet FMCMC fit of the HARPS [33] data is shown in Fig. 7.16.

We carried out a Bayesian re-analysis of the HARPS [33] and HIRES data [34]. Our analysis of the HARPS data found significant evidence for 5 planets, the four reported by [33] and a 399^{+14}_{-16}d period ($6.6^{+2.0}_{-2.7}M_\oplus$) planet similar to the 433 ± 13d period reported by Vogt et al. [34]. Fig. 7.17 shows the 5 planet Kepler periodogram results for the HARPS data. The best set of parameters from the 4 planet fit were used as start parameters. The starting period for the fifth period was set $= 300$d and the most probable period found to be ~ 400d. As illustrated in this example, the parallel tempering feature identifies not only the strongest peak but other potential interesting ones as well. The fifth period parameter (orange points) shows 3 peaks but the 400d period has a maximum value of prior $\times$ likelihood that is almost 1000 times larger than the next strongest which is a second harmonic at 200d. A much

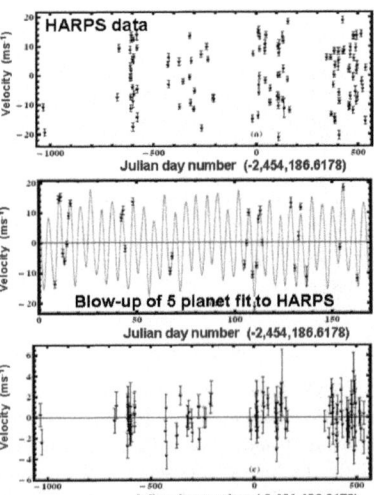

1) 2005 to 2009, European HARPS spectrometer
    Planet e is 1.9 Earth mass
    Planet b is 16 Earth mass
    Planet c is 5 Earth mass
    Planet d is 7 Earth mass   (Habitable zone)
    Latest paper: M. Mayor et al., A&A, 507, p. 487, 2009

2) 2010, combined HARPS and Keck HIRES analysis
    Planet f is 7 Earth mass
    Planet g is 3.1 Earth mass (Habitable zone)
    Vogt et al., ApJ, 723, p. 954, 2011, assumes circular orbits.

3) 2011, Gregory, P. C., MNRAS, 415, 2523
    Conclusions:
    1) Evidence insufficient to claim planet 581g.
    2) Find evidence that Keck HIRES uncertainties are much larger than the quoted values by an extra 1.8 ms⁻¹ added in quadrature.

Fig. 7.16 Brief history of Gliese 581 radial velocity fits.

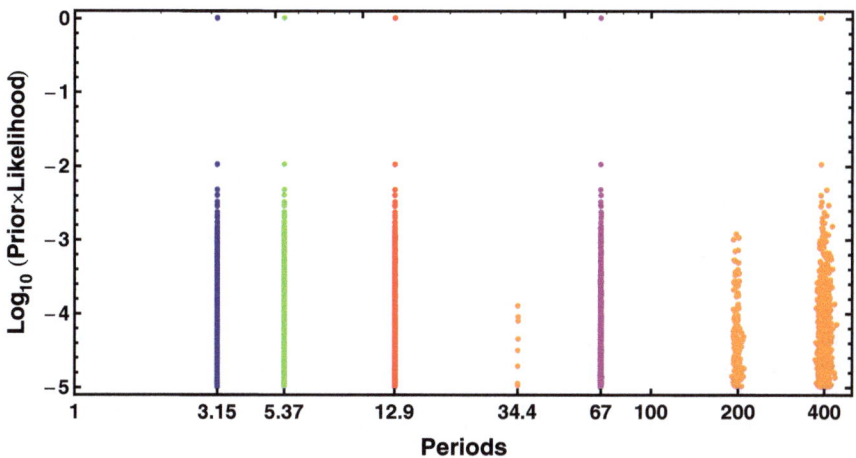

Fig. 7.17 A plot of the 5 period parameter values versus a normalized value of $\text{Log}_{10}[\text{Prior} \times \text{Likelihood}]$ for the 5 planet FMCMC Kepler periodogram of the HARPS data for Gliese 581. The fifth period parameter points are shown in orange.

weaker peak shows up with a 34.4d period. In a Bayesian analysis the relative importance of the three peaks is in proportion to the number of MCMC samples in each peak which is in the ratio of 1.0:0.35:0.04 for the 400, 200, and 34.4d peaks, respectively.

Fig. 7.18 shows a plot of a subset of the FMCMC parameter marginal distributions for the 5 planet fit of the HARPS data after filtering out the post burn-in FMCMC iterations that correspond to the 5 dominant period peaks at 3.15, 5.37, 12.9, 66.9, and 400d. The median value of the extra noise parameter $s = 1.16 \text{m s}^{-1}$.

A six planet model fit found multiple periods for a sixth planet candidate, the strongest of which had a period of 34.4 ± 0.1d but our Bayesian false alarm probability (discussed in Section 7.4) was much too high to consider it significant. Vogt et al. found a period of 36.562 ± 0.052d for Gl 581g.

The analysis of the HIRES data set yielded a reliable detection of only the strongest 5.37 and 12.9 day periods. For a two planet fit the value of the extra noise term s (stellar jitter) was higher for the HIRES data than for the HARPS data suggesting the possibility the HIRES measurement uncertainties were underestimated. The analysis of the combined HIRES/HARPS data again only reliably detected the 5.37 and 12.9d periods. Detection of 4 planetary signals with periods of 3.15, 5.37, 12.9, and 66.9d was only achieved by including an additional unknown but parameterized Gaussian error term added in quadrature to the HIRES quoted errors. The marginal probability density of the sigma for this additional HIRES Gaussian noise term has a well defined peak at $1.84^{+0.35}_{-0.33} \text{m s}^{-1}$.

The two conclusions of our GL 581 analysis are: (1) The current evidence is insufficient to warrant a claim for Gl 581g. (2) The quoted errors for the HIRES data

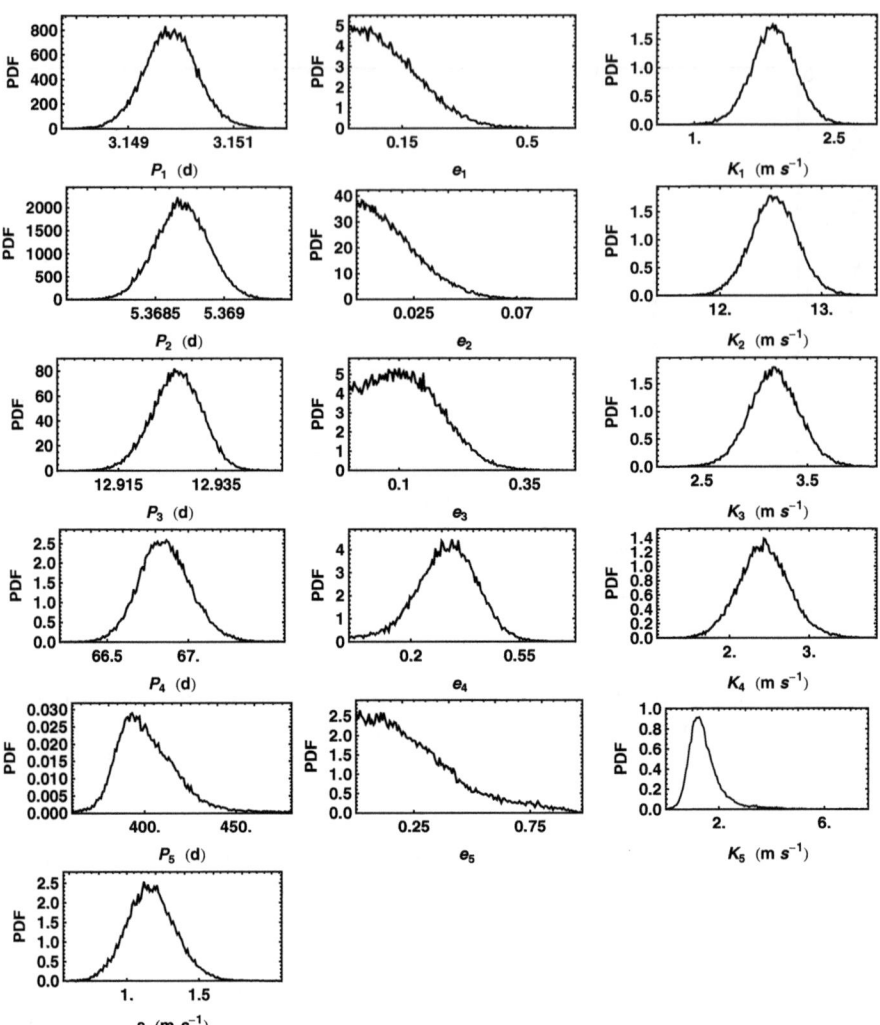

Fig. 7.18 A plot of a subset of the FMCMC parameter marginal distributions for a 5 planet fit of the HARPS data for Gliese 581.

are significantly underestimated equivalent to an effective $1.8^{0.35}_{0.33}$ ms^{-1} Gaussian error added in quadrature with the quoted uncertainties.

7.4 Model Selection

One of the great strengths of Bayesian analysis is the built-in Occam's razor. More complicated models contain larger numbers of parameters and thus incur a larger

Occam penalty, which is automatically incorporated in a Bayesian model selection analysis in a quantitative fashion (see for example, Gregory [10], p. 45). The analysis yields the relative probability of each of the models explored.

To compare the posterior probability of the ith planet model to the four planet model we need to evaluate the odds ratio, $O_{i4} = p(M_i|D, I)/p(M_4|D, I)$, the ratio of the posterior probability of model M_i to model M_4. Application of Bayes' theorem leads to,

$$O_{i4} = \frac{p(M_i|I)}{p(M_4|I)} \frac{p(D|M_i, I)}{p(D|M_4, I)} \equiv \frac{p(M_i|I)}{p(M_4|I)} B_{i4} \quad (7.6)$$

where the first factor is the prior odds ratio, and the second factor is called the *Bayes factor*, B_{i4}. The Bayes factor is the ratio of the marginal (global) likelihoods of the models. The marginal likelihood for model M_i is given by

$$p(D|M_i, I) = \int d\vec{X} p(\vec{X}|M_i, I) \times p(D|\vec{X}, M_i, I). \quad (7.7)$$

Thus Bayesian model selection relies on the ratio of marginal likelihoods, not maximum likelihoods. The marginal likelihood is the weighted average of the conditional likelihood, weighted by the prior probability distribution of the model parameters and s. This procedure is referred to as marginalization.

The marginal likelihood can be expressed as the product of the maximum likelihood and the Occam penalty (e.g., see Gregory [10], page 48). The Bayes factor will favor the more complicated model only if the maximum likelihood ratio is large enough to overcome this penalty. In the simple case of a single parameter with a uniform prior of width ΔX, and a centrally peaked likelihood function with characteristic width δX, the Occam factor is $\approx \delta X / \Delta X$. If the data is useful then generally $\delta X \ll \Delta X$. For a model with m parameters, each parameter will contribute a term to the overall Occam penalty. The Occam penalty depends not only on the number of parameters but also on the prior range of each parameter (prior to the current data set, D), as symbolized in this simplified discussion by ΔX. If two models have some parameters in common then the prior ranges for these parameters will cancel in the calculation of the Bayes factor. To make good use of Bayesian model selection, we need to fully specify priors that are independent of the current data D. The sensitivity of the marginal likelihood to the prior range depends on the shape of the prior and is much greater for a uniform prior than a scale invariant prior (e.g., see Gregory [10], page 61). In most instances we are not particularly interested in the Occam factor itself, but only in the relative probabilities of the competing models as expressed by the Bayes factors. Because the Occam factor arises automatically in the marginalization procedure, its effect will be present in any model selection calculation. Note: no Occam factors arise in parameter estimation problems. Parameter estimation can be viewed as model selection where the competing models have the same complexity so the Occam penalties are identical and cancel out.

The MCMC algorithm produces samples which are in proportion to the posterior probability distribution which is fine for parameter estimation but one needs the proportionality constant for estimating the model marginal likelihood. Clyde et al. [35] reviewed the state of techniques for model selection from a statistical perspective and Ford and Gregory [14] have evaluated the performance of a variety of marginal likelihood estimators in the exoplanet context. Other techniques that have recently been proposed include: Nested Restricted Monte Carlo (NRMC) [24], MultiNest [36], Annealing Adaptive Importance Sampling (AAIS) [37], and Reversible Jump Monte Carlo using a kD-tree [38]. For one planet models with 7 parameters, a wide range of techniques perform satisfactorily. The challenge is to find techniques that handle high dimensions. A six planet model has 32 parameters and one needs to develop and test methods of handling at least 8 planets with 42 parameters. At present there is no widely accepted method to deal with this challenge.

For the 47UMa and Gliese 581 data discussed in Sections 7.3.1 and 7.3.2 the author employed Nested Restricted Monte Carlo (NRMC) to estimate the marginal likelihoods. Monte Carlo (MC) integration can be very inefficient in exploring the whole prior parameter range because it randomly samples the whole volume. The fraction of the prior volume of parameter space containing significant probability rapidly declines as the number of dimensions increase. For example, if the fractional volume with significant probability is 0.1 in one dimension then in 32 dimensions the fraction might be of order 10^{-32}. In restricted MC integration (RMC) this problem is reduced because the volume of parameter space sampled is greatly restricted to a region delineated by the outer borders of the marginal distributions of the parameters for the particular model. However, in high dimensions most of the MC samples will fall near the outer boundaries of that volume and so the sampling could easily under sample interior regions of high probability.

In NRMC integration, multiple boundaries are constructed based on credible regions ranging from 30% to ≥99%, as needed. We are then able to compute the contribution to the total integral from each nested interval and sum these contributions. For example, for the interval between the 30% and 60% credible regions, we generate random parameter samples within the 60% region and reject any sample that falls within the 30% region. Using the remaining samples we can compute the contribution to the NRMC integral from that interval.

The left panel of Fig. 7.19 shows the NRMC contributions to the marginal likelihood from the individual intervals for five repeats of a 3 planet fit to the HARPS data [33]. The right panel shows the summation of the individual contributions versus the volume of the credible region. The credible region listed as 9995% is defined as follows. Let X_{U99} and X_{L99} correspond to the upper and lower boundaries of the 99% credible region, respectively, for any of the parameters. Similarly, X_{U95} and X_{L95} are the upper and lower boundaries of the 95% credible region for the parameter. Then $X_{U9995} = X_{U99} + (X_{U99} - X_{U95})$ and $X_{L9995} = X_{L99} + (X_{L99} - X_{L95})$. Similarly[7], $X_{U9984} = X_{U99} + (X_{U99} - X_{U84})$. For the 3 planet fit the spread in results is within ±23% of the mean. For each credible region interval approximately 320,000 MC samples were used. The mean value of the prior × likelihood within the 30% credible region is a factor of 2×10^5 larger than the mean in the shell between the 97 and 99% credible regions. However, the volume of parameter space

---

[7] Test that the extended credible region (like 9930) for the period parameter does not overlap the credible region of an adjacent period parameter in a multiple planet fit.

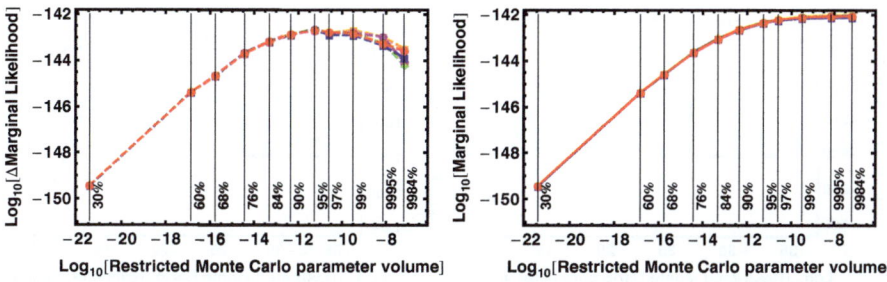

Fig. 7.19 Left panel shows the contribution of the individual nested intervals to the NRMC marginal likelihood for the 3 planet model for five repeats. The right panel shows the sum of these contributions versus the parameter volume of the credible region.

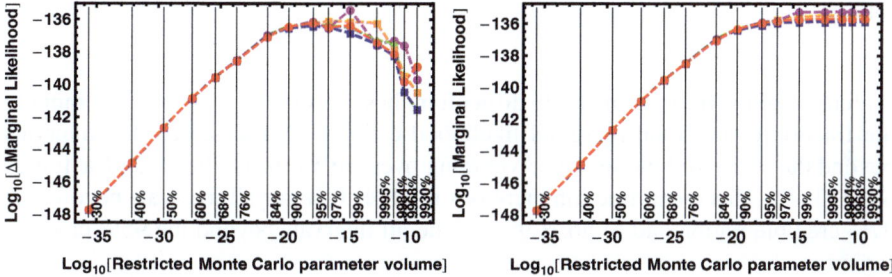

Fig. 7.20 Left panel shows the contribution of the individual nested intervals to the NRMC marginal likelihood for the 5 planet model for five repeats. The right panel shows the integral of these contributions versus the parameter volume of the credible region.

in the shell between the 97 and 99% credible regions is a factor of 8×10^{11} larger than the volume within the 30% credible region so the contribution from the latter to the marginal likelihood is negligible.

The left panel of Fig. 7.20 shows the contributions from the individual credible region intervals for five repeats of the NRMC marginal likelihood estimate for a 5 planet fit to the HARPS data [33]. The right panel shows the summation of the individual contributions versus the volume of the credible region. In this case the spread in five NRMC marginal likelihood estimates extends from 2.1× the mean to 0.65× the mean.

The biggest contribution to the spread in NRMC marginal likelihood estimates for the 5 planet fit comes from the outer credible region intervals starting around 99%. The reason for the increased scatter in the $Log_{10}[\Delta$ Marginal Likelihood] is apparent when we examine the MC samples for one of the five repeats. Fig 7.21 shows plots of the maximum and mean values (Left) and maximum, mean, and minimum values (Right) of the MC samples of the $Log_{10}[prior \times likelihood]$ for each interval of credible region, versus the parameter volume. The range of $Log_{10}[prior \times likelihood]$ values increase rapidly with increasing parameter volume starting around the 99% credible region boundary. This makes the Monte Carlo

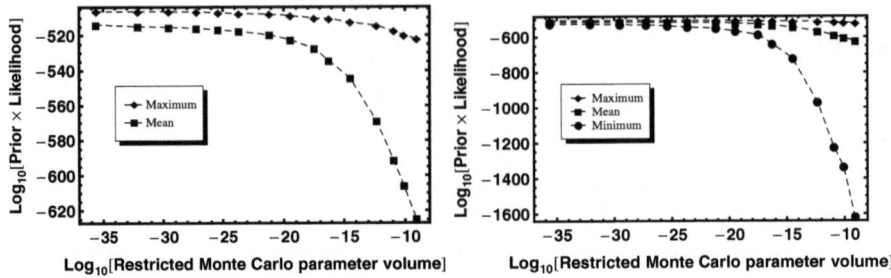

Fig. 7.21 Left panel shows the maximum and mean values of the $\text{Log}_{10}[\text{prior} \times \text{likelihood}]$ for each interval of credible region versus parameter volume for the 5 planet fit. The right panel shows the maximum, mean, and minimum values of the $\text{Log}_{10}[\text{prior} \times \text{likelihood}]$ versus the parameter volume.

evaluation of the mean value more difficult in these outer intervals. We can compute the fraction of our total marginal likelihood estimate that arises for the intervals beyond the 99% credible region. Averaging over the 5 repeats the mean fraction is $6.6 \pm 3.4\%$. This mean fraction increases to $36 \pm 12\%$ for all intervals beyond the 97% credible region and to $56 \pm 9\%$ for the all intervals beyond the 95% credible region.

The NRMC method is expected to underestimate the marginal likelihood in higher dimensions and this underestimate is expected to become worse the larger the number of model parameters, i.e., increasing number of planets [39]. When we conclude, as we do, that the NRMC computed odds in favor of a five planet model for Gliese 581 compared to the four planet model is $\sim 10^2$ (see Table 7.2), we mean that the true odds is $\geq 10^2$. Thus the NRMC method is conservative. One indication of the break down of the NRMC method is the increased spread in the results for repeated evaluations that was discussed above. A recent comparison by Gregory [40] of the NRMC method with a second method, the Ratio Estimator (RE), described in Ford and Gregory [14], indicates the two methods agree within 25% for a 3 planet model fit (17 parameters) to three different exoplanet data sets including Gliese 581. However, the agreement between the two methods breaks down at the 4 planet fit level (22 parameters) with the value of the RE estimate 62 times the NRMC estimate for the Gliese 581 HARPS data. Unlike the NRMC method, the RE method has the potential to pay too much attention to the mode and is expected to overestimate the marginal likelihood [41] at sufficiently high dimensions.

We can readily convert the Bayes factors to a Bayesian False Alarm Probability (FAP) which we define in equation 7.8. For example, in the context of claiming the detection of m planets the FAP_m is the probability that there are actually fewer than m planets, i.e., $m - 1$ or less.

$$\text{FAP}_m = \sum_{i=0}^{m-1} (\text{prob. of i planets}) \qquad (7.8)$$

Table 7.2 Marginal likelihood estimates [17], Bayes factors relative to model 4, and false alarm probabilities.

Model	Periods (d)	Marginal Likelihood	Bayes factor nominal	False Alarm Probability
M_0		6.10×10^{-197}	2.0×10^{-59}	
M_1	(5.37)	$(4.221 \pm 0.003) \times 10^{-155}$	1.4×10^{-17}	1.4×10^{-42}
M_2	(5.37, 12.9)	$(1.94 \pm 0.01) \times 10^{-145}$	6.5×10^{-8}	2.2×10^{-10}
M_3	(5.37, 12.9, 66.9)	$(3.0^{+0.7}_{-0.5}) \times 10^{-142}$	10^{-4}	6.5×10^{-4}
M_4	(3.15, 5.37, 12.9, 66.9)	$(3.0^{+1.1}_{-0.6}) \times 10^{-138}$	1.0	10^{-4}
M_5	(3.15, 5.37, 12.9, 66.9, 399)	$(3.0^{\times 2.1}_{\times 0.65}) \times 10^{-136}$	10^2	0.01
M_6	(3.15, 5.37, 12.9, 34.4, 66.9, 399)	$(6.7^{\times 2.4}_{\times 1/3}) \times 10^{-141}$	2.2×10^{-3}	0.999978

If we assume *a priori* (absence of the data) that all models under consideration are equally likely, then the probability of each model is related to the Bayes factors by

$$p(M_i \mid D, I) = \frac{B_{i4}}{\sum_{j=0}^{N} B_{j4}} \quad (7.9)$$

where N is the maximum number of planets in the hypothesis space under consideration, and of course $B_{44} = 1$. For the purpose of computing FAP_m we set $N = m$. Substituting Bayes factors, given in Table 7.2, into equation 7.8 gives

$$\text{FAP}_5 = \frac{(B_{04} + B_{14} + B_{24} + B_{34} + B_{44})}{\sum_{j=0}^{5} B_{j4}} \approx 10^{-2} \quad (7.10)$$

For the 5 planet model we obtain a low FAP $\approx 10^{-2}$.

Table 7.2 gives the NRMC marginal likelihood estimates[8], Bayes factors and false alarm probabilities for 0, 1, 2, 3, 4, 5 and 6 planet models which are designated $M_0, \cdots, M_6$. For each model the NRMC calculation was repeated 5 times and the quoted errors give the spread in the results, not the standard deviation. The Bayes factors that appear in the third column are all calculated relative to model 4. Based on the HARPS [33] data, the Bayes factor favors a 5 planet model.

7.5 Conclusions

The main focus of this chapter has been on a new fusion MCMC approach to Bayesian nonlinear model fitting. In fusion MCMC the goal has been to develop an

---

[8] Table 7.1 gives two different choices of prior for the eccentricity parameter. The marginal likelihoods listed in the Table 7.2 correspond to the eccentricity noise bias prior. Marginal likelihood values assuming a uniform eccentricity prior were systematically lower. For example, for a 3 planet fit using the uniform eccentricity prior the marginal likelihood was a factor of 3 smaller.

automated MCMC algorithm which is well suited to exploring multi-modal probability distributions such as those that occur in the arena of exoplanet research. This has been accomplished by the fusion of a number of different statistical tools. At the heart of this development is a sophisticated control system that automates the selection of efficient MCMC proposal distributions (including for highly correlated parameters) in a parallel tempering environment. It also adapts to any new significant parameter set that is detected in any of the parallel chains or is bred by a genetic crossover operation. This controlled statistical fusion approach has the potential to integrate other relevant statistical tools as required. A future goal is to automate the selection of an efficient set of β values used in the parallel tempering.

For some special applications it is possible to develop a faster more specialized MCMC algorithm, perhaps for dealing with real time analysis situations. In the development of fusion MCMC, the primary concern has not been speed but rather to see how powerful a general purpose MCMC algorithm we could develop and automate. In real life applications to challenging multi-modal exoplanet data fusion MCMC is proving to be a powerful tool. One can anticipate that this approach might also allow for the joint analysis of different types of data (e.g., radial velocity, astrometry, and transit information) giving rise to statistical fusion and data fusion algorithms.

On the Bayesian model selection front, a wide variety of marginal likelihood estimators perform satisfactorily for models with ~ 7 parameters and the author has achieved satisfactory agreement between two different Bayesian estimators at the 17 parameter level. One of these, the nested restricted Monte Carlo (NRMC) method, has been investigated for up to 32 model parameters. More research aimed at the development and testing of reliable and efficient estimators that work in higher dimensions is an important priority.

Acknowledgements The author would like to thank Wolfram Research for providing a complementary license to run gridMathematica.

References

1. Campbell, B., Walker, G.A.H., Yang, S.: A Search for Substellar Companions to Solar-type Stars. Astrophys. J. **331**, 902–921 (1988)
2. Wolszczan, A., Frail, D.: A Planetary System Around the Millisecond Pulsar PSR1257 + 12. Nature **355**, 145–147 (1992)
3. Mayor M., Queloz D.: A Jupiter-mass companion to a solar-type star. Nature **378**, 355–359 (1995)
4. Marcy G.W., Butler R.P.: A Planetary Companion to 70 Virginis. Astrophys. J. **464**, L147–151 (1996)
5. Udry, S., Bonfils, X., Delfosse, X., Forveille, T., Mayor, M., Perrier, C., Bouchy, F., Lovis, C., Pepe, F., Queloz, D., Bertaux, J.-L.: The HARPS search for southern extra-solar planets. XI. Super-Earths (5 and 8 $M_\oplus$) in a 3-planet system. Astron. Astrophys. **469**, L43–L47 (2007)
6. Lovis, C., Ségransan, D., Mayor, M., Udry, S., Benz, W., Bertaux, J.-L., Bouchy, F., Correia, A.C.M., Laskar, J., Lo Curto, G., Mordasini, C., Pepe, F., Queloz, D., Santos, N.C.: The HARPS search for southern extra-solar planets. XXVIII. Up to seven planets orbiting HD 10180: probing the architecture of low-mass planetary systems. Astron. Astrophys. **528**, 112L (2011)

7. Loredo, T.L., Chernoff, D.: Bayesian Adaptive Exploration. In: Feigelson, E.D., Babu, G.J. (eds.) Statistical Challenges in Modern Astronomy III, pp. 57–69. Springer-Verlag, New York (2003)
8. Loredo, T.: Bayesian Adaptive Exploration. In: Erickson, G.J., Zhai, Y. (eds.) Bayesian Inference And Maximum Entropy Methods in Science and Engineering: 23rd International Workshop, pp. 330–346. AIP Conf. Proc., Vol. 707 (2004)
9. Cumming, A.: Detectability of extrasolar planets in radial velocity surveys. Mon. Not. R. Astron. Soc., **354**, 1165–1176 (2004)
10. Gregory, P.C.: Bayesian Logical Data Analysis for the Physical Sciences: A Comparative Approach with *Mathematica* Support, Cambridge University Press (2005)
11. Gregory, P.C.: A Bayesian Analysis of Extra-solar Planet Data for HD 73526. Astrophys. J. **631**, 1198–1214 (2005)
12. Ford, E.B.: Quantifying the Uncertainty in the Orbits of Extrasolar Planets. Astron. J. **129**, 1706–1717 (2005)
13. Ford, E.B.: Improving the Efficiency of Markov Chain Monte Carlo for Analzing the Orbits of Extrasolar Planets. Astrophys. J. **642**, 505–522 (2006)
14. Ford, E.B., Gregory, P.C.: Bayesian Model Selection and Extrasolar Planet Detection. In: Babu, G.J., Feigelson, E.D. (eds.) Statistical Challenges in Modern Astronomy IV, pp. 189–204. ASP Conf. Ser., Vol. 371 (2007)
15. Cumming, A., Dragomir, D.: An Integrated Analysis of Radial Velocities in Planet Searches. Mon. Not. R. Astron. Soc., **401**, 1029–1042 (2010)
16. Dawson, R.I., Fabrycky, D.C.: Radial Velocity Planets De-aliased: A New, Short Period for Super-Earth 55 Cnc e. Astrophys. J. **722**, 937–953 (2010)
17. Gregory, P.C.: Bayesian Re-analysis of the Gliese 581 Exoplanet System. Mon. Not. R. Astron. Soc. **415**, 2523–2545 (2011)
18. Jaynes, E.T.: Bayesian Spectrum and Chirp Analysis. In: Smith, C.R., Erickson, G.L. (eds.) Maximum Entropy and Bayesian Spectral Analysis and Estimation Problems, pp. 1-37. D. Reidel, Dordrecht (1987)
19. Bretthorst, G.L.: Bayesian Spectrum Analysis and Parameter Estimation, Springer-Verlag, New York (1988)
20. Geyer, C.J.: Markov Chain Monte Carlo. In: Keramidas, E.M. (ed.) Computing Science and Statistics, pp. 156–163. Proceedings of the 23rd Symposium on the Interface, Interface Foundation, Fairfax Station (1991)
21. Hukushima, K., Nemoto, K.: Exchange Monte Carlo Method and Application to Spin Glass Simulations. Journal of the Physical Society of Japan **65**(4), 1604–1608 (1996)
22. Atchadé, Y.F., Roberts, G.O., Rosenthal, J.S.: Towards optimal scaling of metropolis-coupled Markov chain Monte Carlo. Stat. Comput. **21**(4), 555–568 (2011)
23. Gregory, P.C.: A Bayesian Re-analysis of HD 11964: Evidence for Three Planets. In: Knuth, K.H., Caticha, A., Center, J.L., Giffin, A., Rodrguez, C.C. (eds.) Bayesian Inference and Maximum Entropy Methods in Science and Engineering: 27th International Workshop, pp. 307–314. AIP Conference Proceedings, Vol. 954 (2007)
24. Gregory, P.C., Fischer, D.A.: A Bayesian Periodogram Finds Evidence for Three Planets in 47 Ursae Majoris. Mon. Not. R. Astron. Soc. **403**, 731–747 (2010)
25. Roberts, G.O., Gelman, A., Gilks, W.R.: Weak convergence and optimal scaling of random walk Metropolis algorithms. Ann. Appl. Probab. **7**, 110–120 (1997)
26. Ter Braak, C.J.F.: A Markov Chain Monte Carlo version of the genetic algorithm Differential Evolution: easy Bayesian computing for real parameter spaces. Stat. Comput. **16**, 239–249 (2006)
27. Gregory, P.C.: Bayesian Exoplanet Tests of a New Method for MCMC Sampling in Highly Correlated Parameter Spaces. Mon. Not. R. Astron. Soc. **410**, 94–110 (2011)
28. Fischer, D.A., Laughlin, G.L., Butler, R.P., Marcy, G.W., Johnson, J., Henry, G.,Valenti, J., Vogt, S.S., Ammons, M., Robinson, S., Spear, G., Strader, J., Driscoll, P., Fuller, A., Johnson, T., Manrao, E., McCarthy, C., Munõz, M., Tah, K.L., Wright, J., Ida, S., Sato, B., Toyota, E.,

Minniyi, D.: The N2K Consortium. I. A Hot Saturn Planet Orbiting HD 88133. Astrophys. J. **620**, 481–486 (2005)
29. Gregory, P.C.: A Bayesian Kepler Periodogram Detects a Second Planet in HD 208487. Mon. Not. R. Astron. Soc. **374**, 1321–1333 (2007)
30. Jaynes, E.T.: How Does the Brain Do Plausible Reasoning?, Stanford University Microwave Laboratory Report 421, 1957, Reprinted in Erickson, G.J., Smith, C.R. (eds.) Maximum Entropy and Bayesian Methods in Science and Engineering, pp. 1-29. Kluwer Academic Press, Dordrecht (1988)
31. Tinney, C.G., Butler, R.P., Marcy, G.W., Jones, H.R.A., Penny, A.J., McCarthy, C., Carter, B.D., Fischer, D.A.: Three Low-Mass Planets from the Anglo-Australian Planet Search, Astrophys. J. **623**, 1171–1179 (2005)
32. Wittenmyer, R.A., Endl, M., Cochran, W.D., Levison, H.F., Henry, G.W.: A Search for Multi-Planet Systems Using the Hobby-Eberly Telescope. Astrophys. J. Suppl. **182**, 97–119 (2009)
33. Mayor, M., Bonfils, X., Forveille, T., Delfosse, X, Udry, S., Bertaux, J.-L., Beust, H., Bouchy, F., Lovis, C., Pepe, F., Perrier, C., Queloz, D., Santos, N.C.: The HARPS search for southern extra-solar planets. XVIII. An earth-mass planet in the GJ 581 planetary system. Astron. Astrophys. **507**, 487–494 (2009)
34. Vogt, S.S, Butler, R.P., Rivera, E.J., Haghighipour, N., Henry, G.W., Williamson, M.H.: The Lick-Carnegie Exoplanet Survey: A 3.1 Earth Mass Planet in the Habitable Zone of the Nearby M3V Star Gliese 581. Astrophys. J. **723**, 954–965 (2010)
35. Clyde, M.A., Berger, J.O., Bullard, F., Ford, E.B., Jeffreys, W.H., Luo, R., Paulo, R., Loredo, T.: Current Challenges in Bayesian Model Choice. In: Babu, G.J., Feigelson, E.D. (eds.) Statistical Challenges in Modern Astronomy IV, pp. 224–240. ASP Conf. Ser., Vol. 371 (2007)
36. Feroz, F., Balan, S.T., Hobson, M.P.: Detecting extrasolar planets from stellar radial velocities using Bayesian evidence. Mon. Not. R. Astron. Soc. **415**, 3462–3472 (2011)
37. Loredo, T.L., Berger, J.O., Chernoff, D.F., Clyde, M.A., Liu, B.: Bayesian Methods for Analysis and Adaptive Scheduling of Exoplanet Observations. Stat. Meth. **9**, 101–114 (2011)
38. Farr, W.M., Sravan, N., Cantrell, A., Kreidberg, L., Bailyn, C.D., Mandel, I., Kalogera, V.: The Mass Distribution of Stellar-Mass Black Holes. Astrophys. J. **741**, 103–122 (2011)
39. Gregory, P.C.: A Bayesian Periodogram Finds Evidence for Three Planets in HD 11964. Mon. Not. R. Astron. Soc. **381**, 1607–1619 (2007)
40. Gregory, P.C.: Discussion on paper by Martin Weinberg regarding Bayesian Model Selection and Parameter Estimation. In: Feigelson, E.D., Babu, G.J. (eds.) Statistical Challenges in Modern Astronomy V, Springer-Verlag (2012, in press)
41. Jefferys, W.H.: Discusion on "Current Challenges in Bayesian Model Choice" by Clyde et al. In: Babu, G.J., Feigelson, E.D. (eds.) Statistical Challenges in Modern Astronomy IV, pp. 241–244. ASP Conf. Ser., Vol. 371, (2007)

Chapter 8
Classification and Anomaly Detection for Astronomical Survey Data

Marc Henrion, Daniel J. Mortlock, David J. Hand, and Axel Gandy

Abstract We present two statistical techniques for astronomical problems: a star-galaxy separator for the UKIRT Infrared Deep Sky Survey (UKIDSS) and a novel anomaly detection method for cross-matched astronomical datasets. The star-galaxy separator is a statistical classification method which outputs class membership probabilities rather than class labels and allows the use of prior knowledge about the source populations. Deep Sloan Digital Sky Survey (SDSS) data from the multiply imaged Stripe 82 region are used to check the results from our classifier, which compares favourably with the UKIDSS pipeline classification algorithm. The anomaly detection method addresses the problem posed by objects having different sets of recorded variables in cross-matched datasets. This prevents the use of methods unable to handle missing values and makes direct comparison between objects difficult. For each source, our method computes anomaly scores in subspaces of the observed feature space and combines them to an overall anomaly score. The proposed technique is very general and can easily be used in applications other than astronomy. The properties and performance of our method are investigated using both real and simulated datasets.

8.1 Introduction

Astronomy has greatly profited from recent advances in telescope and detector technologies and data storage capabilities. The situation is now such that astronomers are facing a veritable "data avalanche" [1]. The generated datasets are incredibly rich and their analysis requires automated and efficient methods.

In this chapter we will present two tools developed for two types of analysis: a star-galaxy classification method and an algorithm for detecting anomalous sources.

Marc Henrion
Department of Mathematics, Imperial College London, London SW7 2AZ, U.K.,
e-mail: marc.henrion03@imperial.ac.uk

The former is a common task, preliminary to many analyses of astronomical data. The particular classifier that we will describe is a parameterised model that can incorporate prior information about source populations. Though the classifier has been developed for a particular survey, its basic principles can be easily applied to other surveys. Anomaly detection is a more specialised and exploratory task, facilitating the discovery of novel types of objects, though follow-up observations of any set of candidate anomalies must be made. The anomaly detection algorithm that we will describe is, unlike the star-galaxy separator, totally data-driven.

We will summarise the two surveys that we will use throughout this chapter in Sect. 8.2. The star-galaxy separation method is presented in Sect. 8.3 and the anomaly detection algorithm is described in Sect. 8.4.

8.2 The Sloan Digital Sky Survey (SDSS) and the UKIRT Infrared Deep Sky Survey (UKIDSS)

8.2.1 The Sloan Digital Sky Survey (SDSS)

Operations on SDSS [2] started in 1998 and have been continuously ongoing until the present day. So far, there have been 3 phases to SDSS: SDSS-I (2000-2005), SDSS-II (2005-2008) and SDSS-III (2008-ongoing). SDSS-III is currently scheduled to run until 2014.

Observations are made with a dedicated 2.5m telescope [3] at Apache Point Observatory in New Mexico, USA. Sources are observed in 5 optical filters (u, g, r, i, z; [4]) and, for a subset of sources, full spectra are measured using a pair of spectrographs. The SDSS magnitudes are on the AB system.

SDSS uses asinh magnitudes, rather than logarithmic magnitudes. For sources with signal-to-noise ratios exceeding 5, these are essentially identical to logarithmic magnitudes [5]. The advantage of such magnitudes is that it is possible to compute magnitudes for objects with negative background-subtracted flux values. This allows SDSS to include the measured fluxes of undetected sources in their database. However, the additional magnitudes obtained in this way will have low signal-to-noise ratios.

SDSS-I and II have surveyed $\sim 1.16 \cdot 10^4$ deg^2 with single observations, to depths of $u \simeq 22.0$, $g \simeq 22.2$, $r \simeq 22.2$, $i \simeq 21.3$ and $z \simeq 20.5$. Spectra have been obtained for $\sim 1.6 \cdot 10^6$ sources.

The SDSS has also taken repeat measurements in an area along the celestial equator (covering the right ascension (ra) range $\alpha \leq 59$ deg and $\alpha \geq 300$ deg and declinations (dec) of $|\delta| \leq 1.25$ deg), known as Stripe 82. This data was released along with DR7 and reaches depths of $u \simeq 23.6$, $g \simeq 24.5$, $r \simeq 24.2$, $i \simeq 23.8$ and $z \simeq 22.1$.

8.2.2 The UKIRT Infrared Deep Sky Survey (UKIDSS)

UKIDSS [6] is a suite of five separate near-infrared surveys. A detailed technical description of the survey is given by [7], although there have been several improvements in the time since [8]. Observations began in May 2005 and are scheduled to continue until 2012. UKIDSS aims to survey ~ 7500 deg^2 of the Northern Sky. Observations are made using the Wide Field Camera (WFCAM; [9]) on the 3.8m United Kingdom Infrared Telescope (UKIRT) on Mouna Kea, Hawaii.

The UKIDSS surveys include imaging in five near-infrared bands, Z, Y, J, H and K (defined in [10]). One of the surveys, the Large Area Survey (LAS), includes imaging in only four of these bands (Y, J, H and K). The UKIDSS magnitudes are on the Vega system, with the offsets to the AB system provided in [10].

The five UKIDSS surveys are: the Large Area Survey (LAS); the Galactic Plane Survey (GPS); the Galactic Clusters Survey (GCS); the Deep Extragalactic Survey (DXS) and the Ultra Deep Survey (UDS). LAS, DXS and UDS are extra-Galactic surveys, whereas GPS and GCS are Galactic surveys. LAS surveys the widest ($\sim 4,000$ deg^2) and shallowest (up to $K \simeq 18.2$) area, whereas UDS surveys the smallest area (~ 0.77 deg^2), but this to a far greater depth ($K \simeq 23.0$). The UKIDSS LAS survey has also observed the SDSS Stripe 82 region.

For the remainder of this chapter, all photometry is given in the native system of the telescope in question. Thus SDSS u, g, r, i, z photometry is on the AB system, whereas UKIDSS Y, J, H and K photometry is Vega-based.

8.3 A Bayesian Approach to Star-Galaxy classification

Astronomical surveys now gather data on huge numbers of astronomical objects: the 2 Micron All Sky Survey (2MASS; [11]), SDSS and the UKIDSS have all identified hundreds of millions of distinct sources. The scale of these projects immediately necessitates an automated approach to data analysis (although an intriguing alternative is The Galaxy Zoo project described by [12]). Considerable effort has been put into developing algorithms which can decompose an image into a smooth background and a catalogue of discrete objects, the properties of which must be characterised as well. Source positions, fluxes and shapes can all be estimated reliably by using fairly simple moment-based approaches (e.g. [13, 14]), but the separation of point-like stars from more extended galaxies generally requires at least some external astrophysical information be included. As such, the problem of star-galaxy classification is well suited to methods which, by use if Bayes's theorem, combine the measurements of a given source with prior knowledge of the astro-physical populations of which the source might be a member. Here, we will develop a practical formalism for one such method.

The star-galaxy separation method we present below has been published previously in [15]. We will not repeat the contents of that article here, but we will give a more statistical formulation of this particular classification method. We will also make a few additional points, in particular regarding other commonly used methods to separate stars and galaxies.

In Sect. 8.3.1 we will make a few general remarks about existing star-galaxy classification techniques. The general formalism of our star-galaxy separation method is developed in Sect. 8.3.2.1, and specialised to UKIDSS in Sect. 8.3.2.3. Real UKIDSS data are analysed – and the results compared to the classifications from deeper SDSS data – in Sect. 8.3.3. The relative merits of the formalism we have chosen are summarised in Sect. 8.3.3.4.

8.3.1 Star-galaxy Separation Methods in Use

8.3.1.1 Morphology vs. Spectral/Colour Information

Deciding on whether to use morphology, colour or spectral variables to separate stars and galaxies appears to be mainly a feature selection problem. However, different sets of variables correspond to making use of different physical properties of sources to differentiate between stars and galaxies.

The starting point for all morphology-based methods of star-galaxy classification is that stars and galaxies appear differently, the latter being more extended (at a given flux level) and also exhibiting more variety. For bright sources these differences are easily distinguished by the human eye (as demonstrated so well by the Galaxy Zoo project; [12]), but for faint sources this visual classification is much more difficult – and time-consuming. Further, the challenge is to develop automatic algorithms that can perform the same task from measured image properties. To characterise an object's morphology, astronomers typically use areal profiles (i.e. the area of the source's image above certain light intensity thresholds), or curves of growth (i.e. the amount of light contained within certain radii around the object centre). Sometimes, however, the (calibrated) pixel intensities are used directly to assess object types [16, 17].

Galaxies and stars are physically different objects and this difference is reflected in their spectral profiles: they emit different amounts of light at different parts of the electromagnetic spectrum, cf. Fig. 1 from [18]. A spectral approach can usually differentiate more easily between stars and other sources which appear point-like (e.g. quasars). Spectral classification usually proceeds by determining a best-fit template for each source and assigning the source to the class of the template, e.g. [19]. However, measuring spectra is time-consuming and expensive. As a result most surveys do not measure full spectra, but only measure the light emitted by sources over broad ranges of wavelengths, called filter passbands. For this reason we will not consider spectral source classification further in this work.

It is possible to use colour information (i.e. the difference in light emitted in different filters) to classify stars and galaxies. As galaxies typically emit more light at the longer wavelengths, such a classification is equivalent to classifying sources into red (typically galaxies) and blue (typically stars) objects. This approach is often preferred for quasar selection [20, 21], but can also be used for star-galaxy separation [18, 22].

8.3.1.2 Multi-band Surveys: Single-band and Global Classifications; Missing Detections

The vast majority of surveys nowadays obtain data in multiple filter passbands. As far as star-galaxy separation is concerned, this raises two issues: first there is the problem of whether classifications should be computed in individual bands or globally, in a multi-band approach and, secondly, there is the problem of what should be done with sources with missing detections in one or several bands.

Multi-dimensional methods making use of the full multi-band data can be used to compute global object classifications ([18, 20–25]). In these articles only a global classification is computed and single-band classifications are not made available. However, for some science uses, it might be more useful to have a band-specific classification, rather than a multi-band one.

One approach to obtaining both single-band classifications and a global classification consists in classifying sources in each band and then combining these band-specific classifications to obtain a global, multi-band, classification. The UKIDSS pipeline classifier [26] obtains a global classification in two different ways (one yielding a class label, the other a posterior class membership probability). By thresholding on a morphology statistic, the classifier assigns class labels to objects in each individual band. A multi-band class label is obtained by combining the band-specific morphology statistics to a multi-band morphology statistic and applying the same thresholding rule to this combined statistic. To obtain multi-band posterior class probabilities, each band-specific class label is assigned approximate probabilities that objects assigned to this class are truly of that type or one of the other types (e.g. for a given band, an object assigned the class label "star", "probable star", "probable galaxy" or "galaxy" is considered to actually be a star with probability 0.9, 0.7, 0.25 and 0.05 respectively in that band). These approximate probabilities are then used to compute a global classification. While this scheme has the flexibility of providing both single-band and multi-band classifications, its heuristic nature is unsatisfactory.

Missing detections in multi-band surveys constitute another problem: for various reasons (emission profiles of sources, sensitivities of the different filters, ...) some objects are not observed in all the bands of the survey. Ideally sources with missing detections in some bands should also be classified using the data from the bands they have been observed in.

In [21, 23, 25], the authors discard objects with missing detections or non-physical values in some bands as their classification methods require data with no missing values. [27] compute classifications in each of three bands and then use a system of priority to select a global classification from these. The three bands are ranked according to the classifier's performance on test data and, if an object has been observed in more than one band, the classification from the best performing band is used, otherwise the classification from the only available band is used.

One way around the problem posed by non-detections is given by the SDSS survey: rather than use logarithmic magnitudes, SDSS uses asinh magnitudes [5]. This allows SDSS to measure magnitudes in all bands even for objects which have not been detected in some bands. The classification methods from [20, 23, 24] thus avoid the non-detection problem. However, the data from objects with missing values will have low signal-to-noise ratios in the missing bands and the measured colours of such objects can be affected by the magnitude measurements in the missing bands, which will be close to the zero flux magnitude in those bands. This in turn can affect the quality of the classifications.

8.3.1.3 Classifier Training and Testing: Labelled Data

To train or test a given classifier we need labelled data. However, we do not know the true type of most objects in the sky. In particular, faint sources are key to testing a classifier as classifying bright sources is usually straightforward. However, getting good quality labelled data of faint sources can be problematic.

Classification by visual inspection is usually reliable, but is not perfect, particularly for faint sources, and suffers from the subjective bias due to the expert. It is also very time-consuming and thus visually classified training / testing datasets tend to be small in size.

Taking a survey and cross-matching its sources with those from a deeper or high-resolution spectroscopic survey is one way of getting labelled data. The assumption here is that any classifier, however simple, used on the deeper or spectroscopic survey will be able to classify sources correctly, as these data will be of higher resolution. However, a deeper and overlapping survey does not always exist.

Some authors (e.g. [14]) choose to use simulated training data. Another strategy consists in using hybrid real/simulated training data by taking medium bright objects, which are easy to classify, and simulating their appearance at fainter magnitudes or simply adding noise (e.g. [28] test the effect of noisy data on their classifier performance by degrading labelled images). However, simulated data will always reflect the authors' understanding of the processes affecting the visual appearance of sources.

While the problem of testing the classifier remains, training / fitting the classifier can be done with un- or only partially labelled data. For instance, one can develop a more robust classifier, able to deal with unlabelled and even misclassified data, as long as correctly classified data form the vast majority of the training data. [29], for instance, propose a semi-supervised clustering technique, which also allows defining classes of unknown object types if some clusters contain only unlabelled data. Another approach consists in using prior knowledge about source populations and extrapolating the classification rules from brighter sources to fainter regimes. Cut-based techniques, such as [26] and [30], use this approach, as do more sophisticated techniques, e.g. [31].

Having made these general remarks, we will now develop a general formalism for star-galaxy separation in UKIDSS. For a more thorough review of commonly used source classification techniques, see [15].

8.3.2 The Model

Before developing our model, we should note that, in the astronomy literature the term "Bayesian" is taken to include any method which uses both information on the source populations (to obtain prior probabilities) and the observed data (to form the likelihood terms) to compute posterior class membership probabilities. Referring to such methods, of which our classifier is an example, simply as Bayes classifiers might be more correct, but here we will adopt the convention used in the astronomical community.

Let T be a random variable giving the object type of an astronomical source and let X be the random vector giving the available data. Throughout this chapter, for notational ease, we have replaced the more formal $\Pr(T = t | X = x)$ with the less cumbersome, if occasionally ambiguous, $p(t|x)$. Also, again for notational ease, we will not differentiate between probabilities, probability mass functions and probability density functions: we will write $p(.)$ for each of these. Thus, when we write $p(x)$, we could mean the probability $p(X = x)$ or the probability density function of the random variable X (or even the density of a random variable Y) evaluated at the point x. What we mean in each case, however, will always be clear from the context.

8.3.2.1 General Formalism for Classifying Astronomical Sources

Suppose a noisy, seeing-smeared and pixelated image of a source has been measured. We want to express our confidence that the source is of a given type. Suppose there are N_t different types of astronomical objects: $\{t_1, t_2, \ldots, t_{N_t}\}$. Given data $x = (x_1, x_2, \ldots, x_{N_x})$, where N_x is the number of measured variables, we wish to calculate the posterior class membership probabilities, $p(t|x)$, for each object type t. This is achieved by using Bayes' theorem:

$$p(t|x) = \frac{p(t)p(x|t)}{\sum_{t'=1}^{N_t} p(t')p(x|t')}, \qquad (8.1)$$

where $p(t)$ is the prior probability that the source is of type t and $p(x|t)$ is the probability density of getting the observed data under the hypothesis that the source is of type t.

8.3.2.2 Star-galaxy Classification

At this point we will start making simplifying assumptions, notably:

- Every object is either a star or a galaxy, i.e. we assume, for this work, that T can take one of two values: s for stars and g for galaxies. While there are many other populations of astronomical objects, the vast majority of sources will either be Galactic stars or galaxies. The next most common type of objects are quasars, but as their name suggest, they appear as point-sources in the optical and near-infrared bands, and hence can be included with the stars in the context of morphological classification.
- The morphological information contained in an image of a source can be compressed into a single statistic. The potentially large parameter space is greatly reduced by the use of a single morphology statistic, c. c simply encodes the degree to which a given source is not point-like. There is great freedom in specifying c and, even, what the null distribution of c for stars should be. For UKIDSS, this assumption is motivated by the existence of an excellent morphology statistic: `ClassStat`, see Sect. 8.3.2.3.
- The source flux is sufficiently well measured so that the uncertainty in the photometry can be ignored.

The first of these assumptions simplifies Bayes' theorem, Eq. (8.1):

$$P_s = p(s|x) = \frac{p(s)p(x|s)}{p(s)p(x|s) + p(g)p(x|g)}. \tag{8.2}$$

Thus we need to compute a single number: P_s.

The second assumption means that we assume that the intrinsic properties of a given source can be summarised by the parameter vector $\{m_1, m_2, \ldots, m_{N_b}, c\}$, where m_i is the true apparent magnitude of the source in band i, c is its intrinsic morphology statistic and N_b is the number of bands of the survey. The notion of a true morphology statistic is somewhat artificial, given that c is generally defined in terms of image properties such as pixel values; however it is taken to be the value of the morphology statistic that would have been measured if the source was observed without photometric noise, but with the smearing of the observational point-spread function (PSF). As such c is not actually an intrinsic property of the source.

The data consist of the measured magnitudes, $\{\hat{m}_1, \hat{m}_2, \ldots, \hat{m}_{N_b}\}$, and the measured morphology statistics, $\{\hat{c}_1, \hat{c}_2, \ldots, \hat{c}_{N_b}\}$, in each of the N_b bands. Ideally, our model should encode both the photometric and morphology measurement uncertainties as well as correlations between measurements in different bands. However, for the particular task at hand, we will, not unreasonably, assume that inter-band photometric noise correlations are negligible and that the photometric part of the model likelihood is Gaussian in magnitude units. This latter approximation will break down for faint sources [32], but here all sources are unambiguously detected.

We need, however, to include the survey incompleteness, expressed here as the probability that a source is detected in at least one band (or, more specifically, in a reference band). The detection probability is assumed to drop from unity to zero over a magnitude range Δm_b around the nominal detection limit of the survey, $m_{\lim,b}$. The specific form adopted for the incompleteness is

$$p(\det|m_b) = \frac{1}{2}\text{erfc}\left(\frac{m_b - m_{\lim,b}}{\Delta m_b}\right), \tag{8.3}$$

where $\text{erfc}(x) = 2\int_{\sqrt{2}x}^{\infty} \varphi(x'; 0, 1)\,dx'$ is the complementary error function, and where $\varphi(x; \mu, \sigma) = \exp\{-1/2[(x-\mu)/\sigma]^2\}/[(2\pi)^{1/2}\sigma]$ is the unit-normalised Gaussian probability density with mean μ and variance σ^2. By convention, the fainter an object is, the higher its corresponding magnitude measurement will be; hence the detection probability decreases with increasing magnitude.

Note that $p(\det|m_b)$ depends only on the true magnitude m_b and not the measured $\hat{m}_b$. Since the tail of the detection limit distribution is significantly longer for stars (as, being more centrally concentrated, there is a greater chance of faint stars meeting the detection criteria of most surveys). Hence we condition this probability on t as well: $p(\det|m_b, t)$. A somewhat subtle result of this is that the majority of the very faintest sources in a sample generated in this way are stars, even for surveys that are sufficiently deep that galaxies are intrinsically much more numerous at such faint fluxes.

Finally, for computing multi-band posterior probabilities, we will assume that the true morphology, c, depends only on the object type and on the apparent brightness of the source in a reference band, m.

Taken together, the above assumptions allow us to reduce the data vector for each source in the survey to $x = (\hat{c}, \hat{m}) = (\hat{c}_1, \hat{c}_2, \ldots, \hat{c}_{N_b}, \hat{m}_1, \hat{m}_2, \ldots, \hat{m}_{N_b})$.

8.3.2.3 Star-galaxy Separation in UKIDSS

The classifier which we will develop is geared towards the UKIDSS LAS survey. We will use deep SDSS Stripe 82 data to verify our classifications. Hence we will quickly review the morphology statistics that are recorded, respectively that can be derived from other measurements, in each survey.

8.3.2.4 Morphology Statistic in UKIDSS

Aside from basic image parameters (e.g. positions, counts, ...) the UKIDSS catalogues include a number of derived statistics, including an "extendedness" statistic in each band. The data variables we use are the magnitude variables in all four UKIDSS LAS bands (Y, J, H, K) and the above mentioned extendedness statistic, ClassStat, in each band.

Thus, for our classifier, the measured data vector will be $x = (\hat{c}, \hat{m}) = (\hat{c}_Y, \hat{c}_J, \hat{c}_H, \hat{c}_K, \hat{m}_Y, \hat{m}_J, \hat{m}_H, \hat{m}_K)$.

The extendedness statistic c, as defined in [26], is based on the fact that all the unsaturated stars in each field have the same average curve of growth (i.e. fraction of their total flux as a function of angular radius). This average can be measured empirically, and a mismatch statistic calculated for each source. In a given magnitude range the statistic is scaled so that, for stars, it has zero mean, unit variance and is approximately Gaussian distributed; this scaled mismatch statistic is referred to as ClassStat in the WSA. Extended galaxies (and blended pairs of sources) have positive ClassStat values, whereas most noise sources (e.g. cosmic rays), being more compact than the PSF, have negative ClassStat values. ClassStat encodes much of the important morphological information, even in faint images, and is a superb morphology statistic. However because it is a statistic based solely on the image data (i.e. it does not include prior information about a source's nature) it cannot encode all the information about a source (as distinct from the image of it). Moreover, there is no well-motivated method of combining the ClassStat values obtained from multiple measurements of a source. (In UKIDSS, combined source probabilities and ClassStat values are reported, but these are heuristic in nature, and do not retain all the information present in the band-specific ClassStat values.)

8.3.2.5 Morphology Statistic in SDSS

The SDSS approach to star-galaxy classification is based on the use of model magnitudes, each detected source being fit as both a point-source (i.e. the measured point-spread function) and a galaxy (i.e. a Sérsic profile [33] with one of two different exponents). The difference between the two different magnitudes, termed the concentration, c, is then used as a morphology statistic [34]. The basic classification is done by designating sources with measured concentration $\hat{c} \leq 0.145$ as stars and sources with $\hat{c} > 0.145$ as galaxies. Whilst this scheme is very effective, it is also important to note that the classifications of up to a third of sources contradict in different bands [34].

The Stripe 82 data are significantly deeper than the UKIDSS LAS (in the sense that all but the reddest sources are detected with a greater signal-to-noise ratio in Stripe 82 than in the LAS, and an average UKIDSS-selected source has $\sigma_r \simeq 0.1\,\sigma_Y$). Even though the SDSS optical imaging has a significantly larger seeing ($\sim$1.2 arcsec) than the UKIDSS NIR data ($\sim$0.8 arcsec), the SDSS Stripe 82 data of the morphologically ambiguous sources near the LAS detection limit is able to separate point and extended sources reliably (for more information, cf. [15]).

By limiting ourselves to sources with $16 \leq r \leq 20.5$ (thus avoiding both saturated and very faint sources) we assume the SDSS class labels to be correct. In particular, for $Y \simeq 20$, the SDSS r-band class labels misclassify only $\sim$4% of sources (this number is obtained by fitting a Gaussian distribution to the star population and a log-normal to the galaxy population for the SDSS concentration data). This is a very good result when compared to the UKIDSS ClassStat data which, at this faintness regime, no longer allow a separation into two populations of sources [15].

Hence, for the purpose of star-galaxy separation, we treat the SDSS Stripe 82 data as definitive classifications against which our LAS classifications can be tested.

8.3.2.6 Test Sample

Our starting point is a sample of 121,902 UKIDSS sources in a 14.38 deg^2 area defined by right ascensions of either $\alpha \leq 60$ deg or $\alpha \geq 300$ deg and declinations of $|\delta| \leq 0.1$. This area is entirely within the SDSS Stripe 82 region, and has been covered by UKIDSS in the Y, J, H and K bands. Our main aim is to classify these sources and compare the results to the SDSS Stripe 82 classifications. But to do so requires the preliminary task of generating the magnitude-dependent prior distributions of ClassStat, along with the star and galaxy number counts. This is not part of the actual classification process (i.e. it is independent of any single source), and so is considered separately from the results.

8.3.2.7 Model Specification

We have noted above that we assume the ClassStat statistic c to depend only on object type and the apparent magnitude in a reference band. For UKIDSS LAS, we have chosen the Y-band as the reference band. As not all sources have been observed in all four UKIDSS bands, we choose $\hat{m}$ to be the average of the measured magnitudes $\hat{m}_b$ in which a given source has been observed. To convert all of these magnitudes onto the scale of the reference band, we have added the average colours $Y - J$, $Y - H$ and $Y - K$ to the respective magnitudes $\hat{m}_J$, $\hat{m}_H$, $\hat{m}_K$. Provided that the typical value of c for an object of type t does not vary rapidly with its magnitude, the average colour relationship for each population is a reasonable approximation.

So, finally, we have reduced the data vector to $x = (\hat{c}, \hat{m}) = (\hat{c}_Y, \hat{c}_J, \hat{c}_H, \hat{c}_K, \hat{m})$. Together with the assumption that there are only two object types, s and g, and keeping conditioning on $\hat{m}$, Eq. (8.1) becomes

$$p(t|\hat{c},\hat{m}) = \frac{p(t|\hat{m})p(\hat{c}|t,\hat{m})}{p(s|\hat{m})p(\hat{c}|s,\hat{m}) + p(g|\hat{m})p(\hat{c}|g,\hat{m})}. \tag{8.4}$$

For given $\hat{m}$, we assume that the random variable T follows a Bernoulli distribution with parameter $p(s|\hat{m})$:

$$T|\hat{m} \sim \text{Bernoulli}[p(s|\hat{m})], \tag{8.5}$$

i.e. given $\hat{m}$,

$$T = \begin{cases} s & \text{with probability } p(s|\hat{m}) \\ g & \text{with probability } p(g|\hat{m}) = 1 - p(s|\hat{m}) \end{cases}, \tag{8.6}$$

where $p(s|\hat{m})$ incorporates both prior knowledge or belief about population numbers and the incompleteness model discussed above in Eq. (8.3).

The observed Y-band counts of stars and galaxies (identified here by using our model with number counts obtained by binning the data by magnitude and

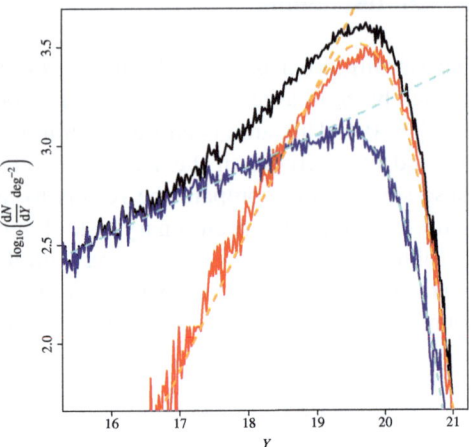

Fig. 8.1 Differential number counts of all sources (black), stars (blue) and galaxies (red) from UKIDSS observations. Classifications are obtained by using our model with number counts obtained by binning the data into equal-sized magnitude bins and fitting simple mixture models to the c_Y data in each bin. Also shown as dashed lines are the model fits (see Eq. (8.7)), both with and without a correction for incompleteness. The latter are those used as priors in our model. (Reproduced from [15], with permission.)

interpolating the parameters) from the test sample described above are shown in Fig. 8.1. Both exhibit exponential counts down to $Y \simeq 19$, beyond which the survey incompleteness dominates (as expected, given the average UKIDSS LAS limit of $Y \simeq 20.2$). For both stars and galaxies the intrinsic number counts are taken to be of the form

$$\frac{dN_t}{dY} = \alpha_t \ln(10) 10^{\alpha_t(Y-\beta_t)}, \tag{8.7}$$

where α_t is the type-dependent logarithmic slope and β_t is a scale parameter.

In order to fit these parameters, however, it is necessary to account for the incompleteness in each band, denoted here as $p(\det|Y)$, which was introduced in equation (8.3). The magnitude limit $m_{\lim,b}$ and incompleteness range Δm_b are fitted in the Y, J, H and K bands for both stars and galaxies. Although there are some discrepancies, the key point is that the relative numbers of stars and galaxies at a given magnitude will give far more accurate prior probabilities than, say, an uninformative prior (i.e. $p(s) = p(g) = 0.5$ for all sources).

As we have remarked above, due to the difference in detection distribution for point and extended sources, near the detection limit of a given survey, the majority of sources will be stars. For this reason we will account for the incompleteness in the magnitude dependent prior star probability $p(s|\hat{m})$. Thus, with the above models for number counts, Eq. (8.7), and incompleteness, Eq. (8.3), $p(s|\hat{m})$ can be written as

$$p(s|\hat{m}) = \frac{10^{\alpha_s(\hat{m}-\beta_s)} \frac{1}{2}\mathrm{erfc}\left(\frac{\hat{m}-m_{\lim,Y,t}}{\Delta m_{Y,t}}\right)}{10^{\alpha_s(\hat{m}-\beta_s)} \frac{1}{2}\mathrm{erfc}\left(\frac{\hat{m}-m_{\lim,Y,t}}{\Delta m_{Y,t}}\right) + 10^{\alpha_g(\hat{m}-\beta_g)} \frac{1}{2}\mathrm{erfc}\left(\frac{\hat{m}-m_{\lim,Y,t}}{\Delta m_{Y,t}}\right)}. \tag{8.8}$$

`ClassStat` is constructed so that, on average, $c = 0$ for stars and $c > 0$ for extended sources. We observe $\hat{c}$ however, the distribution of which, for isolated stars, should be normal, with zero mean and unit variance, again by construction. For galaxies, the distribution of c is more complicated: galaxies are intrinsically more varied, and the definition of the morphology statistic is essentially independent of galaxies' properties. For the UKIDSS sample an empirical function was sought which could represent the distribution of galaxies' c values as a function of magnitude. Particular care was taken to ensure a good fit close to the survey's limit, for which there is minimal morphological information and $c \to 0$, even for galaxies. We found a log-normal distribution to be a good choice for the ClassStat distribution of galaxies. Therefore the distribution of the variables representing the true ClassStat value is assumed to be

$$C|t,\hat{m} \sim \begin{cases} \delta_0 & \text{if } t = s \\ \Phi_{\log}[\mu(\hat{m}), \sigma(\hat{m})] & \text{if } t = g \end{cases}, \quad (8.9)$$

where δ_0 is the Dirac delta and $\Phi_{\log}(\mu, \sigma)$ a log-normal distribution with parameters μ and σ. The probability density function $\varphi_{\log}(x; \mu, \sigma)$ of the log-normal distribution is given by

$$\varphi_{\log}(x; \mu, \sigma) = \frac{1}{x\sqrt{2\pi\sigma^2}} \exp\left\{-\frac{[\ln(x) - \mu]^2}{2\sigma^2}\right\}. \quad (8.10)$$

As we have already stated, we will ignore inter-band noise correlations. We will also assume that the random variables representing the measured ClassStat values, $\hat{C}_b$, are identically distributed for stars and galaxies, with $\hat{C}_b$ given by

$$\hat{C}_b = C + \epsilon_b, \quad b = Y, J, H, K, \quad (8.11)$$

where the ϵ_b are independently and identically distributed random variables.

By construction, the distribution of ϵ_b should be a zero mean, unit variance normal. However the observed ClassStat distribution of bright stars appears to be significantly non-Gaussian, [15]. Therefore, for the observed ClassStat distribution we have instead adopted a Gaussian mixture model of the form

$$p(\hat{c}_b|c) = a\varphi(\hat{c}_b - c; \mu_1, 1) + (1 - a)\varphi(\hat{c}_b - c; \mu_2, \sigma_2), \quad (8.12)$$

where, for stars, $c = 0$, and μ_1, μ_2 and σ_2 are free parameters to be fit. These were fit using a simple maximum likelihood (ML) approach in each of the four UKIDSS bands.

Rather than specifying the functions $\mu(m)$ and $\sigma(m)$ of the standard parameterisation of the log-normal distribution, as given in Eq. (8.10), we model the mean $\mu'(m)$ and standard deviation $\sigma'(m)$ of the log-normal distribution by the empirical functions below,

$$\mu'(m) = \left(1 - \frac{m}{m_{\max}}\right) \times \left\{[v_1 m^2 + v_2 m + v_3]^{v_4} + v_5\right\}, \quad (8.13)$$

$$\sigma'^2(m) = 10^{\eta_1(m - \eta_2)}, \quad (8.14)$$

where $m_{\max}$ is the upper detection limit in the reference band and $\nu_1, \nu_2, \nu_3, \nu_4, \nu_5$, η_1 and η_2 are free parameters fitted by a simple least-squares (LS) procedure.

Note that if μ' and σ' are the mean and standard deviation of a random variable the logarithm of which is normally distributed with mean μ and standard deviation σ, then these parameters are related via a standard distributional result: $\mu' = e^{\mu + \sigma^2/2}$ and $\sigma'^2 = (e^{\sigma^2} - 1) e^{2\mu + \sigma^2}$.

The stellar and galactic densities implied by our models are shown as contours in Fig. 8.2, along with the sample from which the fit was derived. (The H-band, rather than the Y-band, was chosen as it has the highest number of saturated sources, thus emphasising an aspect of the data that is not included in the model.) The fit is not perfect (e.g. the true density is underestimated at the bright end and slightly overestimated in two regions near the faint end), but is very good. Also, the bright UKIDSS stars (with $H \lesssim 12.5$) have significantly positive ClassStat values, as they are saturated; we do not attempt to include this phenomenon as essentially all sources bright enough to be saturated in UKIDSS images can be classified as stars on the basis of prior information.

Hence $p(\hat{c}|t, \hat{m})$ is given by,

$$p(\hat{c}|t, \hat{m}) = \int_{-\infty}^{\infty} p(c|t, \hat{m}) \prod_{b=Y,J,H,K} p(\hat{c}_b|c) dc \qquad (8.15)$$

$$= \begin{cases} \prod_{b=Y,J,H,K} p(\hat{c}_b|C = 0) & \text{if } t = s \\ \int_{-\infty}^{\infty} \varphi_{\log}[c; \mu(\hat{m}), \sigma(\hat{m})] \prod_{b=Y,J,H,K} p(c_b|c) dc & \text{if } t = g \end{cases}, \qquad (8.16)$$

where $p(c_b|c)$ is given by Eq. (8.12).

All the terms in Bayes's Theorem, Eq. (8.4), are now completely specified.

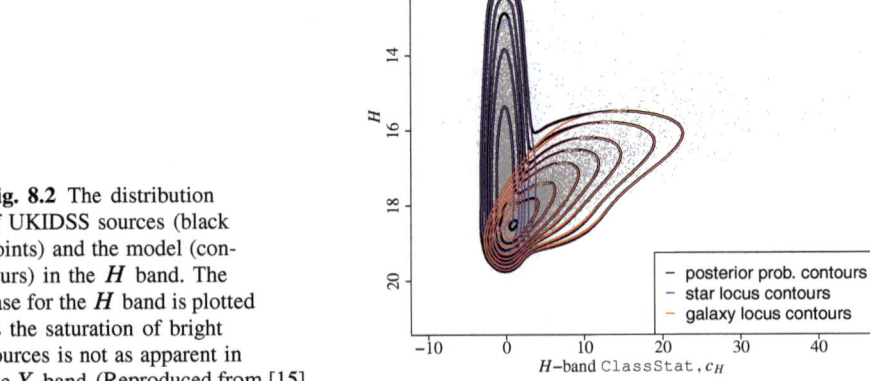

Fig. 8.2 The distribution of UKIDSS sources (black points) and the model (contours) in the H band. The case for the H band is plotted as the saturation of bright sources is not as apparent in the Y-band. (Reproduced from [15], with permission.)

8.3.3 Results

In this presentation the goal was to give an alternative, more statistical, formulation of the star-galaxy classification method from [15], and to make a few additional points regarding existing star-galaxy separation methods. The performance of the particular classifier that we have described is evaluated to great detail in [15], and below we limit ourselves to summarise some of the main results.

8.3.3.1 Importance of Prior Knowledge

Figure 8.3 (right) shows the posterior stellar probabilities in the c_Y vs. Y plane (the choice of band is unimportant, as the J, H and K band plots are similar). It is clear that for the overwhelming majority of objects, in particular those with either $Y \lesssim 18$ or $c_Y \gtrsim 5$, the Bayesian (in the sense defined earlier in the text) classifier gives very definite classifications (i.e. values close to either 0 or 1). Unsurprisingly, the region where the classifier is most often confounded (i.e. where $P_s \simeq 0.5$) is where the star and galaxy loci merge. Indeed, as the two loci overlap completely at the faint end, there is very little information regarding object class to be extracted from the measured `ClassStat` values, and the prior knowledge drives the classification.

8.3.3.2 Single- and Multi-band Classifications

Our classifier can compute both single-band classifications and combined, multi-band posterior star probabilities. Figure 8.3 shows both the Y-band-only probabilities and the joint, multi-band posterior star probabilities from our model.

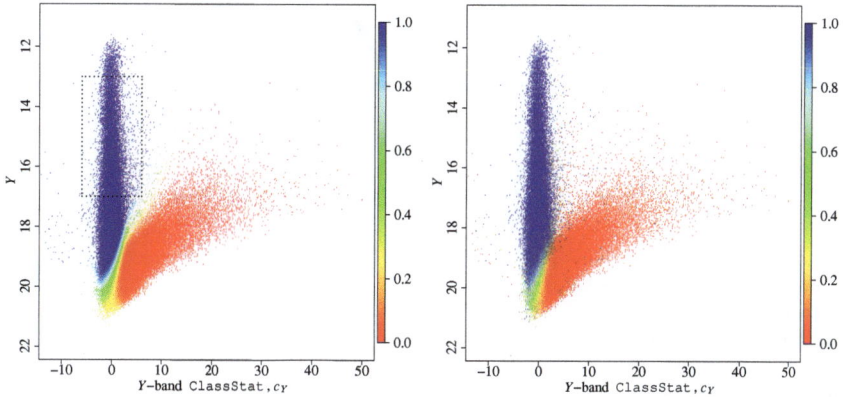

Fig. 8.3 Single- (left) and multi-band (right) star probabilities derived from our Bayesian method as a function of the measured Y-band `ClassStat` and magnitude. The dotted box indicates the selection region for bright stars, used to fit the parameters from Eq. (8.12). (Reproduced from [15], with permission.)

The most notable difference is that for the latter there are fewer sources which confound the classifier. In fact, compared to the single-band model, there is a decrease of at least 50% in the number of classifier-confounding sources for the combined model [15]. While a reduction in the classifier-confounding region is not always desirable, here this decrease translates the fact that the classifier will be at a loss only when the data from different bands are contradictory, or when a source's type is unclear in all the bands in which it was detected.

8.3.3.3 Comparison to SDSS Stripe 82 and UKIDSS Pipeline Classifications

Figure 8.4 shows the posterior star probabilities from our model as a function of SDSS concentration and r-band magnitude. The dotted line indicates the threshold concentration value (0.145) for SDSS star/galaxy labels. Overall there is good agreement with most sources with low P_s lying to the left of the line and sources with high P_s lying to the right.

Comparing our classifier and the UKIDSS pipeline to the Stripe 82 data, Fig. 8.5 shows the mismatch rates of both classifiers, taking the SDSS r-band classifications as a reference. To do this we have converted the posterior probabilities into class labels; an object is labelled as a star if $P_s \geq 0.5$, otherwise as a galaxy. We have limited the sources to those with $16 < r < 20.5$ so as to avoid saturated sources ($r \lesssim 16$) and sources for which the uncertainty of the SDSS labels is non-negligible ($r \gtrsim 20.5$). It is clear that our classifier is more accurate than the UKIDSS pipeline classifier; even though the difference in performance decreases for fainter magnitudes. For sources with $16.6 \leq Y < 17.4$, our classifier achieves a mismatch rate of 0.0154, compared to 0.0314 for the UKIDSS pipeline. At the faint end ($Y > 20$), the mismatch rates are 0.0679 (our classifier) and 0.0751 (UKIDSS pipeline). For all

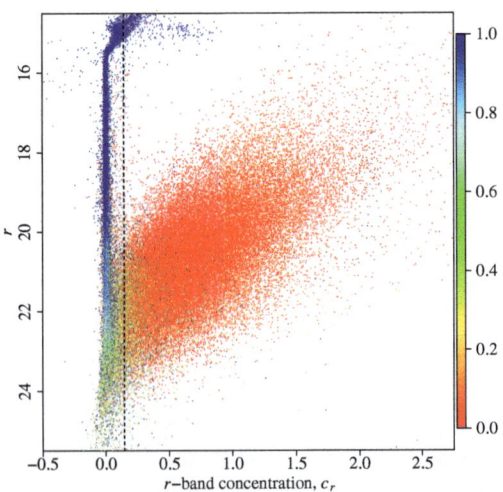

Fig. 8.4 UKIDSS posterior star probabilities shown as a function of the measured SDSS Stripe 82 concentration vs. r-band magnitude. Sources to the left/right of the dotted line (with concentration = 0.145) are classified as stars/galaxies in SDSS. (Reproduced from [15], with permission.)

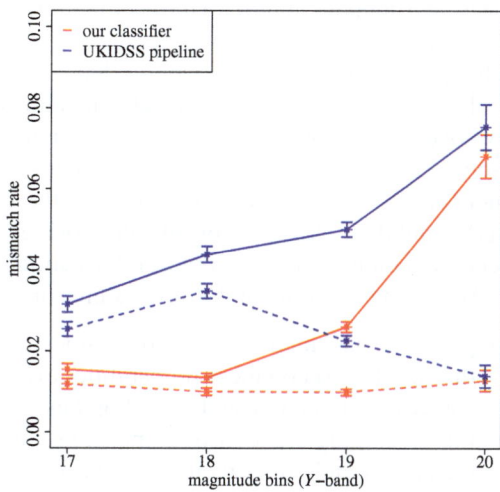

Fig. 8.5 Mismatch rates between the SDSS r-band class labels and labels based on our classifier (red) and the UKIDSS pipeline (blue). Mismatch rates are shown both for all sources (with $16 < r < 20.5$; solid lines) and for those sources for which our classifier outputs very definite classifications ($P_s < 0.1$ or $P_s > 0.9$; dashed lines). The magnitude values on the horizontal axis are the mid-range values of the bins used to compute the rates. Also shown are the standard errors of the mismatch rates. (Reproduced from [15], with permission.)

sources with $16 < r < 20.5$, the mismatch rate for the UKIDSS pipeline (0.0440) is more than double that of our classifier (0.0218).

8.3.3.4 Value of our Approach

A practical application of our method would be to look at the amount of telescope time that would be required to follow-up a morphologically-selected sample of targets. If one imagines a spectroscopic survey of faint stars, and one was to trust star-galaxy separators such as the ones used by UKIDSS or SDSS versus selecting sources with $P_s > 0.9$ from our method, then a certain proportion of telescope time would be spent observing compact/faint galaxies that were misclassified. While there will certainly also be misclassified sources when selecting objects by basing the selection on P_s, in [15] we show that their proportion can be greatly reduced.

8.3.4 Conclusion

We have developed a Bayesian formalism for star-galaxy classification in optical and/or NIR surveys that combines the morphological properties of an object (as measured in multiple passbands) with prior knowledge of the star and galaxy populations.

We have demonstrated our method on data from the UKIDSS LAS, combining morphology statistics measured in the Y, J, H and K bands (or whatever subset of these a source was detected in). The specific morphology statistic used, ClassStat [26], represents a powerful means of data compression from the full image, and contains almost all the useful information for the faint sources (which are the main motivation for the development of sophisticated star-galaxy classification techniques). However, the existing UKIDSS data products include only heuristic combinations of the band-specific classifications, and the application of the method developed here makes it possible to extract all the useful UKIDSS information on a source's morphology in as self-consistent a manner as is possible without using colour information as well. In particular, the use of prior information avoids the overly-confident classification of faint sources, for which the available measurements contain little morphological information.

We summarise some of the key features of our classifier below:

- posterior class membership probabilities: astronomers can select themselves the thresholds to set on these probabilities to assign class labels, or to decide which sources are to be rejected as ambiguous, respectively flagged for further study; this essentially allows astronomers to set the levels of contamination and completeness required for their specific needs
- through the use of Bayes' theorem, prior information can be fed into the model
- probabilistic, parametric model: the classification of any object can be easily interpreted; the model is also less prone to overfitting than unparametric, purely data-driven models
- morphological classification: class labels are unresolved (primarily stars) and resolved sources (primarily galaxies)
- based on the ClassStat statistic [26], which very efficiently describes the shape of an object
- ability to produce single- and multi-band classifications
- sources can be classified independently of the number of bands they have been detected in

Our test sample of UKIDSS LAS sources was chosen to lie in the multiply-scanned SDSS Stripe 82 region, giving us independent and almost totally reliable classifications of all our sources. (This is a rare situation outside simulations, and an opportunity that could be used for a number of similar testing schemes.) Converting the posterior probabilities into class labels, the Bayesian classifier achieves an error rate of 0.068 at the UKIDSS detection limit, compared to 0.075 for the UKIDSS pipeline. For all non-saturated sources, the error rate for our model lies at 0.022, compared to 0.044 for the UKIDSS pipeline.

The Bayesian model used to separate stars and galaxies described here can be very easily applied to other surveys with similar statistics measuring the extendedness of sources. The multiple advantages of such a classifier (posterior probabilities, use of prior knowledge, rigorous computation of multi-band classifications, ability to cope with missing detections) and its good performance exhibited for the UKIDSS data provide a strong argument in favour of a wider use of this methodology. In particular the use of our method can improve the efficiency of telescope time.

8.4 CASOS: a Subspace Method for Anomaly Detection in High Dimensional Astronomical Databases

In this section we describe a tool for detecting anomalies in cross-matched survey data. This work will be published in [35]. Here we will, once again, focus on the reasons behind developing such an anomaly technique, with less emphasis on the empirical evaluation of the particular method that is presented.

8.4.1 Introduction

Depending on the purpose of a given astronomical survey, it can be designed to record flux in Gamma-ray, X-ray, ultraviolet, optical, infrared, microwave or radio passbands. The number of completed or ongoing surveys is very large, and the surveys differ widely in regions of the sky that are mapped, the filter passbands used, the detection limits (survey depth), etc. This is due to different scientific aims of the different surveys and, maybe more fundamentally, technological limitations. But many surveys overlap, i.e. a given source can be observed in different surveys, depending on which region in the sky it lies in, how bright it is and in which parts of the light spectrum it radiates.

This overlap can be exploited for data analysis purposes by cross-matching surveys: identical sources in multiple surveys are identified (i.e. matched) and their data aggregated.

8.4.1.1 Problem description and motivation

In this work, using cross-matched survey data, we want to detect astronomical objects with strange physical properties. Finding such objects (and then studying them more closely with follow-up observations) is one of the main aims of astronomical surveys. As such sources (e.g. quasars, brown dwarfs,..., but also potentially unknown types of objects) are typically rare, the task of finding them can be formulated as an anomaly detection problem.

For detecting interesting unusual objects, automatic outlier detection methods are only the first step, to be followed by human examination. In the case of astronomy, anomaly detection methods can be used to select a set of candidate objects, with potentially highly unusual measured properties, for which detailed follow-up observations will be made.

Different methods are effective for detecting different kinds of anomalies in different situations, and it would be naive to expect a single method always to be best. Our approach is intended to be complementary to other methods, with properties described in the sections below.

Surveys in themselves can be large and high-dimensional (thousands to hundreds of millions of objects; a handful to hundreds of variables). Hence a database compiled by cross-matching surveys can be large and high-dimensional. While the gain of information about source populations achieved by cross-matching surveys is highly desirable, the resulting, potentially massive, datasets can pose various computational and methodological problems (e.g. sparsity of high-dimensional feature spaces, feasibility problems of algorithms scaling non-linearly with sample size and/or the number of variables, etc.).

Another property of cross-matched catalogues is that they contain many missing values. Different objects will be observed in different surveys: a given object might have been observed in surveys A, B and C, but not in surveys D and E, while another source is observed in C and E but not A, B and D. Further, within each survey there can be missing values as the different bands have different sensitivities and thus not all bands will detect a faint source.

The method we propose below essentially reduces the problem of working in a high-dimensional space to working in many lower-dimensional subspaces. While the reasons for taking this approach are given by the problem above, the specific reasons for working in lower-dimensional data subspaces are five-fold:

- Data in high-dimensional spaces are sparse [36]. A first consequence of this fact is that the local density around every object is low. Since anomalies are typically defined as objects that lie in low-density regions of the data-space, the very concept of an anomaly makes less sense in higher dimensions. A second consequence is that the notion of distance becomes less meaningful in high-dimensional spaces. [37] show that the discrimination between the nearest and furthest neighbour of any given point becomes poor in high-dimensional datasets. This particularly affects nearest neighbour based anomaly detection techniques.
- In Fig. 8.6 there is one anomaly only apparent when all three dimensions are considered jointly: the red triangle, sitting inside a sphere around which other points are lying. This illustrates that unless there is a relationship between *all* the variables in a dataset, anomalies are apparent in subspaces of the data. The more variables there are, the more complex such a relationship will be. Also, the more variables are collected, the higher the chances of some being independent of each other. For these two reasons, we think, such a complex relationship is increasingly unlikely as the dimensionality increases and that lower-dimensional approaches can be successful.
- Figure 8.6 also illustrates that anomalies can be anomalous in only a subset of variables: three anomalies are anomalous in only two of three variables, and two anomalies are outlying in one variable only. In a full-dimensional approach the combined anomaly scores of such objects will be less extreme because of the contributions from the variables they are not anomalous in, and thus these anomalies can go undetected. For example, in a 100-dimensional data space, suppose there is an object which is anomalous in three variables only. When the anomaly score is computed, the three exceptional contributions are averaged out by the other 97.

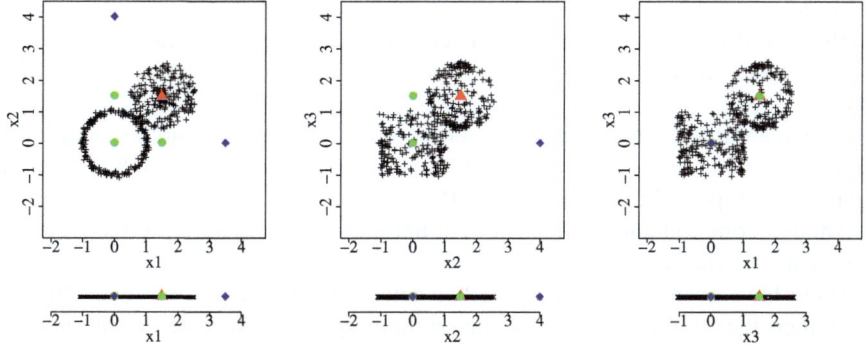

Fig. 8.6 Two- and one-dimensional plots of a dataset that will serve as a motivating example. There are two groups of ordinary objects: a first group lying (subject to additional noise) on a sphere with radius 1 and centre $(1.5, 1.5, 1.5)$ and those lying (again subject to additional noise) on a cylinder with radius 1, height 2 and centre at the origin $(0, 0, 0)$. There are three types of anomalies: those that can be discerned in either one- (blue diamonds), two- (green circles) or three-dimensional (red triangle) subsets of the original data variables. Note that the three-dimensional anomaly (red triangle) is overprinted by a two-dimensional anomaly (green circle) on some of the graphs.

- The need for imputing missing data, or discarding objects with missing values can be eliminated by restricting oneself to the variables in which a particular object has been observed in. Thus, a lower-dimensional approach will allow a more efficient and rigorous use of the available data.
- Understanding why a given object has been declared anomalous will be easier the less variables have been used to declare it anomalous. Thus, lower-dimensional approaches can result in increased interpretability.

8.4.1.2 Existing and Similar Work

A comprehensive review of anomaly detection methods similar to the method presented here can be found in [35]. We will however review methods which have been used specifically for astronomical datasets.

Our method reduces the problem of detecting anomalies over a multi-dimensional feature space to one of detecting anomalies in many lower-dimensional subspaces. Further, our method will use a local density, nearest neighbour based anomaly score calculation algorithm (but this could be replaced by any algorithm computing a numerical anomaly score). In particular we will use the Local Outlier Factor (LOF; [38]). LOF generalises distance-based outliers (DB-outliers), introduced by [39]. The DB-outliers technique computes the number of neighbours within a certain radius of a given object. If that number is less than a threshold, the object is flagged as anomalous. Alternatively the inverse of the number of neighbours within a chosen radius of an object can be used as anomaly score. LOF looks at the local density around an object. The LOF score is essentially the average of the ratios of the average distance to the k nearest neighbours of the k nearest neighbours of a given object and the average distance to the k nearest neighbours of this object (though there is some smoothing for small distances involved as well).

To find anomalies in astronomical datasets, object-specific search strategies are usually adopted, e.g. searching for high redshift quasars [32] or finding outlying light curves of periodic variable stars [40].

Another common approach consists in finding outliers as a by-product of object classification [18, 29].

There have been more general outlier detection approaches, however. In [41], the outlier detection method of the data mining system Distributed Exploration of Massive Astronomy Catalogs (DEMACS) is described. As the name suggests this is an anomaly detection method tailored to the need of virtual observatories where datasets from different surveys can be located at different sites.

Like many methods that reduce the dimensionality before checking for anomalies, the DEMACS anomaly detector uses principal component analysis (PCA; [42]). PCA projects the data onto directions of decreasing variance so that the first principal component corresponds to the direction of maximum variance. The DEMACS method defines anomalies as the objects for which the projections onto the last principal component deviate most (i.e. the objects with largest residuals, given the overall correlation structure of the dataset).

[43] take a similar approach: they also describe an anomaly detection method for distributed astronomical datasets and their method is also PCA based. They take however a slightly different view than [41] and restrict the search for anomalies to the top few principal components.

However, PCA cannot be performed if the data contain missing values. As we wish to analyse datasets containing missing values, PCA based techniques are not applicable.

8.4.2 Combining Anomaly Scores from Observed Subspaces (CASOS)

The proposed method to address the anomaly detection problem in datasets obtained by cross-matching astronomical surveys can be summarised in a few easy steps. Our main idea consists in looking for anomalies not over the full-dimensional dataset, but in lower-dimensional subspaces of the data. We could, in theory, compute any lower-dimensional projection of the data. But, computing projections of data with missing values will introduce further missing values. Also, we would need to determine a set of subspaces that are 'best' for anomaly detection; if there are different types of anomalies, such a 'best' set of subspaces might not exist (but see [44]). Therefore, we will limit ourselves to the subspaces that are given by subsets of the original data variables.

```
1. for i in 1 : D
2.        for j in 1 : (N_x choose i)
3.              compute ASs for objects with no MVs in jth
                i-dimensional subspace
4.              store the AS vector for this subspace
5.        end for j
6.        combine the AS vectors for all i-dimensional subspaces
7. end for i
8. output D AS vectors or D lists of anomaly candidates
```

Algorithm 1. Proposed approach; if only D-dimensional subspaces are of interest, then the loop on i needs only to run once for $i = D$ and only one AS vector is output.

Our approach is summarised by Algorithm 1, but let us first define some notation:

- n - the number of objects (rows) in the dataset
- N_x - the number of variables (columns) of the dataset
- D - the maximum dimensionality of the subspaces($1 \leq D \leq N_x$)
- AS - anomaly score (we assume the more anomalous an object is, the higher its AS)
- MV - missing value
- NA - the encoding of a missing value in the data

Our method is not a novel AS computation algorithm, but attempts to use an AS calculator designed for low-dimensional data on high-dimensional data whilst avoiding the curse of dimensionality. In practice, any AS computation algorithm can be used with our approach. For this work we have used the Local Outlier Factor (LOF; [38]).

8.4.2.1 Combination Functions and Required Properties

The key step in our approach is step 6 in Algorithm 1 above. It is by – sensibly – combining, for each object, the ASs of the subspaces the object has been observed in, that we can directly compare the anomalousness of objects with different sets of observed variables. While one could use any function to combine the different ASs, there are a few obvious choices, such as averaging the observed ASs. As we have stated, our key aim is to be able to directly compare the resulting combined ASs of different objects – regardless of which variables these objects have been observed in. Not all candidate combination functions that come to mind will allow us to do this. If one were to, say, sum all the ASs from the observed subspaces for each object, then objects with many observed variables are more likely to have higher ASs than objects with few observed variables and objects are not directly comparable. Hence restrictions on what constitutes a valid combination function need to be imposed.

In [35] we give mathematical descriptions of properties needed to guarantee the comparability of ASs. Below we state, in words, the two properties that any combination function needs to satisfy. In [35] we list three additional properties that characterise the behaviour of combination functions.

8.4.2.2 Property 1

Objects with different numbers of MVs have ASs on the same scale.

8.4.2.3 Property 2

Objects with at least one non-missing AS, have a non-missing combined AS.

8.4.3 Examples of Combination Functions

We will use the following notation:

$N_D = \binom{N_x}{D}$, the number of distinct subspaces of dimension D in an N_x-dimensional dataset

$X = (AS_{i,j})_{\substack{1 \leq i \leq n \\ 1 \leq j \leq N_D}}$, a matrix of ASs, with $AS_{ij} \in \mathbb{R} \cup \{NA\}$, X_i the ith row of X

$\mathcal{G} = \{X \mid X \text{ an } n \times N_D \text{ AS matrix}\}$, the set of all $n \times N_D$ AS matrices

We define a *combination function* to be a function $\rho : \mathcal{G} \rightarrow (\mathbb{R} \cup \{NA\})^n$ which satisfies Props. 1 and 2 above. If, in addition, a combination function satisfies Props. 3-5 from [35], it is termed *well-behaved*.

Let $S_i = \{X_{i,j} | j \in \{1, \ldots, N_D\} \text{ and } X_{i,j} \neq NA\}, i = 1, \ldots, n$.

- Selecting the highest AS:

$$\rho^{(ext)}_i(X) = \max S_i \quad \forall i \in \{1, \ldots, n\}.$$

- Averaging the ASs:

$$\rho^{(avg)}_i(X) = \frac{1}{|S_i|} \sum_{X \in S_i} X \quad \forall i \in \{1, \ldots, n\}.$$

- Averaging the top N ASs:

$$\rho^{(topN)}_i(X) = \frac{1}{N} \sum_{j=0}^{N-1} X_{i,(|S_i|-j)} \quad \forall i \in \{1, \ldots, n\},$$

where $X_{i,(j)}$ is the jth order statistic of the ASs of object i. (N.B. if an object has less than N ASs, the combined AS is the average of all the available ASs.)

- Sum of the excess above a certain quantile:
  For each $j \in 1, \ldots, N_D$ let $q_j^{(1-\alpha)}$ be the $(1 - \alpha)$ quantile of the ASs recorded for subspace j. For all j, we subtract $q_j^{(1-\alpha)}$ from the AS for that subspace. Finally, for each object, we sum the non-negative values.

$$\rho^{(topquant)}{}_i(X) = \sum_{X \in S_i} \left(X - q_j^{(1-\alpha)}\right)^+ \quad \forall i \in \{1, \ldots, n\},$$

where $(X)^+ \equiv \max(X, 0)$.

- Sum of the excess above a certain quantile and below another one:
  We choose $0 \leq \alpha_2 < \alpha_1 \leq 1$ and compute, for each j, $q_j^{(1-\alpha_1)}$ and $q_j^{(1-\alpha_2)}$. For all j, we set all those ASs exceeding $q_j^{(1-\alpha_2)}$ equal to $q_j^{(1-\alpha_2)}$ and then subtract the amount by which they exceed $q_j^{(1-\alpha_2)}$, i.e. for all i so that $X_{i,j} > q_j^{(1-\alpha_2)}$ we set $X_{i,j} = q_j^{(1-\alpha_2)} - (X_{i,j} - q_j^{(1-\alpha_2)})$. Then, for all j, we subtract $q_j^{(1-\alpha)}$ from all the ASs for that subspace. Finally, for each object, we sum the non-negative values.

$$\rho^{(midquant)}{}_i(X) = \sum_{X \in S_i} \left[\left(X - q_j^{(1-\alpha_1)}\right)^+ - 2\left(X - q_j^{(1-\alpha_2)}\right)^+\right]^+$$

All of the above are valid combination functions. The first four are also well-behaved (for details see [35]).

8.4.4 Properties of CASOS

Having defined CASOS in Sect. 8.4.2, we can now look at further properties of our approach. For an evaluation of the computational complexity of CASOS, the reader is referred to [35].

8.4.4.1 Flexibility

Through the choice of combination function, our method can very easily be adapted to specific needs. For instance $\rho^{(avg)}$ will find objects which either have very large ASs in some subspaces, or which have consistently high ASs. However, $\rho^{(avg)}$ can be affected by the masking effect from irrelevant features since it averages ASs over all available subspaces. $\rho^{(ext)}$ will be less affected by irrelevant features, as it is enough for an object to have a high AS in a single subspace to have a high combined AS. However, if a dataset contains both noise objects (e.g. cosmic rays in astronomical datasets) and objects which are physically anomalous, both $\rho^{(ext)}$ and $\rho^{(avg)}$ will result in high combined ASs for noise sources, as their measured values are often highly outlying. This can, in turn, affect the detection of true, non-noise anomalies. $\rho^{(midquant)}$ can adjust for this, as it essentially ignores too extreme ASs during the combination step. Thus one can design a combination function which will suit whatever beliefs one might hold about a particular datasets. All combination functions should be checked, however, to see that they satisfy Props. 1 and 2.

8.4.4.2 Transition Between One-variable-at-a-time and Full-dimensional Approach

The case when $D = 1$ corresponds to a one-variable-at-a-time approach, whereas $D = N_x$ is equivalent to the full-dimensional approach. Thus our approach can be regarded as a generalisation of these two, including them as special cases.

In the astronomy setting, if we only use magnitude variables (i.e. measures of brightness), then the one-dimensional AS vectors will be of little use as they will merely flag up very bright and/or very faint objects. Indeed, the individual magnitude variables have low-density regions only at the upper and lower ends of their range. CASOS can be used with $D = 1$ to check the dataset for objects with physically impossible values, but such a quality-control step should, ideally, have been performed prior to the actual data analysis.

8.4.4.3 Analysis of the Motivating Example

We can now return to the motivating example from Fig. 8.6. This dataset features different kinds of outliers and it also contains missing values. The anomalies are $o1, o2, o3, o4, o5$ and $o6$. $o1$ appears anomalous only when all three data variables are considered jointly. $o2, o3$ and $o4$ are two-dimensional anomalies, but $o3$ has a missing value in one of the data variables. Finally $o5$ and $o6$ can be seen to be anomalous by considering only one variable, but for $o5$ only two attributes have been recorded.

We apply CASOS with $D = 1, 2, 3$, the latter being equivalent to the full-dimensional LOF approach [38], Local Density Factor (LDF, [45]), Local Outlier Correlation Integral (LOCI, [46]), the method described in [44] and Angle-Based Outlier Detection (ABOD and fastABOD, [47]) to this dataset and list the results in Table 8.1. For CASOS we have used $\rho^{(avg)}$ as combination function. ABOD was too slow even for this small, low-dimensional dataset, hence why we have used fastABOD instead.

Note that CASOS and the method from [44] have been applied to the full dataset, whereas LDF, LOCI and fastABOD have only been applied to the reduced dataset of 373 objects with no missing values.

A full description of the results from this experiment can be found in [35], here we only give the overall result, summarised in Table 8.1. We note that of the six methods compared in Table 8.1, CASOS with $D = 2$ works best, which shows that lower-dimensional approaches can outperform their full-dimensional counterparts.

8 Classification and Anomaly Detection for Astronomical Survey Data

Table 8.1 Various anomaly detection methods applied to the dataset from Fig. 8.6. Checkmarks indicate successful detection.

anomaly type	o1 3D	o2 2D	o3 2D + MV	o4 2D	o5 1D + MV	o6 1D
[a] CASOS, $D = 1$					✓	✓
[a] CASOS, $D = 2$		✓	✓	✓	✓	✓
[b] CASOS, $D = 3$ / LOF	✓	✓				✓
[b] LDF						✓
[c] LOCI	✓	✓				✓
[c] [44]			✓	✓	✓	✓
[b] fastABOD	✓	✓		✓		✓

[a] objects with the 6 highest ASs have been flagged as anomalies
[b] objects with the 4 highest ASs (resp. smallest ASs, for fastABOD) have been flagged as anomalies
[c] for binary anomaly / non-anomaly labels, we report those objects flagged as anomalies; note that LOCI gave 1 false-positive result whereas the method from [44] gave 41 false-positives

We conclude that this example shows the advantages (ability to handle data with missing values, ability – for some datasets at least – to outperform full-dimensional approaches, ability to avoid the masking effect from irrelevant features) and limitations (inability to detect full-dimensional anomalies) of our approach.

8.4.5 Empirical Evaluation

Here we present the empirical evaluation of CASOS on two real, astronomical datasets. For a more thorough evaluation of CASOS, in particular on simulated datasets where the distribution of anomalies can be controlled, the reader is referred to [35].

8.4.5.1 Cross-matched SDSS-UKIDSS Data

The cross-matched SDSS-UKIDSS data that we will use contain seventeen measured variables: eight colour variables ($u - g$, $g - r$, $r - i$, $i - z$, $z - Y$, $Y - J$, $J - H$ and $H - K$) and nine morphology statistics (concentrations in five SDSS bands and ClassStat statistics in four UKIDSS bands). As in Sect. 8.3, we use data from the SDSS Stripe 82 region (more specifically we have selected sources with $0 <$ ra < 10 and $0 <$ dec < 1), for which there is a good overlap between SDSS and UKIDSS, and for which SDSS made multiple observations, thus providing much more information on the detected sources.

The extracted data contained 170,413 sources. However, an initial run of our algorithm on the data returned many noisy sources and artifacts (e.g. cosmic rays, saturated sources, sources affected by bad sky background estimation). We have used the UKIDSS data flags to remove such sources and, after applying these various data quality filters, our final dataset consists of 109,368 sources.

In UKIDSS, if a source is too faint in a given band to be detected, it will have missing values for the variables from that band. In SDSS however, if a source is detected in any of the band, the SDSS data processing pipeline will re-extract the measurements for any band in which the source has not been detected. Thus the SDSS data contain no missing values. However we wish to test CASOS in the more usual setting where such non-detections result in missing values. Further, the re-extracted measurements contain little useful information. We have therefore re-introduced missing values in the SDSS data by setting those detections to missing for which the signal-to-noise ratio is less than 5.

Due to the missing values contained in this dataset it has not been possible to apply full-dimensional approaches such as LOF, LDF or fastABOD to this dataset. All results are for CASOS (with $k = 100$, $D = 2$ and $\rho^{(avg)}$).

We have flagged the 109 sources with highest combined ASs (i.e. top 1% of sources) as potential anomalies. Looking at these sources we can identify several types of objects. A first type consists of sources for which the SDSS pipeline has re-extracted photometry, despite these objects having negative (after sky background subtraction) measured fluxes. Some of the SDSS magnitudes for these sources are extremely faint compared to the other measured magnitudes. These are in fact the re-extracted measurements whenever a source was too faint to be detected in a given band. While we have tried to re-instate the original missing values, we have not been able to do so for all sources. The physical properties of the majority of such sources will not exhibit any true anomalousness. However any anomaly detection method should flag sources with such measured data, as they are anomalous in the feature space that has been analysed. Fig. 8.7 shows low-resolution spectra (obtained by using the measured SDSS and UKIDSS magnitudes) of two such sources. The right-hand-side source has a measured i-band magnitude clearly fainter than the SDSS detection limit ($i \simeq 24.36$), and the left-hand-side source has a measured u-band magnitude much fainter than the u-band limit ($u \simeq 24.64$).

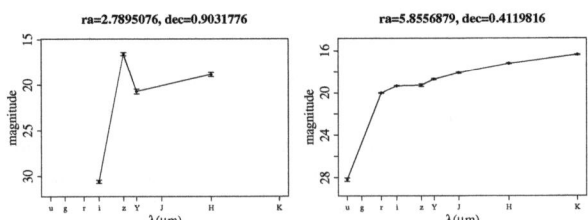

Fig. 8.7 Low-resolution spectra, obtained by using the measured SDSS and UKIDSS magnitudes, for sources for which one of the SDSS bands has been re-extracted despite the measured flux not being above the detection limit. Note that bright sources have low magnitude values and faint sources have large values. The error bars represent the errors reported by the SDSS and UKIDSS data processing pipelines.

8 Classification and Anomaly Detection for Astronomical Survey Data

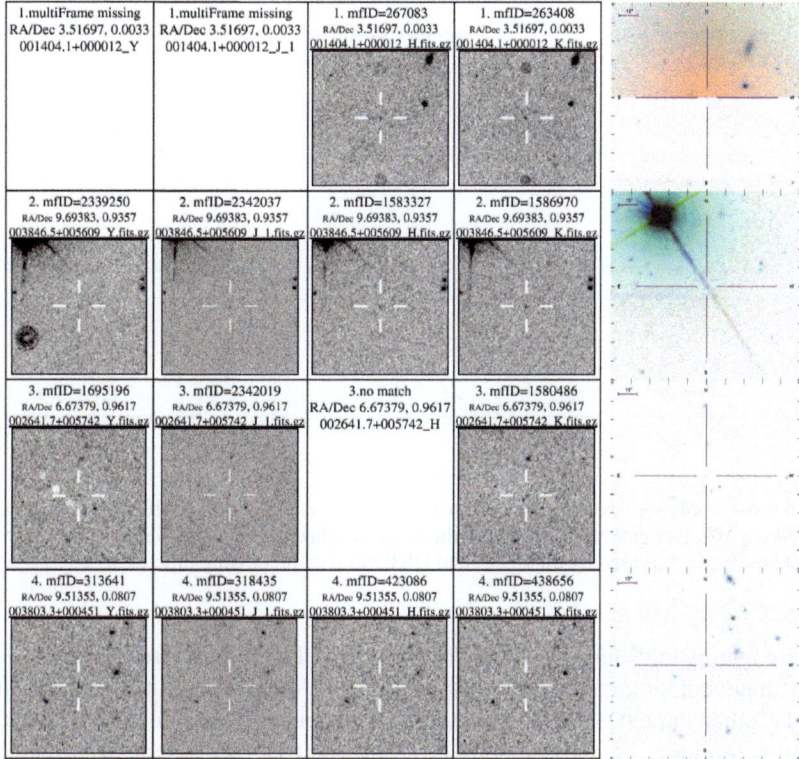

Fig. 8.8 Sources which are either noise artifacts or the photometry of which has been affected by nearby bright sources as they appear in the four UKIDSS bands and in SDSS. UKIDSS images have been obtained from the WFCAM Science Archive ([48]; http://surveys.roe.ac.uk/wsa/) and SDSS images have been obtained using the SDSS Image List Tool from the SDSS SkyServer (http://cas.sdss.org/dr7/en/tools/chart/list.asp).

Another type of anomaly candidate consists of sources which are either noise artifacts or the photometry of which has been affected by nearby bright sources. Again, the inherent physical properties of such sources are unlikely to be anomalous, but we would expect an anomaly detection technique to detect them as anomalies due to their unusual measured values. Figure 8.8 shows images in each of the four UKIDSS bands (Y, J, H, K; the four leftmost columns) as well as inverted false-colour images from SDSS (the far-right column). Each row corresponds to one source. For the top row object, the SDSS photometry is clearly bad. However, there are also some artifacts to be seen on the H- and K-band UKIDSS images. The source in the second row is near a very bright star and, in SDSS, a diffraction spike from this bright star affects the photometry. For the source in the third row, some artifacts can be seen in the Y-band UKIDSS image, which have probably affected the photometry in that band. Finally, the bottom row source can be seen to be just noise in the UKIDSS Y-band, whereas it is not detected in the remaining three UKIDSS band. In SDSS the object has been detected.

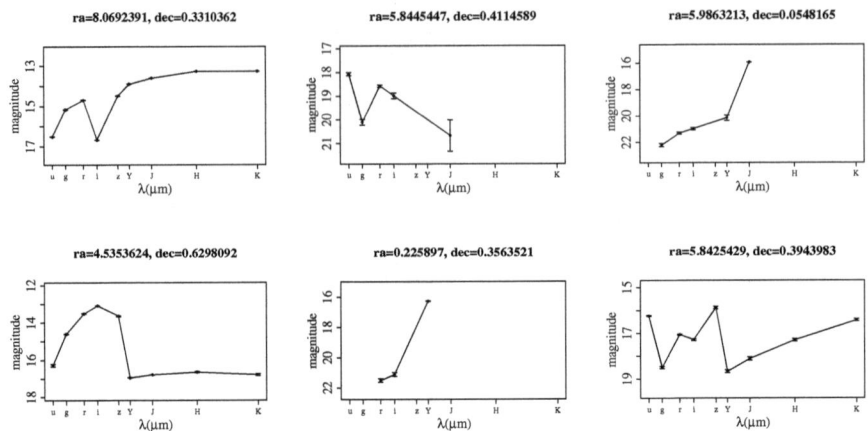

Fig. 8.9 Low-resolution spectra (made up from the original SDSS and UKIDSS survey data) for six of the top 109 anomaly candidates from the cross-matched SDSS-UKIDSS data. The error bars represent the errors reported by the SDSS and UKIDSS data processing pipelines.

More interesting sources are shown in Fig. 8.9. The three sources in the top row have strange colours: the source on the left has an i-band magnitude much fainter than the other optical SDSS bands, the middle source is similar, but with a faint g-band magnitude, finally, the source on the right is very bright in the J-band, resulting in an extreme $Y - J$ colour. The two left-most sources in the bottom row have apparently contradictory SDSS and UKIDSS measurements: the first source is bright in the SDSS bands, but much fainter in the UKIDSS band, whereas the second source is very bright in the UKIDSS Y-band, when compared to its measurements in the SDSS r and i bands. Finally, the bottom-right corner source has an odd spectrum in the SDSS bands, alternating widely in brightness in the different bands.

One type of source for which it is relatively straightforward to establish why their data appear to be contradictory are sources which lie very close to each other on the sky. For these sources there is the possibility that they can be detected as two, or three separate sources in one survey, but as one blended source in the other survey. Since SDSS has observed Stripe 82 multiple times, it is usually in SDSS that such sources can be clearly separated, whereas in UKIDSS they appear as one blended source. Figure 8.10 shows UKIDSS and SDSS images of three such blended sources. In fact some UKIDSS sources have been cross-matched to multiple sources.

We conclude that CASOS does show some potential on cross-matched survey data, as evidenced by the properties of the top 1% anomaly candidates flagged by CASOS. But, as we stated in Sect. 8.4.1, any sources flagged by CASOS would need to be observed to greater detail with follow-up observations. It will require much more focused input from astronomers to get meaningful anomaly candidate lists, as otherwise CASOS will flag up noisy sources, or sources with weird, but uninteresting measurements (such as, e.g. the artificially low SDSS magnitudes).

8 Classification and Anomaly Detection for Astronomical Survey Data 179

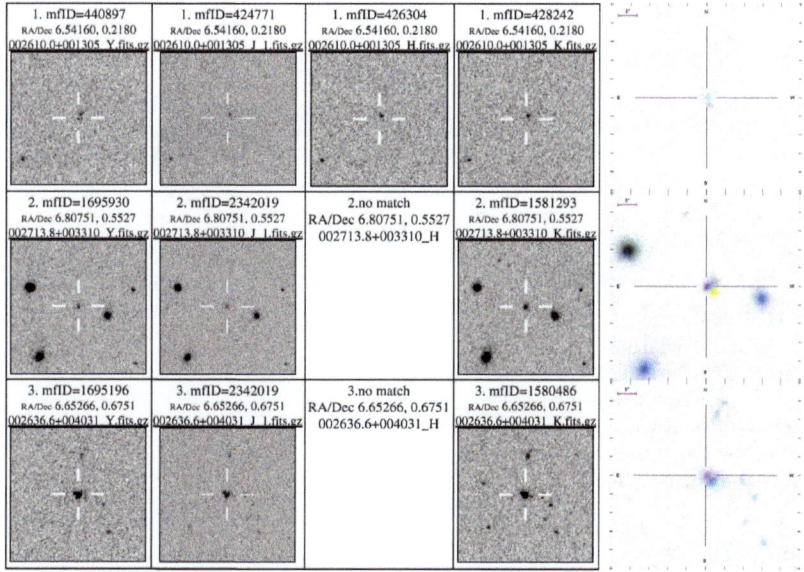

Fig. 8.10 Sources close to each other on the sky, which appear as one blended source in UKIDSS, but as several, distinct sources in SDSS. Each row corresponds to one source, the first four columns are the four UKIDSS band images and the last column are SDSS inverted false-colour images. UKIDSS images have been obtained from the WFCAM Science Archive ([48]; http://surveys.roe.ac.uk/wsa/) and SDSS images have been obtained using the SDSS Image List Tool from the SDSS SkyServer (http://cas.sdss.org/dr7/en/tools/chart/list.asp).

8.4.5.2 Quasar Candidates Datasets

We have also applied CASOS to a set of 12,074 pre-selected high-redshift quasar candidates. This dataset is described in greater detail in [32]. The dataset contains 5 colour variables: $i - z, z - Y, Y - J, J - H$ and $H - K$. Objects have been pre-selected to lie in a certain region in $i - Y$ vs. $Y - J$ space (shown on Fig. 8.11).

The dataset contains 7 confirmed high-redshift quasars. The aim was to see if CASOS would be able to detect these. We have used CASOS with $D = 2$ and $\rho^{(avg)}$. Figure 8.11 shows the top 120 anomaly candidates from CASOS.

Flagging the top 1% of sources as anomalous (121 sources), CASOS detects one of the seven high-redshift quasars. The probability of flagging one or more of the seven sources by chance is $\sum_{i=1}^{7} \binom{7}{i} \cdot 0.01^i \cdot 0.99^{7-i} = 0.0679$. However when the top 10% of sources (1,207 sources) are flagged as anomalous, then CASOS is able to detect five of the seven quasars. Now the probability of flagging five or more of the seven sources by chance is $\sum_{i=5}^{7} \binom{7}{i} \cdot 0.1^i \cdot 0.9^{7-i} = 1.765 \cdot 10^{-4}$.

So this shows that there is some potential in CASOS. However, again, more guided input from astronomers will be needed, as the efficiencies (1/121, respectively 5/1207) are very low. Still, without using any additional information, CASOS managed to reduce the problem of finding 7 high-redshift quasars in 12,074 sources to finding 5 quasars in 1,207 sources. Comparing this to the result from [32], which

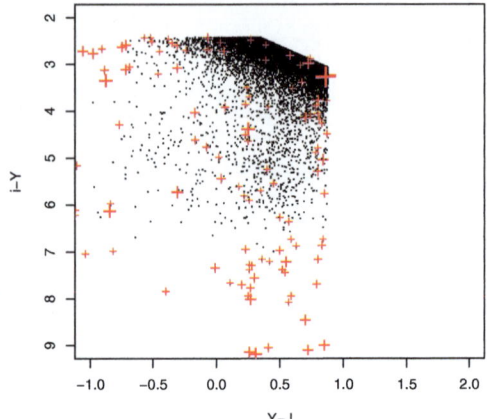

Fig. 8.11 12,074 preselected quasar candidates (compare to Fig. 1 from [32]). Red crosses indicate the top 120 anomaly candidates, with the size of the red crosses proportional to the combined anomaly score.

was obtained by using Bayesian model comparison, CASOS is impressively competitive.

8.4.6 Conclusion

We have introduced a novel algorithm, CASOS, which performs anomaly detection in lower-dimensional subspaces of the data. The advantages of this algorithm are multiple:

- ability to directly use data with missing values
- addresses some of the problems of high-dimensional data spaces (such as the breakdown of the notion of anomaly and distance)
- less susceptible to the masking effect from irrelevant features
- the choice of combination function adds flexibility to adapt the method to the requirements of a particular dataset
- better interpretability

We should, however, also note that CASOS has the disadvantage that it will not be able to detect outliers which are only apparent in multivariate spaces with significant numbers of variables. But we believe such situations are rare, and normally outliers will be apparent in lower-dimensional spaces.

CASOS is not intended to be a universal solution to anomaly detection problems, but rather is a new method with complementary properties to other techniques, so that it provides a useful addition to the armoury of anomaly detection methods.

We have applied CASOS to three real datasets, in particular a set of cross-matched SDSS-UKIDSS data. While the results for the astronomy datasets look promising, CASOS needs to be supervised more closely by astronomers in order to get meaningful results, which would justify the costs of follow-up observations.

We have implemented CASOS in R, a freely available software environment for statistical programming (http://www.r-project.org/).

Full source code for CASOS and for generating the various datasets described in this chapter is available from MH (marc.henrion03@imperial.ac.uk).

8.5 Conclusion

In this chapter we have described two new statistical tools for astronomical datasets: a probabilistic star-galaxy separation method for the UKIDSS LAS survey and a novel anomaly detection algorithm for cross-matched survey data. In addition to describing the details of each technique, we have also presented our reasoning for developing them and have reviewed similar work. We have evaluated both methods on real datasets.

While the star-galaxy separation work is targeted for morphological classification of astronomical sources, and in particular for the UKIDSS survey, the anomaly detection algorithm has a wide applicability and can be used on any dataset with numerical attributes.

Both tools are designed to be a first data analysis step: the star-galaxy separator allows astronomers to collect samples of sources according to desired levels of completeness or purity and CASOS can be used to pre-select strange astronomical sources for deeper follow-up observations. As such, both are designed to facilitate the work of astronomers.

Acknowledgements The results presented here would not have been possible without the efforts of the many people involved in the SDSS and UKIDSS projects.

Marc Henrion was supported by an EPSRC research studentship, and David Hand was partially supported by a Royal Society Wolfson Research Merit Award.

References

1. Borne, K.D.: Data-Driven Discovery through e-Science Technologies. In: 2nd IEEE International Conference on Space Mission Challenges for Information Technology (SMC-IT'06), pp. 251–256. IEEE Computer Society (2006)
2. York, D.G., et al.: The Sloan Digital Sky Survey: technical summary. Astron. J. **120**, 1579–1587 (2000)
3. Gunn, J.E., et al.: The 2.5 m telescope of the Sloan Digital Sky Survey. Astrophys. J. **131**, 2332–2359 (2006)
4. Fukugita, M., Ichikawa, T., Gunn, J.E., Doi, M., Shimasaku, K., Schneider, D.P.: The Sloan Digital Sky Survey photometric system. Astron. J. **111**, 1748–1756 (1996)

5. Lupton, R.H., Gunn, J.E., Szalay, A.S.: A modified magnitude system that produces well-behaved magnitudes, colors, and errors even for low signal-to-noise ratio measurements. Astron. J. **118**, 1406–1410 (1999)
6. Lawrence, A., Warren, S.J., Almaini, O., Edge, A.C., Hambly, N.C., Jameson, R.F., Lucas, P., Casali, M., Adamson, A., Dye, S., Emerson, J.P., Foucaud, S., Hewett, P., Hirst, P., Hodgkin, S.T., Irwin, M.J., Lodieu, N., McMahon, R.G., Simpson, C., Smail, I., Mortlock, D., Folger, M.: The UKIRT Infrared Deep Sky Survey (UKIDSS). Mon. Not. R. Astron. Soc. **379**, 1599–1617(19) (2007)
7. Dye, S., et al.: The UKIRT Infrared Deep Sky Survey early data release. Mon. Not. R. Astron. Soc. **372**, 1227–1252 (2006)
8. Warren, S.J., et al.: The United Kingdom Infrared Telescope Infrared Deep Sky Survey first data release. Mon. Not. R. Astron. Soc. **375**, 213–226 (2007)
9. Casali, M., et al.: The UKIRT wide-field camera. Astron. Astrophys. **467**, 777–784 (2007)
10. Hewett, P.C., Warren, S.J., Leggett, S.K., Hodgkin, S.T.: The UKIRT Infrared Deep Sky Survey ZY JHK photometric system: passbands and synthetic colours. Mon. Not. R. Astron. Soc. **367**, 454–468 (2006)
11. Skrutskie, M.F., Cutri, R.M., Stiening, R., Weinberg, M.D., Schneider, S., Carpenter, J.M., Beichman, C., Capps, R., Chester, T., Elias, J., Huchra, J., Liebert, J., Lonsdale, C., Monet, D.G., Price, S., Seitzer, P., Jarrett, T., Kirkpatrick, J.D., Gizis, J.E., Howard, E., Evans, T., Fowler, J., Fullmer, L., Hurt, R., Light, R., Kopan, E.L., Marsh, K.A., McCallon, H.L., Tam, R., Van Dyk, S., Wheelock, S.: The Two Micron All Sky Survey (2MASS). Astron. J. **131**, 1163–1183 (2006)
12. Lintott, C.J., et al.: Galaxy Zoo: morphologies derived from visual inspection of galaxies from the Sloan Digital Sky Survey. Mon. Not. R. Astron. Soc. **389**, 1179–1189 (2008)
13. Irwin, M.J.: Automatic analysis of crowded fields. Mon. Not. R. Astron. Soc. **214**, 575–604 (1985)
14. Bertin, E., Arnouts, S.: SExtractor: software for source extraction. Astron. Astrophys. Suppl. Ser. **117**, 393–404 (1996)
15. Henrion, M., Mortlock, D.J., Hand, D.J., Gandy, A.: A Bayesian approach to star-galaxy classification. Mon. Not. R. Astron. Soc. **412**, 2286–2302 (2011)
16. Bazell, D., Peng, Y.: A comparison of neural network algorithms and preprocessing methods for star-galaxy discrimination. Astrophys. J. Suppl. Ser. **116**, 47–55 (1998)
17. Cortiglioni, F., Mähönen, P., Hakala, P., Frantti, T.: Automated star-galaxy discrimination for large surveys. Astrophys. J. **556**, 937–943 (2001)
18. Wolf, C., Meisenheimer, K., Röser, H.J.: Object classification in astronomical multi-color surveys. Astron. Astrophys. **365**(3), 660–680 (2001)
19. Aihara, H., et al.: The eighth data release of the Sloan Digital Sky Survey: first data from SDSS-III. Astrophys. J. Suppl. Ser. **193**, 29–45 (2011)
20. Richards, G.T., Nichol, R.C., Gray, A.G., Brunner, R.J., Lupton, R.H., Vanden Berk, D.E., Chong, S.S., Weinstein, M.A., Schneider, D.P., Anderson, S.F., Munn, J.A., Harris, H.C., Strauss, M.A., Fan, X., Gunn, J.E., Ivezić, Ž., York, D.G., Brinkmann, J., Moore, A.W.: Efficient photometric selection of quasars from the Sloan Digital Sky Survey: 100,000 $z < 3$ quasars from data release one. Astrophys. J. Suppl. Ser. **155**, 257–269 (2004)
21. Richards, G.T., Deo, R.P., Lacy, M., Myers, A.D., Nichol, R.C., Zakamska, N.L., Brunner, R.J., Brandt, W.N., Gray, A.G., Parejko, J.K., Ptak, A., Schneider, D.P., Storrie-Lombardi, L.J., Szalay, A.S.: Eight-dimensional mid-infrared/optical Bayesian quasar selection. Astron. J. **137**, 3884–3899 (2009)
22. Wolf, C., Meisenheimer, K., Röser, H.J., Beckwith, S.V.W., Chaffee Jr., F.H., Fried, J., Hippelein, H., Huang, J.S., Kümmel, M., von Kuhlmann, B., Maier, C., Phleps, S., Rix, H.W., Thommes, E., Thompson, D.: Multi-color classification in the Calar Alto Deep Imaging Survey. Astron. Astrophys. **365**, 681–698 (2001)
23. Bazell, D., Miller, D.J.: Class discovery in galaxy classification. Astrophys. J. **618**, 723–732 (2005)

24. Suchkov, A.A., Hanisch, R.J., Margon, B.: A census of object types and redshift estimates in the SDSS photometric catalog from a trained decision tree classifier. Astron. J. **130**, 2439–2452 (2005)
25. Ball, N.M., Brunner, R.J., Myers, A.D., Tcheng, D.: Robust machine learning applied to astronomical data sets. I. Star-galaxy classification of the Sloan Digital Sky Survey DR3 using decision trees. Astrophys. J. **650**, 497–509 (2006)
26. Irwin, M., Lewis, J., Riello, M., Hodgkin, S., Gonzales-Solares, E., Wyn Evans, D., Bunclark, P.: Pipeline processing of wide-field near-infrared data from WFCAM (in preparation)
27. Odewahn, S.C., de Carvalho, R.R., Gal, R.R., Djorgovski, S.G., Brunner, R., Mahabal, A., Lopes, P.A.A., Moreira, J.L.K., Stalder, B.: The Digitized Second Palomar Observatory Sky Survey (DPOSS). III. Star-galaxy separation. Astron. J. **128**, 3092–3107 (2004)
28. Philip, N.S., Wadadekar, Y., Kembhavi, A., Joseph, K.B.: A difference boosting neural network for automated star-galaxy classification. Astron. Astrophys. **385**, 1119–1126 (2002)
29. Miller, D.J., Browning, J.: A mixture model and EM-based algorithm for class discovery, robust classification, and outlier rejection in mixed labeled/unlabeled data sets. IEEE T. Pattern. Anal. **25**, 1468–1483 (2003)
30. Bardeau, S., Kneib, J.P., Czoske, O., Soucail, G., Smail, I., Ebeling, H., Smith, G.P.: A CFH12k lensing survey of X-ray luminous galaxy clusters. I. Weak lensing methodology. Astron. Astrophys. **434**, 433–448 (2005)
31. Scranton, R., Johnston, D., Dodelson, S., Frieman, J.A., Connolly, A., Eisenstein, D.J., Gunn, J.E., Hui, L., Jain, B., Kent, S., Loveday, J., Narayanan, V., Nichol, R.C., O'Connell, L., Scoccimarro, R., Sheth, R.K., Stebbins, A., Strauss, M.A., Szalay, A.S., Szapudi, I., Tegmark, M., Vogeley, M., Zehavi, I., Annis, J., Bahcall, N.A., Brinkman, J., Csabai, I., Hindsley, R., Ivezic, Z., Kim, R.S.J., Knapp, G.R., Lamb, D.Q., Lee, B.C., Lupton, R.H., McKay, T., Munn, J., Peoples, J., Pier, J., Richards, G.T., Rockosi, C., Schlegel, D., Schneider, D.P., Stoughton, C., Tucker, D.L., Yanny, B., York, D.G.: Analysis of systematic effects and statistical uncertainties in angular clustering of galaxies from early Sloan Digital Sky Survey data. Astrophys. J. **579**(1), 48–75 (2002)
32. Mortlock, D.J., Patel, M., Warren, S.J., Hewett, P.C., Venemans, B.P., McMahon, R.G., Simpson, C.: Probabilistic selection of high-redshift quasars. Mon. Not. R. Astron. Soc. **419**, 390–410 (2012)
33. Sérsic, J.L.: Influence of the atmospheric and instrumental dispersion on the brightness distribution in a galaxy. La Plata Bol **6**, 41 (1963)
34. Yasuda, N., Fukugita, M., Narayanan, V.K., Lupton, R.H., Strateva, I., Strauss, M.A., Ivezić, Z., Kim, R.S.J., Hogg, D.W., Weinberg, D.H., Shimasaku, K., Loveday, J., Annis, J., Bahcall, N.A., Blanton, M., Brinkmann, J., Brunner, R.J., Connolly, A.J., Csabai, I., Doi, M., Hamabe, M., Ichikawa, S.I., Ichikawa, T., Johnston, D.E., Knapp G. R. andKunszt, P.Z., Lamb, D.Q., McKay, T.A., Munn, J.A., Nichol, R.C., Okamura, S., Schneider, D.P., Szokoly, G.P., Vogeley, M.S., Watanabe, M., York, D.G.: Galaxy number counts from the Sloan Digital Sky Survey commissioning data. Astron. J. **122**, 1104–1124 (2001)
35. Henrion, M., Hand, D.J., Gandy, A., Mortlock, D.: CASOS: a Subspace Method for Anomaly Detection in High Dimensional Astronomical Databases (2011). Submitted
36. Aggarwal, C.C., Hinneburg, A., Keim, D.A.: On the surprising behavior of distance metrics in high himensional space. In: Van den Bussche, J., Vianu, V. (eds.) Database Theory - ICDT 2001, *Lecture Notes in Computer Science*, vol. 1973, pp. 420–434. Springer (2001)
37. Beyer, K., Goldstein, J., Ramakrishnan, R., Shaft, U.: When is "nearest neighbor" meaningful? In: Beeri, C., Buneman, P. (eds.) Database Theory - ICDT '99, *Lecture Notes in Computer Science*, vol. 1540, pp. 217–235. Springer (1999)
38. Breunig, M.M., Kriegel, H.P., Ng, R.T., Sander, J.: LOF: identifying density-based local outliers. Proceedings of the 2000 ACM SIGMOD International Conference on Management of Data **29**(2), 93–104 (2000)
39. Knorr, E.M., Ng, R.T., Tucakov, V.: Distance-based outliers: algorithms and applications. VLDB J. **8**, 237–253 (2000)

40. Rebbapragada, U., Protopapas, P., Brodley, C., Alcock, C.: Finding anomalous periodic time series. Mach. Learn. **74**, 281–313 (2009). 10.1007/s10994-008-5093-3
41. Dutta, H., Gianella, C., Borne, K., Kargupta, h.: Distributed top-K outlier detection in astronomy catalogs using the DEMAC system. In: Proceedings of the 2007 SIAM International Conference on Data Mining, pp. 473–478 (2007)
42. Jolliffe, I.T.: Principal Component Analysis, 2nd edn. Springer Series in Statistics. Springer (2002)
43. Mahule, T., Borne, K., Dey, S., Arora, S., Kargupta, H.: PADMINI: a peer-to-peer distributed astronomy data mining system and a case study. In: Proceedings of the Conference on Intelligent Data Understanding 2010 (2010)
44. Aggarwal, C.C., Yu, P.S.: An effective and efficient algorithm for high-dimensional outlier detection. VLDB J. **14**(2), 211–221 (2005)
45. Latecki, L., Lazarevic, A., Pokrajac, D.: Outlier detection with kernel density functions. In: Perner, P. (ed.) Machine Learning and Data Mining in Pattern Recognition, *Lecture Notes in Computer Science*, vol. 4571, pp. 61–75. Springer (2007)
46. Papadimitriou, S., Kitagawa, H., Gibbons, P.B., Faloutsos, C.: LOCI: fast outlier detection using the local correlation integral. In: Proceedings of the IEEE 19th International Conference on Data Engineering (ICDE'03). IEEE Computer Society (2003)
47. Kriegel, H.P., Schubert, M., Zimek, A.: Angle-based outlier detection in high-dimensional data. In: Proceedings of the 14th ACM SIGKDD International Conference on Knowledge Discovery and Data Mining (KDD'08) (2008)
48. Hambly, N.C., et al.: The WFCAM Science Archive. Mon. Not. R. Astron. Soc. **384**, 637–662 (2008)

Chapter 9
Independent Component Analysis for Dimension Reduction Classification: Hough Transform and CASH Algorithm

Asis Kumar Chattopadhyay[1], Tanuka Chattyopadhyay[2], Tuli De[1], and Saptarshi Mondal[1]

Abstract Classification of galaxies has been carried out by using two recently developed methods, viz., Independent Component Analysis (ICA) with K-means clustering and Clustering in Arbitrary Subspace based on Hough Transform (CASH) for different data sets. The first two sets are consisting of dwarf galaxies and their globular clusters whose distributions are non Gaussian in nature. The third one is a larger one containing a wider range of galaxies consisting of dwarfs to giants in 56 clusters of galaxies. Morphological classification of galaxies are subjective in nature and as a result can not properly explain the formation mechanism and other related issues under the influence of different correlated variables through a proper scientific approach. Hence objective classification by using the above mentioned methods are preferred to overcome the loopholes.

9.1 Introduction

In Statistics, very often we face empirical and large datasets to analyze. One of the most recent powerful statistical techniques for analyzing such datasets is Independent Component Analysis (ICA). Such datasets are generally multivariate in nature. The common problem is to find a suitable representation of the multivariate data. For the sake of computational and conceptual simplicity such representation is sought as a linear transformation of the original data. Principal Component Analysis, Factor Analysis, Projection Pursuit are some popular methods for linear transformation. But ICA is different from other methods, because it looks for the components in the representation that are both statistically independent and non Gaussian. In essence, ICA separates statistically independent component data, which is the original source

---

[1] Department of Statistics, Calcutta University, Kolkata, 35 Ballygunge Circular Road, Kolkata 700019, India, e-mail: akcstat@caluniv.ac.in

[2] Department of Applied Mathematics, Calcutta University, Kolkata, 35 Ballygunge Circular Road, Kolkata 700019, India

data, from an observed set of data mixtures. All information in the multivariate datasets are not equally important. We need to extract the most useful information. Independent Component Analysis extracts and reveals useful hidden factors from the whole datasets. ICA defines a generative model for the observed multivariate data, which is typically given as a large database of samples. ICA can be applied in various fields like speech processing, brain imaging, stock predictions, signal separation, telecommunications, econometrics, etc.

Clustering is a technique used to place data elements into related groups without advance knowledge of the group definitions. Clustering algorithms are attractive for the task of identification in coherent groups for existing data sets under consideration. However, clustering algorithm needs the following requirements when applied to large data sets:

1. Minimal requirements of domain knowledge to determine the input parameters.
2. Discovery of clusters with arbitrary shape and good efficiency on large databases.
3. Automatic determination of the optimum number of homogeneous classes.

Popular clustering techniques such as the K-Means Clustering and Expectation Maximization (EM) Clustering fail to give solution to the combination of these requirements. Thus keeping in view the above considerations some new approaches have been developed known as Density Based Clustering Techniques and Subspace Clustering Techniques.

In the Density based approach the main reason why a cluster is recognized is that within each cluster there is a typical density of points which is considerably higher than outside the cluster. Furthermore, the density within the areas of noise is lower than the density in any of the clusters. In other words, the clusters and consequently the classes are easily and readily identifiable because they have an increased density with respect to the points they possess. The single points scattered around the database are outliers, which means they do not belong to any clusters as a result of being in an area with relatively low concentration. Here discussions have been focused on Subspace Clustering Techniques which is a data mining task.

Clustering seeks to find groups of similar objects based on the values of their attributes. Traditional clustering algorithms, i.e., the Full Space algorithms use distance on the whole dataspace to measure similarity between objects. As the number of dimensions in a dataset increases, distance measures become increasingly meaningless. For very high dimensional datasets, the objects are almost equidistant from each other. This is known as the curse of high dimensionality.

The concept of subspace clustering has been proposed to cope with this problem by discovering clusters embedded in the subspaces of high dimensional datasets. Subspace Clustering is the task of detecting all clusters in all subspaces. This means that a point might be a member of multiple clusters, each existing in a different subspace.

Subspaces can either be axis parallel or arbitrarily oriented affine subspaces. The two approaches towards clustering differ in how they interpret the overall goal, which is finding clusters in data sets with high dimensionality.

In both cases, the data objects which are grouped into a common subspace cluster, are very dense (i.e., the variance is small) when projected onto the hyperplane which is perpendicular to the subspace of the cluster (called the perpendicular space plane). The objects may form a completely arbitrary shape with a high variance when projected onto the hyperplane of the subspace in which the cluster resides (called the cluster subspace plane). This means that the objects of the subspace cluster are all close to the cluster subspace plane. The knowledge that all data objects of a cluster are close to the cluster subspace plane is valuable for many applications.

If the plane is axis-parallel, this means that the values of some of the attributes are more or less constant for all cluster members. The whole group is characterized by this constant attribute value, an item of information which can definitely be important for the interpretation of the cluster. This property may also be used to perform a dedicated dimensionality reduction for the objects of the cluster and may be useful for data compression (because only the higher-variance attributes must be stored at high precision individually for each cluster member) and similarity search (because only the high-variance attributes need to be individually considered for the search) and an index needs only to be constructed for the high-variance attributes.

If the cluster subspace plane is arbitrarily oriented, the knowledge is even more valuable. In this case, it is known that the attributes which define the cluster subspace plane have a complex dependency among each other. This dependency defines a rule, which again characterizes the cluster and which is potentially useful for cluster interpretation.

Many subspace clustering algorithms use a grid based approach to find dense regions. They partition the data space into non-overlapping rectangular cells by discretizing each dimension into a number of bins. A cell is dense if the fraction of total objects contained in the cell is greater than a threshold. Dense cells in all subspaces are identified using a bottom-up strategy and connected dense cells are merged together to form clusters. In the grid based approach, objects around the boundaries of the bins have similar values, but they are put into different bins. As a result, a cluster may be divided into several small clusters.

These methods are popular due to two main reasons. Firstly, conventional (full space) clustering algorithms often fail to find useful clusters when applied to data sets of higher dimensionality, because typically many of the attributes are noisy, some attributes may exhibit high correlations with others and only few of the attributes really contribute to the cluster structure. Secondly, the knowledge gained from a subspace clustering algorithm is much richer than that of a conventional clustering algorithm because it can be used for interpretation, data compression, similarity search, etc.

Arbitrarily Oriented Clustering assumes that the cluster structure is significantly dense in the local neighborhood of the cluster centers or other points that participate in the cluster.

In the context of high-dimensional data, this "locality assumption" is rather optimistic. Theoretical considerations show that concepts like "local neighbourhood" are not meaningful in high-dimensional spaces because distances can no longer be used to differentiate between points. This is a consequence of the well-known curse of dimensionality.

In the present work the concept of a arbitrarily oriented subspace clustering technique developed by [1] has been applied.

9.2 Data Sets for ICA

To demonstrate Independent Component Analysis with subsequent K-means clustering we have used two data sets of dwarf galaxies and their globular clusters (GCs) in the Local Volume(LV).

9.2.1 Data set 1

This consists of 60 dwarf galaxies taken from a data set of 104 dwarf galaxies [2]. The reduced set has been constructed such that its members have most of their parameters without missing values. The parameters considered from [2] are distance modulus (μ_0, in mag), morphological index (T), mean metallicity of the Red Giant Branches ([Fe/H], in dex), effective color ($(V - I)_e$, in mag), logarithm of projected major axis (log(Diam), in Kpc) from [3], logarithm of limiting diameter (log(Dlim), in Kpc), limiting V and I magnitudes (M_V, M_I, in mag) within the diameter Dlim, limiting V and I surface brightness (SBVL, SBIL, in mag arcsec^{-2}) taken at the distance Dlim/2 from centre of the host galaxy, effective surface brightness in V band (SBVe, in mag arcsec^{-2}), logarithm of effective radius (log(R_e) in Kpc), logarithm of model exponential scale length (logh, in Kpc), best exponential fitting central surface brightness in V and I bands (SBVC, SBIC in mag arcsec^{-2}) respectively. Among these T and μ_0 are not directly used in the analysis. The parameters used from [3] are HI rotational velocity (V_m in Km s^{-1}), HI mass to luminosity ratio (M_{HI}/L in solar units) and tidal index (Θ). The scaling parameters used from [4] are total stellar mass ($M_{*,V}$ in $10^7 M_\odot$) and HI mass of the host galaxy (M_{HI} in $10^7 M_\odot$) respectively. Among all these parameters only 13 parameters from [2] (excluding T and μ_0) together with (Θ) are directly used for ICA with K-means clustering as the sample is complete without any missing values with respect to these 14 parameters. The remaining parameters are used to study the properties of the groups found as a result of the final classification as these parameters have missing values more than 5% of the sample size.

9.2.2 Data set 2

This consists of 103 GCs in the Local Volume dwarf galaxies [5]. The parameter set consists of logarithm of half light radius (log(r_h) in parsec), apparent axial ratio (e), integrated absolute magnitude (V0, in mag) corrected for extinction, integrated absolute $(V-I)_0$ color (corrected for Galactic extinction, in mag), projected distance from the host galaxy (d_{proj}, in Kpc), central surface brightness in V and I bands (μ_{V0}, μ_{I0} in mag arsec^{-2}), logarithm of King core radius and tidal radius ($\log(r_c)$, $\log(r_t)$ in parsec).The values of the parameters Age (in Gyr), Z/H (in dex) and α/H (in dex) are taken from [6].

9.2.3 Test for Normality

Before applying ICA we have tested Gaussianity of the data sets. Here the null hypothesis was that the entire data set follows multivariate normal distribution. For testing this we performed multivariate Shapiro-Wilk test. We found that the p-value for data set 1 is 0.001456 whereas for data set 2 it is 6.462×10^{-7}, which are too small. Thus the null hypotheses have been rejected and we could conclude that the data sets do not follow multivariate normal distribution.

9.3 Independent Component Analysis

Independent Component Analysis (ICA) was most clearly stated by [7]. Formally, the classical ICA model is of the form:

$$X = AS, \tag{9.1}$$

where $X = [X_1, \ldots, X_m]'$ is a random vector of observations, $S = [S_1, \ldots, S_m]'$ is a random vector of hidden sources whose components are mutually independent and A is nonsingular mixing matrix. So A^{-1} is the unmixing matrix. Let us have n independently and identically distributed (i.i.d.) samples of X, say $\{X(j) : 1 \leq j \leq n\}$. The main goal of ICA is to estimate the unmixing matrix A^{-1} and thus to recover hidden source using $S_k = A_k^{-1} X$, where A_k^{-1} is the kth row of A^{-1}.

In the model, it is assumed that the data variables are linear or non-linear mixtures of some latent variables and the mixing system is also unknown. The latent variables are assumed non-Gaussian and mutually independent and they are called the independent components of the observed data.

Suppose n random variables $X_1, \ldots, X_n$ are expressed as linear combinations of n random variables $S_1, \ldots, S_n$. Equation (9.1) can be written as:

$$X_i = a_{i1} S_1 + a_{i2} S_2 + \cdots + a_{in} S_n, \quad i = 1, 2, \ldots, n \tag{9.2}$$

The S_i's are statistically mutually independent, where a_{ij}'s are the entries of the nonsingular matrix A. All we observe are the random variables X_i, and we have to estimate both the mixing coefficients a_{ij} and the independent components S_i, using the X_i.

There are many computer algorithms for performing ICA. A first step in those algorithms is to whiten (sphere) the data. This means that any correlations in the data are removed, i.e., the data are forced to be uncorrelated. Mathematically speaking, we need a linear transformation V such that $Z = VX$, where $E(ZZ') = I$. This can be easily accomplished by choosing $V = C^{-1/2}$, where $C = E(XX')$.

After sphering, the separated data can be found by an orthogonal transformation on the whitened data Z.

9.4 ICA by Maximization of Non-Gaussianity

In ICA estimation, non-Gaussianity is very important. Without non-Gaussianity the estimation is not possible. Non-Gaussianity is motivated by the central limit theorem. Under certain conditions, the statistical distribution of a sum of independent random variables tends toward a Gaussian distribution. A sum of two independent random variables usually has a distribution that is closer to Gaussian than any of the two original random variables. Here,

$$X = AS, \quad VX = VAS$$
$$\Rightarrow Z = (VA)S, \tag{9.3}$$

which implies that Z_i is closer to Gaussian than S_i. S_i is estimated by Z_i through maximization of non-Gaussianity. From Eq. (9.3) we can write

$$S = WZ, \tag{9.4}$$

where $W = (VA)^{-1}$.

We can measure non-Gaussianity by *Negentropy* [8]. The entropy of a discrete variable is defined as the sum of the products of probability of each observation and the log of those probabilities. On the other hand, for a continuous function the entropy is called differential entropy which is given by the integral of the function times the log of the function. Negentropy is the difference between the differential entropy of a source S from the differential entropy of a Gaussian source with the same covariance of S. It is denoted by $J(S)$ and defined as follows:

$$J(S) = H(S_{\text{Gauss}}) - H(S), \tag{9.5}$$

where

$$H(S) = -\int p_S(\eta) \log p_S(\eta) d\eta, \tag{9.6}$$

$p_S(\eta)$ is the density function of S. Negentropy is always non-negative, and it is zero if and only if S has a Gaussian distribution. Negentropy has an interesting property that it is invariant for invertible linear transformation. It is also a robust measure of non-Gaussianity. Here we estimate S by maximizing the distance of its entropy from Gaussian entropy as the noises are assumed to be Gaussian and if the signals are non-Gaussian only then can they be separated from the noise. If the signals are Gaussian, then ICA will not work.

9.5 Approximation of Negentropy

One drawback of negentropy is that it is very difficult to compute. That is why it needs to be approximated [8]. The approximation is given by:

$$J(S) \propto (E[G(S)] - E[G(S_{\text{Gauss}})])^2, \quad (9.7)$$

where S_{Gauss} is a Gaussian random variable, G is a non-quadratic function. In particular, G should be so chosen that it does not grow too fast. Two popular choices of G are:

$$G_1(S) = \frac{1}{a} \log \cosh aS \quad (9.8)$$

$$G_2(S) = -e^{-S^2/2},$$

where $1 \leq a \leq 2$ is some suitable constant, which is often taken equal to 1.

9.6 The FastICA Algorithm

In this method the independent components are estimated one by one. This algorithm converges very fast and is very reliable. This algorithm is also very easy to use. We follow the following steps to perform the algorithm:

1. We center the data such that its mean becomes zero.
2. We whiten the data and denote it by Z.
3. We choose the number of Independent Components to be estimated and set $k = 1$.
4. We take an initial value of unit norm for W_k randomly, i.e., we initialize W_k, where W_k is the kth row of W.
5. We set $W_k = E\{Zg(W_k'Z)\} - E\{g'(W_k'Z)\}W_k$, where g is derivative of G and G is defined as in (9.8).
6. We orthogonalize as:

$$W_k = W_k - \sum_{j=1}^{k-1} (W_k' W_j) W_j.$$

7. We set $W_k = W_k / \|W_k\|$.
8. If W_k does not converge, then we go back to step 5.
9. We set $k = k + 1$. If k does not exceed the number of independent components to be estimated, we go back to step 4.

Thus we find the estimated independent components.

9.7 Properties of the Groups of Data Set 1

Two groups, G1 and G2 of dwarf galaxies in the local volume have been found by using ICA along with K-means clustering on the basis independent components irrespective of their morphological classification (viz., T). In order to find the the optimum value of K (viz., 2) we have used the method developed by [9] which is discussed under section 9.13. In G1, 71% are dwarf irregulars, 20% are dwarf spirals and 3% are of transition type whereas in G2, 80% are dwarf irregulars, 15% are dwarf spirals and 5% are of transition type. G1 contains brighter galaxies having larger sizes and high degree of rotation whereas G2 consists of fainter galaxies of smaller sizes having insignificant rotation. A luminosity-metallicity (Fig. 9.1) relation shows a significant correlation for the galaxies in G2 ($r = 0.44$; $p = .024$) and for a part of G1. This indicates that formation of dwarf galaxies is primarily governed by self enrichment though partial processes also take place during the evolution of stars and interaction of interstellar gas pervading the dwarf galaxies with intergalactic medium [10, 11]. Galactic winds, supernovae explosions, tidal or ram pressure etc are responsible for significant loss of metals from dwarf galaxies [12]. Multiple bursts of of star formation [13, 14] also counts for complex behaviour of luminosity-metallicity relation in G1 and G2. Also, shift in almost two orders of magnitudes from G1 to G2 at the same metallicity and same scale length (see Fig. 9.2) suggests that galaxies of G2 evolved from some galaxies of G1 due to depletion of gas outflows produced by supernovae explosion, ram pressure and tidal striping. Fig. 9.3 shows the scatter plot of tidal index [15] versus logarithm of scale length for the sample galaxies. It is seen that galaxies having tidal indices greater than -0.5, scale length increases with tidal index. This shows that neighbours influence the thickening of the galactic disks ($r(\log h, \Theta) = 0.559$, $p = 0.001$) irrespective of morphological types.

9.8 Properties of the Groups of Data Set 2

Four groups of Globular clusters (GCs), GC1, GC2, GC3, and GC4 are found as a result of K-means clustering through independent components. The optimum vale of K (viz., 4) has been found by the same method. Among the four clusters GCs of GC2 and GC4 are younger, metal rich and dynamically less evolved as is evident from their mean $\log(r_t/r_c)$ values as well as highest eccentricity (viz., $e = 0.16$ from GC4). On the other hand GCs of GC1 and GC3 are older, metal poor, dynamically evolved (hence rounder) and GCs of GC1 are almost round. The metal deficiency shows a tidal depletion due to external origin. So, G1 galaxies might be the formation sites for GC2, GC4 globular clusters whereas G2 galaxies can be considered as the sites for GC1 and GC3 globular clusters.

9 Independent Component Analysis for Dimension Reduction Classification

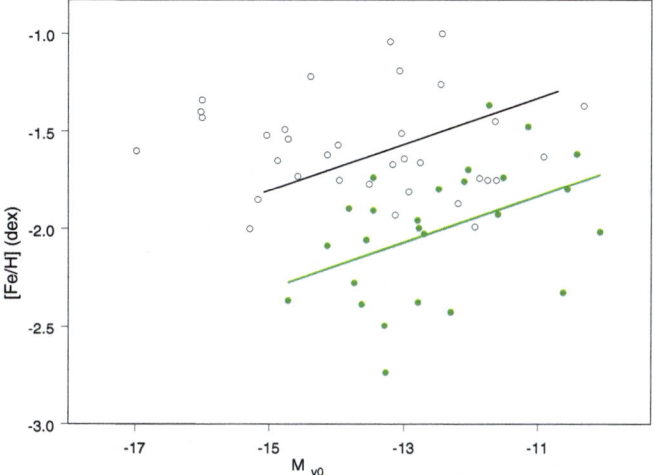

Fig. 9.1 Luminosity (M_{V0}) metallicity ([Fe/H]) diagram for the two groups G1 and G2 of dwarf galaxies found as a result of CA in the LV. The black open circles are for group G1 and the green solid circles are for G2.

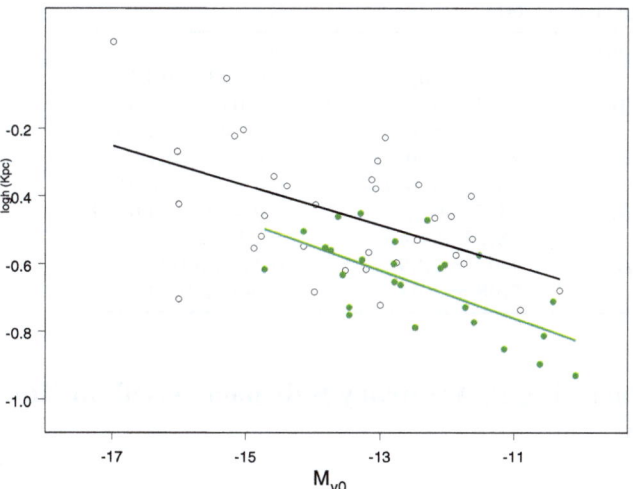

Fig. 9.2 The logarithm of scale length (logh) versus Luminosity (M_{V0}) of two groups G1 and G2. The black open circles are for group G1 and the green solid circles are for G2.

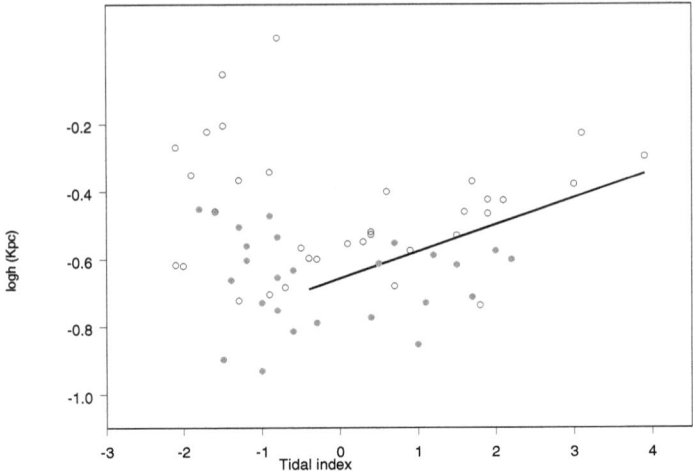

Fig. 9.3 The logarithm of scale length (logh) versus tidal index (⊙) for the groups G1 and G2. The black open circles are for group G1 and the green solid circles are for G2.

Table 9.1 Mean values of the significant parameters of the two groups G1 and G2 with standard errors.

Parameters	G1	G2
Number	34	26
θ	0.129 ± 0.285	-0.173 ± 0.241
[Fe/H]	-1.5806 ± 0.0430	-2.0127 ± 0.0658
M_{V0}	-13.555 ± 0.275	-12.491 ± 0.244
SBV_{e0}	23.321 ± 0.212	23.277 ± 0.145
$(V-I)_{e0}$	0.8813 ± 0.0343	0.7675 ± 0.0405
$\log R_e$	-0.2366 ± 0.0322	-0.4407 ± 0.0261
V_m	28.27 ± 4.46	17.0 ± 2.85
M_{H1}	4.07 ± 1.89	6.42 ± 5.07
M_{*V}	5.08 ± 2.41	0.640 ± 0.340

9.9 CASH: Clustering in Arbitrary Subspace based on Hough Transform

The locality assumption that the clustering structure is dense in the entire feature space and that the Euclidean neighborhood of points in the cluster, or of cluster centers, does not contain noise is a very strict limitation for high-dimensional real-world datasets. In high-dimensional spaces, the distance to the nearest neighbor and the distance to the farthest neighbour converge. As a consequence, distances can no

Table 9.2 Mean values of the significant parameters for the four groups of globular clusters in the Local Volume.

Parameters	GC1	GC2	GC3	GC4
Number	34	36	9	21
V_0	−7.253	−7.298	−6.619	−6.353
μ_{V0}	19.488	20.579	20.406	19.629
$\log(r_h)$	0.8524	1.0616	0.8886	0.7147
$\log(r_t)$	1.6004	1.6802	2.253	1.3405
$\log(r_c)$	0.4562	0.7319	0.4891	0.3530
$\log(r_t/r_c)$	1.1443	0.9484	1.7638	0.9875
e	0.0971	0.1278	1.1556	0.1619
$(V-I)_0$	1.0959	0.8158	0.872	0.5824
d_{proj}	0.897	1.280	1.118	0.650
Age(Gyr)	8.25	2.25	6	4
Z/H	−1.5	−1.425	−1.8	−1.4
α/H	0.2	0.25	0.1	0.1

longer be used to differentiate between points in high dimensional spaces and concepts like the neighborhood of points become meaningless. Usually, although many points share a common hyperplane, they are not close to each other in the original feature space. In those cases, existing approaches will fail to detect meaningful patterns because they cannot learn the correct subspaces of the clusters. In addition, as long as the correct subspaces of the clusters cannot be determined, obviously outliers and noise cannot be removed in a preprocessing step.

In this method, development of an original principle to characterize the subspace holding a cluster is based on the idea of the Hough Transform. This transform charts out the points from a 2-dimensional data space (also known as picture space) of Euclidean co-ordinates (eg., pixel of an image) into a parameter space. It is the parameter space that stands for all possible 1-dimensional lines in the original 2-dimensional data space. In principle, each point of the data space is mapped into an infinite number of points in the parameter space which is, however, not an infinite set but actually a trigonometric function relating to the parameter space. Each function in the parameter space represents all lines in the picture space crossing the corresponding point in data space. The intersection of the dual curves in the parameter space points to a line through the corresponding points alike in the picture space.

The objective of a clustering algorithm is to find intersections of many curves in the parameter space representing lines through many database objects. The key feature of the Hough transform is that the distance of the points in the original data space is not considered any more. Objects can be identified as associated to a common line even if they are far apart in the original feature space. As a consequence, the Hough transform is a promising candidate for developing a principle for subspace analysis that does not require the locality assumption and, thus, enables a global subspace clustering approach.

9.10 Input Parameters

CASH requires the user to specify two input parameters. The first parameter m specifies the minimum number of sinusoidal curves (minpts) that need to intersect a hypercuboid in the parameter space such that this hypercuboid is regarded as a dense area. Obviously, this parameter represents the minimum number of points in a cluster and thus is very intuitive. The second parameter s specifies the maximal number of splits along a search path (split level). CASH is robust with respect to the choice of s. Since CASH does not require parameters that are hard to guess like the number of clusters, the average dimensionality of the subspace clusters, or the size of the Euclidean neighborhood based on which the similarity of the subspace clusters is learned, it is much more usable.

9.11 Data Set 3

In order to evaluate the efficiency of the algorithm CASH, the method has been applied to a data compiled and standardized by [16] for a sample of 56 low-redshift galaxy clusters containing 699 early-type galaxies. After eliminating the missing observations the sample size has been reduced to 528 and the CASH method has been performed using four parameters (variables), viz., the logarithm of the effective Radius(log R_e in Kpc), the surface brightness averaged over half light radius(μ_e in mag arcsec^{-2}), central velocity dispersion(σ in km s^{-1}) and magnesium index (M_{g2} index).

9.11.1 Experimental Evaluation

9.11.1.1 Initial choice of constraints:

Since CASH only needs two constraints, viz., m the minpts and s the number of split levels, the constants have been selected by trial and error method. The jitter has been fixed to a preassigned small value 0.15. The value of m has been taken from 100 to 40 and values of s have been varied from 1 to 3. It is expected that the value for m should not be larger than 100 for a sample of size 528 because it is the minimum number of points to be included per cluster. Also the number of split levels should be moderate for a data set of moderate size.

It is seen that the stability has been achieved by taking $m = 60$ and $s = 2$. After that even a decrease in the value of m has not contributed in the result. With the above mentioned combination of s and m, the number of cluster has been found to be seven.

9.12 Properties of the Groups of Data Set 3

The efficiency of CASH has been checked by several properties.

The average properties of the seven groups are shown in Table 9.3 where N_{gal} represents the size of each clusters and M_{dyn} represents the dynamical mass which can be obtained from the relation

$$M_{dyn} \approx A\sigma^2 R_e/G, \quad \text{where } A \text{ and } G \text{ are constants.}$$

It is well known that Fundamental Plane (FP) is a relationship between the effective radius, average surface brightness and central velocity dispersion of normal elliptical galaxies and Virial Plane (VP) is the parametric plane constituted by effective radius, surface brightness averaged over effective radius and velocity dispersion when a galaxy is in dynamical equilibrium.

The slopes for $\log R_e$ with respect to $\log M_{dyn}$ are shown in Table 9.4 for seven clusters.

From Table 9.4 it is clear that all the slopes are greater than 0.38. So the galaxies are not formed as a result of pure disk mergers [17]. Since the slope of C_4 is more or less close to 0.38, it might be formed due to pure disk merger. For the remaining ones, the slopes are steeper which might be due to merger of non-disky objects or the result of repeated merging of small systems [18].

The Mg2 index more or less increases chronologically. So accordingly, higher Mg2 indicates that the galaxies are dynamically more evolved and lower Mg2 value signifies that the galaxies are dynamically less evolved (see Table 9.3, column 6).

Table 9.3 Average properties of the seven groups of galaxies obtained by CASH Method:

clusters	N_{gal}	$\log_\sigma$	μ_e	$\log R_e$	M_{g2}	$\log M_{dyn}$
C1	21	2.2345 ± 0.021	19.382 ± 0.119	2.7791 ± 0.017	0.26286 ± 0.0052	10.179 ± 0.012
C2	60	2.2575 ± 0.236	20.414 ± 0.231	2.8638 ± 0.016	0.28143 ± 0.0035	10.310 ± 0.021
C3	75	2.2526 ± 0.025	19.461 ± 0.172	2.8694 ± 0.027	0.28239 ± 0.0043	10.306 ± 0.013
C4	167	2.2537 ± 0.071	19.437 ± 0.261	2.8257 ± 0.035	0.28837 ± 0.0036	10.264 ± 0.014
C5	82	2.2895 ± 0.013	20.487 ± 0.266	2.8032 ± 0.013	0.26951 ± 0.0023	10.313 ± 0.035
C6	63	2.2752 ± 0.014	19.667 ± 0.143	2.9244 ± 0.029	0.29371 ± 0.0014	10.406 ± 0.023
C7	60	2.3450 ± 0.023	19.346 ± 0.235	2.9127 ± 0.013	0.30400 ± 0.0028	10.534 ± 0.034

Table 9.4 Slopes of seven different clusters.

clusters	N_{gal}	slope
C1	21	0.447
C2	82	0.443
C3	60	0.587
C4	75	0.417
C5	167	0.471
C6	63	0.663
C7	60	0.662

Table 9.5 The Values of the M_{g2} index, Tilt and Slope of seven clusters.

clusters	N_{gal}	M_{g2}	Tilt	Slope
C1	21	0.26286	0.3487	0.447
C2	82	0.28143	0.744	0.443
C3	60	0.28239	0.645	0.587
C4	75	0.28837	0.649	0.417
C5	167	0.26951	0.728	0.471
C6	63	0.29371	0.749	0.663
C7	60	0.30400	0.732	0.662

The Fundamental Plane (FP) is expressed by the relationship,

$$\log_{10} R_e = a \log \sigma + b\mu_e + c$$

Where a, b and c are constants to be determined.

The virial Plane (VP) is expressed by the relationship,

$$\log_{10} R_e = 2 \log \sigma + 0.4\mu_e$$

The ratio of the slopes of the FP with VP is defined as the tilt. Hence a small value of tilt indicates that the FP is farther from the corresponding VP. The tilt values for the seven groups are shown in Table 9.5.

It is also clear from Table 9.5 that the tilts are almost increasing in parity with the Mg2 index and also the tilts are approximately increasing from C1 to C7 indicating that the galaxies in the later groups are more dynamically evolved (hence closer to their corresponding virial planes). This is also consistent with the fact that the magnesium indices are also increasing for the groups indicating that in a dynamically evolved galaxy the metal content is higher. Since none of the tilt values are close to 1, it can be concluded that these galaxies have been formed by dissipational Mergers [17]. So the groups can be considered as evolutionary tree with respect to FP, VP and M_{g2} indices as

$$C1 \longrightarrow C2 \longrightarrow C3 \longrightarrow C4 \longrightarrow C7 \longrightarrow C6 \quad \text{(excluding C5)}$$

which is irrespective of the scatter of the FP giving rise to several controversial arguments so far.

9.13 Comparison with the K-means Clustering

In order to study the efficiency of CASH, cluster analysis have also been carried out using K-Means [19] method. It is seen that the optimum number of clusters is $k = 7$

9 Independent Component Analysis for Dimension Reduction Classification

which is similar with CASH results. The optimality of K-Means method has been checked by the criteria developed by [9] described as follows,

Let the data set be modeled as a p-dimensional random variable, X, consisting of a mixture distribution of G components with common covariance, Γ. If we let $c_1, \ldots, c_K$ be a set of K cluster centers, with c_X the closest center to a given sample of X, then the minimum average distortion per dimension when fitting the K centers to the data is given by

$$d_K = \frac{1}{p} \min_{c_1, c_2, \ldots, c_k} E[(X - C_X)^T (X - C_X)]$$

Then the jump is calculated as follows

$$j_K = d_K^{-\frac{p}{2}} - d_{K-1}^{-\frac{p}{2}}$$

Then a graph of j_K versus K (the number of clusters) is plotted. Then the value of K for the maximum j_K is obtained is chosen to be the optimum number of clusters. In the present data set, $p = 4$, so that j_K is taken as:

$$j_K = d_K^{-2} - d_{K-1}^{-2}$$

After calculating the jumps it is seen that the optimum number of clusters is $K = 7$, since the highest jump is achieved at $K = 7$. Figs. 9.4 and 9.5 correspond to the distortion curve (d_K versus K) and the Jump Curve (j_K versus K) respectively.

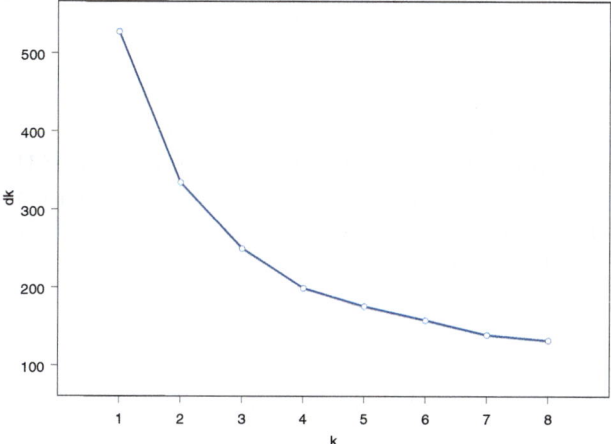

Fig. 9.4 The Distortion Curve.

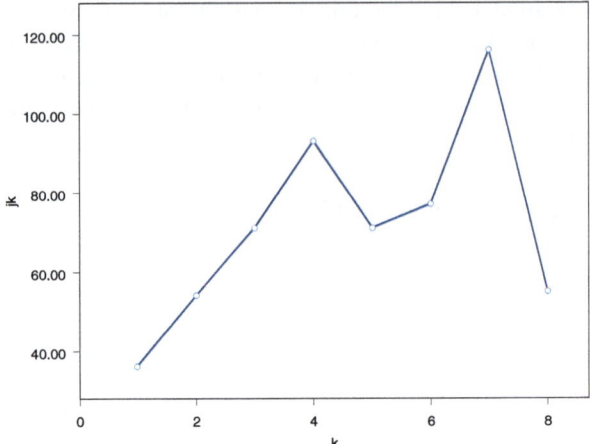

Fig. 9.5 The Jump Curve.

Table 9.6 Average properties of the seven groups of galaxies obtained by K-Means Method

clusters	N_{gal}	$\log_\sigma$	μ_e	$\log R_e$	M_{g2}	$\log M_{dyn}$
C1	110	2.1571 ± 0.006	19.215 ± 0.033	2.6391 ± 0.012	0.26655 ± 0.002	9.8846 ± 0.021
C2	133	2.3663 ± 0.005	19.315 ± 0.027	2.8450 ± 0.009	0.30395 ± 0.001	10.509 ± 0.015
C3	83	2.4090 ± 0.007	19.990 ± 0.037	3.2030 ± 0.012	0.31200 ± 0.002	10.925 ± 0.024
C4	36	2.0105 ± 0.017	20.024 ± 0.101	2.7641 ± 0.028	0.21617 ± 0.004	9.7162 ± 0.047
C5	21	2.3493 ± 0.018	21.305 ± 0.116	3.5794 ± 0.042	0.31076 ± 0.005	11.209 ± 0.061
C6	84	2.2362 ± 0.003	19.924 ± 0.038	2.9729 ± 0.013	0.27133 ± 0.002	10.376 ± 0.027
C7	60	2.2625 ± 0.009	18.314 ± 0.066	2.4214 ± 0.021	0.28669 ± 0.002	9.8776 ± 0.032

Like CASH results, for K-means clustering the average properties are shown in Table 9.6 while magnesium indices (M_{g2}), tilts and slopes are listed in Table 9.7 for the seven groups. It is clear from Tables 9.6 and 9.7, unlike CASH results, the M_{g2} indices, M_{dyn}, tilts and slopes do not increase along the sequence of the groups continuously as in the previous case. Here the trend is to form two sequences of groups of galaxies of which one is

$$C1 \longrightarrow C2 \longrightarrow C3$$

and the other is

$$C4 \longrightarrow C7 \longrightarrow C6 \quad (\text{excluding C5}).$$

The above two sequences correspond to dynamically less and more evolved galaxies respectively. Numerical values of slopes indicates that C6 galaxies can be formed as a result of pure disk merger whereas galaxies of remaining groups might be formed due to repeated merger of small systems. So from the above two

Table 9.7 The values of the M_{g2} index, Tilt and Slope of seven clusters

clusters	N_{gal}	M_{g2}	Tilt	Slope
C1	21	0.26655	0.415	0.444
C2	82	0.30395	0.501	0.453
C3	60	0.31200	0.606	0.444
C4	75	0.21617	0.461	0.435
C5	167	0.31076	0.736	0.559
C6	63	0.27133	0.553	0.418
C7	60	0.28669	0.554	0.528

analyses it is clear that CASH results are physically more interpretable compared to K-means. From the point of view of an unique evolutionary path with respect to FP, CASH gives a stronger footing on the explanation of different degrees of scatter of the respective FP for the seven clusters which is not very clear for the evolutionary path found through K-means clustering. This also indicates that CASH method is more appropriate for a larger data set to extract the hidden features of the objects therein.

References

1. Achtert, E., Böhm, C., Jörn, D., Kröger, P., Zimek, A.: Global Correlation Clustering Based on Hough Transform. Stat. Anal. Data Min. **1**, 111–127 (2008)
2. Sharina, M.E., Karachentsev, I.D., Dolphin, A.E., Karachentseva, V.E., Tully, R.B., Karataeva, G.M., Makarov, D.I., Makarova, L.N., Sakai, S., Shaya, E.J., Nikolaev, E.Y., Kuznetsov, A.N.: Photometric properties of the Local Volume dwarf galaxies. Mon. Not. R. Astron. Soc. **384**, 1544–1562 (2008)
3. Karachentsev, I.D., Karachentseva, V.E., Huchtmeier, W.K., Makarov, D.I.: A Catalog of Neighboring Galaxies. Astron. J. **127**, 2031–2068 (2004)
4. Georgiev, I.Y., Puzia, T.H., Goudfrooij, P., Hilker, M.: Globular cluster systems in nearby dwarf galaxies: III. Formation efficiencies of old globular clusters. Mon. Not. R. Astron. Soc. **406**, 1967–1984 (2010)
5. Sharina, M.E., Puzia, T.H., Makarov, D.I.: Hubble Space Telescope imaging of globular cluster candidates in low surface brightness dwarf galaxies. Astron. Astrophys. **442**, 85–95 (2005)
6. Puzia, T.H., Sharina, M.E.: VLT spectroscopy of globular clusters in low surface brightness dwarf galaxies. Astrophys. J. **674**, 909–926 (2008)
7. Comon, P.: Independent component analysis, A new concept? Signal Process. **36**, 287–314 (1994)
8. Hyvärinen, A., Karhunen, J., Oja, E.: Independent Component Analysis. Wiley, New York (2001)
9. Sugar, A.S., James, G.M.: Finding the number of clusters in a data set: An information theoretic approach. J. Amer. Statist. Assoc. **98**, 750–763 (2003)
10. Grebel, E.K., Gallagher, J.S., Harbeck, D.: The Progenitors of Dwarf Spheroidal Galaxies. Astron. J. **125**, 1926–1939 (2003)
11. Chattopadhyay, T, Sharina, M., Karmakar, P.: Statistical analysis of dwarf galaxies and their globular clusters in the Local Volume. Astrophys. J. **724**, 678–686 (2010)

12. Shaya, E.J., Tully, R.B.: The angular momentum content of galaxies. Astrophys. J. **281**, 56–66 (1984)
13. Carraro, G., Chiosi, C. Girardi, L., Lia, C.: Dwarf elliptical galaxies: structure, star formation and colour-magnitude diagrams. Mon. Not. R. Astron. Soc. **327**, 69–79 (2001)
14. Hirashita, H.: Intermittent Star-Formation Activities of Dwarf Irregular Galaxies. Publ. Aston. Soc. Jpn. **52**, 107–112 (2000)
15. Karachentsev, I.D, Makarov, D.I.: Galaxy Interactions in the Local Volume. In: Barnes, J.E., Sanders, D.B. (eds.) Galaxy interactions at high and low redshifts. Proc. IAU Symposium, vol. 186, pp. 109–118 (1998)
16. Hudson, M.J., Lucey, J.R., Smith, R.J., Schlegel, D.J., Davies, R.L.: Streaming motions of galaxy clusters within 12 000 km s^{-1} - III. A standardized catalogue of Fundamental Plane data. Mon. Not. R. Astron. Soc. **327**, 265–295 (2001)
17. Robertson B., Cox, T.J., Hernquist, L., Franx, M., Hopkins, P.F., Martini, P., Springel, V.: The Fundamental Scaling Relations of Elliptical Galaxies. Astrophys. J. **641**, 21–40 (2006)
18. Shen, S., Mo, H.J., White, S.D.M., Blanton, M.R., Kauffmann, G., Voges, W., Brinkmann, J., Csabai, I.: The size distribution of galaxies in the Sloan Digital Sky Survey. Mon. Not. R. Astron. Soc. **343**, 978–994 (2003)
19. MacQueen, J.B.: Some Methods for Classification and Analysis of Multivariate Observations. In: Cam, L.M. Le, Neyman, J. (eds.) Proceedings of Fifth Berkeley Symposium on Mathematical Statistics and Probability, pp. 281–297. University of California Press (1967)

Chapter 10
Improved Cosmological Constraints from a Bayesian Hierarchical Model of Supernova Type Ia Data

Marisa Cristina March, Roberto Trotta, Pietro Berkes, Glenn Starkman, and Pascal Vaudrevange

Abstract We present a Bayesian hierarchical model for inferring the cosmological parameters from the supernovae type Ia fitted with the SALT-II lightcurve fitter. We demonstrate with simulated data sets that our method delivers tighter statistical constraints on the cosmological parameters over 90% of the time, that it reduces statistical bias typically by a factor $\sim$2–3 and that it has better coverage properties than the usual χ^2 approach. As a further benefit, a full posterior probability distribution for the dispersion of the intrinsic magnitude of SNe is obtained. We apply this method to recent SNIa data, and by combining them with CMB and BAO data we obtain $\Omega_m = 0.28 \pm 0.02$, $\Omega_\Lambda = 0.73 \pm 0.01$ (assuming $w = -1$) and $\Omega_m = 0.28 \pm 0.01$, $w = -0.90 \pm 0.05$ (assuming flatness; statistical uncertainties only). We constrain the intrinsic dispersion of the B-band magnitude of the SNIa population, obtaining $\sigma_\mu^{\text{int}} = 0.13 \pm 0.01$ [mag].

10.1 Supernovae Type Ia and the Accelerating Universe

Observations of the supernovae type Ia played the leading role in the discovery of the the apparent late time acceleration of our Universe, [1, 2], the importance of supernovae Ia as standardizable candles with which to probe the expansion history of the Universe, has not diminished since. There are several possible models to explain the apparent late time acceleration including dark energy, modified gravity, the backreaction and void models. Identifying which of these modes is the best given the observed data is one of the key questions in modern cosmology. One statistical approach to evaluating the relative merits of the various models would be to use Bayesian model selection, but whilst reviewing the suitability of supernovae type Ia (SNe Ia) data for use in Bayesian model selection, two problems became apparent:

Marisa Cristina March
Astrophysics Group, Imperial College, London. e-mail: marisa.march06@imperial.ac.uk
The Astronomy Centre, University of Sussex. e-mail: m.march@sussex.ac.uk

1. The SNe Ia data are affected by an unknown intrinsic dispersion in the SNe Ia absolute magnitudes, increasing the number of SNe Ia observed will not reduce this error.
2. The common method for doing cosmological parameter inference from the SALT-II light curve fitted SNe Ia data uses a procedure which precludes the subsequent use of this data in Bayesian model selection.

In this chapter we describe a different method for Bayesian analysis of SNe Ia data fitted with the SALT-II light curve fitter. This chapter is a review of work first presented in [3] and follows that publication closely. The Bayesian Hierarchical method replaces the second step (parameter inference step) of the SALT-II method. The aims of this new methodology are twofold:

1. To provide a rigorous statistical framework for assessing and understanding the unknown intrinsic dispersion.
2. To provide a fully Bayesian method for cosmological parameter inference from the SNe Ia data in order that the SNe Ia data can be used in Bayesian model selection, and exploited with the full suite of Bayesian methods.

This chapter describes how the new Bayesian method for cosmological parameter inference has been developed and tested, and also describes the beginnings and plans for applications of this method to investigations of evolution of SNe Ia with redshift. Use of this new method in problems of Bayesian model selection is not covered in this work, but will be implemented in future papers.

10.2 Lightcurve Fitting and a Description of the Problem

Several methods are available to fit SNe lightcurves, including the MLCS method, the Δm_{15} method, CMAGIC, [4, 5] SALT, SALT-II and others. Recently, a sophisticated Bayesian hierarchical method to fit optical and infrared lightcurve data has been proposed by [6, 7]. MLCS fits the cosmological parameters at the same time as the parameters controlling the lightcurve fits. The SALT and SALT-II methods, are a two step process, The first step is to fit to the SNe lightcurves three parameters controlling the SN magnitude, the stretch and colour corrections. From those fits, the cosmological parameters are then fitted in a second, separate step. In this chapter we will consider the SALT-II method (although our discussion is equally applicable to SALT), and focus on the second step in the procedure, namely the extraction of cosmological parameters from the fitted lightcurves. We briefly summarize below the lightcurve fitting step, on which our method builds.

The rest-frame flux at wavelength λ and time t is fitted with the expression

$$\frac{dF_{\text{rest}}}{d\lambda}(t,\lambda) = x_0[M_0(t,\lambda) + x_1 M_1(t,\lambda)]\exp(c \cdot CL(\lambda)), \quad (10.1)$$

where M_0, M_1, CL are functions determined from a training process, while the fitted parameters are x_0 (which controls the overall flux normalization), x_1 (the stretch parameter) and c (the colour correction parameter). The B-band apparent magnitude m_B^* is related to x_0 by the expression

$$m_B^* = -2.5\log\left[x_0 \int d\lambda M_0(t=0,\lambda) T^B(\lambda)\right], \quad (10.2)$$

where $T^B(\lambda)$ is the transmission curve for the observer's B-band, and $t = 0$ is by convention the time of peak luminosity. After fitting the SNIa lightcurve data with SALT-II algorithm, e.g. [8] report the best-fit values for m_B^*, x_1, c, the best-fit redshift z, and the covariance matrix.

The objective of the parameter inference step of the SALT-II light curve fitting process is to constrain the cosmological parameters Ω_m, Ω_Λ, Ω_κ, w using the fitted parameters $\hat{m}_{Bi}^*, \hat{x}_{1i}, \hat{c}_i$ resulting from the first step of the process. In this chapter, quantities which are measured are denoted by a circumflex, and we label the set of measured values as

$$D_i = \{\hat{z}_i, \hat{m}_{Bi}^*, \hat{x}_{1i}, \hat{c}_i, \hat{C}_i\}. \quad (10.3)$$

where index i labels each of the N SNe Ia in the dataset and $\hat{C}_i$ is the covariance matrix for the measured values

$$\hat{C}_i = \begin{pmatrix} \sigma_{m_B^*i}^2 & \sigma_{m_B^*i, x_1 i} & \sigma_{m_B^*i, ci} \\ \sigma_{m_B^*i, x_1 i} & \sigma_{x_1 i}^2 & \sigma_{x_1 i, ci} \\ \sigma_{m_B^*i, ci} & \sigma_{x_1 i, ci} & \sigma_{ci}^2 \end{pmatrix}. \quad (10.4)$$

The distance modulus μ_i for each SN (i.e. the difference between its apparent B-band magnitude and its absolute magnitude) is modelled by the SALT-II relation as:

$$\mu_i = m_{Bi}^* - M_i + \alpha \cdot x_{1i} - \beta \cdot c_i \quad (10.5)$$

where M_i is the (unknown) B-band absolute magnitude of the SN, while α, β are global nuisance parameters controlling the stretch and colour correction.

Since dark energy can mimic curvature and vice versa, we conduct this work either in a ΛCDM model of the Universe with w fixed $w \equiv -1$ but allowing non zero curvature, or alternatively in a flat wCDM model of the Universe with $\Omega_\kappa \equiv 0$ but allowing $w \neq 1$ but constant with redshift. We denote the complete set of cosmological parameters as

$$\mathcal{C} = \{\Omega_m, \Omega_\Lambda \text{ or } w, h\} \quad (10.6)$$

h is defined as $H_0 = 100h\,\text{km/s/Mpc}$, where H_0 is the value of the Hubble rate today. H_0 is degenerate with the absolute magnitude of the supernovae M_0 and cannot be determined from the SNe Ia data alone, H_0 is effectively a nuisance parameter in this work.

In a Friedman–Robertson–Walker cosmology defined by the parameters $\mathcal{C}$, the theoretical distance modulus to a SN at redshift z_i is given by

$$\mu_i = \mu(z_i, \mathcal{C}) = 5\log\left[\frac{D_L(z_i, \mathcal{C})}{\text{Mpc}}\right] + 25, \qquad (10.7)$$

where D_L denotes the luminosity distance to the SN. We have defined the dimensionless luminosity distance (with $D_L = c/H_0 d_L$, where c is the speed of light)

$$d_L(z, \Omega_m, \Omega_\Lambda, w) = \frac{(1+z)}{\sqrt{|\Omega_\kappa|}} \text{sinn}\left\{\sqrt{|\Omega_\kappa|}\int_0^z dz'\left[(1+z')^3\Omega_m + \Omega_{\text{de}}(z') + (1+z')^2\Omega_\kappa\right]^{-1/2}\right\} \qquad (10.8)$$

with the dark energy density parameter

$$\Omega_{\text{de}}(z) = \Omega_\Lambda \exp\left(3\int_0^z \frac{1+w(x)}{1+x}dx\right). \qquad (10.9)$$

In the above equation, we have been completely general about the functional form of the dark energy equation of state, $w(z)$. In the rest of this work, however, we will make the further assumption that w is constant with redshift, i.e. $w(z) = w$. We have defined the function $\text{sinn}(x) = x, \sin(x), \sinh(x)$ for a flat Universe ($\Omega_\kappa = 0$), a closed Universe ($\Omega_\kappa < 0$) or an open Universe, respectively.

The problem now is how to infer the cosmological parameters $\mathcal{C}$ given the measured values D. We shall describe the standard χ^2 method for solving this problem in section 10.3 before going on to describe our new Bayesian method in 10.4.

10.3 The Standard χ^2 Method

The most common method found in the literature for estimating the cosmological parameters from SNe Ia data fitted using SALT-II involves some variation of the χ^2 method outlined in this section (e.g. [8–10]). The exact χ^2 method varies between consortia, but the essential elements are common to all, and are outlined below.

The χ^2 statistic is defined as

$$\chi^2_\mu = \sum_{i=1}^N \frac{(\mu_i - \mu_i^{\text{obs}})^2}{\sigma^2_{\mu i}}, \qquad (10.10)$$

where μ_i is the theoretical distance modulus and is a function of redshift and the cosmological parameters $\mathcal{C}$. The 'observed' distance modulus μ_i^{obs} is given by the following equation with the estimated values from step 1 of the light curve fitting process for $\hat{m}_{Bi}^*, \hat{x}_{1i}, \hat{c}_i$

$$\mu_i^{obs} = \hat{m}_{Bi}^* - M_0 + \alpha \cdot \hat{x}_{1i} - \beta \cdot \hat{c}_i \qquad (10.11)$$

The total error on the distance modulus $\sigma_{\mu i}^2$ is the sum of several errors added in quadrature

$$\sigma_{\mu i}^2 = (\sigma_{\mu i}^{fit})^2 + (\sigma_{\mu i}^z)^2 + (\sigma_\mu^{int})^2, \qquad (10.12)$$

The three components of the error are:

1. Fitting error $\sigma_{\mu i}^{fit}$ which is given by

$$\left(\sigma_{\mu i}^{fit}\right)^2 = \underline{\Psi}^T \hat{C}_i \underline{\Psi} \qquad (10.13)$$

where $\Psi = (1, \alpha, -\beta)$ and $\hat{C}_i$ is the covariance matrix given in Eq. (10.4). Here we see that the global fit parameters α, β enter into the denominator as well as the numerator of the χ_μ^2 expression.
2. Redshift error $\sigma_{\mu i}^z$, error in the redshift measurement (given by host galaxy redshift) due to uncertainties in the peculiar velocity of the host galaxy and uncertainties in the spectroscopic measurements.
3. The intrinsic dispersion of the SNe Ia absolute magnitudes, σ_μ^{int} an unknown quantity which must be estimated from the data and parameter estimation process. This number describes the variation in the SNe Ia absolute magnitudes which remain after correction for stretch and colour, the variation may be due to physical differences in the SNe Ia population or differences in the survey and data reduction technique.

Additional errors due to lensing, Milky Way dust extinction, etc. can also be added in at this stage, but we do not consider those errors in this work.

The χ_μ^2 statistic of Eq. (10.10) is minimized by sampling over parameter space and simultaneously fitting for both the cosmological parameters Ω_m, Ω_Λ, Ω_κ, w and the SNeIa global fit parameters α, β, M_0. Variations on this include using an iterative process to update the α, β values in the denominator, e.g. Astier and Guy [9] who point out that Tripp [11] realized that minimizing over α, β directly could result in artificially inflated values of α, β so as to reduce the χ_μ^2 value. The problem with the unknown intrinsic dispersion σ_μ^{int} is dealt with by using an iterative process in which χ_μ^2 is minimized, then σ_μ^{int} is adjusted between minimizations in such a way as to give χ_μ^2 per degree of freedom to be unity, i.e. $\chi_\mu^2/d.o.f \sim 1$.

Although the method described above has been fully tested by the consortia that use this method and has been found to give satisfactory results for cosmological parameter inference, several problems remain:

1. The use of the χ_μ^2 expression is not well motivated statistically, but is based on a heuristic derivation.

2. The global fit parameters α, β act as both range and location parameters, appearing in both the numerator and denominator, hence the errors on these parameters are not Gaussian. The informal test which states that $\chi^2/d.o.f \sim 1$ for a good fit model only holds in the Gaussian case, and its use cannot be justified here.
3. The total error on the distance modulus $\sigma^2_{\mu i}$ is adjusted in a way which assumes that the model under consideration (generally either ΛCDM or flat wCDM) is a good fit for the data. This means that this method of parameter inference, or error bars produced using this method cannot be used to investigate problems of model selection, since the model used to derive σ^{int}_μ will by construction be the favoured model.
4. This method obtains a single value for σ^{int}_μ, giving no indication of the error on that value. A better approach would be to obtain a probability distribution function for σ^{int}_μ so that an indication can be given about the degree of belief about the value obtained for σ^{int}_μ.
5. Because the different parameters are not treated in the same way (e.g. some are updated iteratively, sometimes marginalization is used, other times minimization is used), this method cannot be used with standard MCMC or nested sampling techniques.

The new method for Bayesian parameter inference from the SNe Ia data which we will present in this chapter seeks to address some of these problems with the χ^2_μ method and provide a statistically well motivated framework for parameter inference which can also be used in problems of model selection. This method replaces the second step in the SALT-II light curve fitting process. We first describe the Bayesian Hierarchical Model for the system, and then present the details of the calculation.

10.4 Bayesian Hierarchical Model (BHM): Description

The fitted values of stretch and colour resulting from the first stage of the SALT-II light curve fitting procedure suffer from the particular problem that the errors on these values are large compared with the range of these values. If not correctly treated, this particular problem can result in the reconstruction of biased parameters. A simple linear model illustrating this particular problem and its solution is described by [12].

The method which we present here takes the methodology of [12] and applies it to the SNe Ia case. We use a similar initial set up to the χ^2 approach described by [12], with one important difference. Whereas in the χ^2 approach M_0 appears as a global fit parameter explicitly in Eq. (10.11), where it is the mean absolute magnitude of the SNe Ia population, we use M_i to represent the absolute magnitude of each individual SNe Ia. M_i is different for each SNe Ia because even after correction for colour and stretch some variation remains in the SNe Ia population. (M_0 also appears explicitly in our method as the mean absolute magnitude of the SNe Ia population in Eq. (10.15)). Our version of Eq. (10.5) is:

$$\mu_i = m^*_{Bi} - M_i + \alpha \cdot x_{1i} - \beta \cdot c_i. \tag{10.14}$$

The graphical representation of the Bayesian hierarchical network in Fig. 10.1 shows the dependencies of the parameters within the problem. One can see that each SNe Ia has a true (unobserved) redshift z_i, and a true absolute magnitude M_i. The M_i are drawn from a Gaussian distribution with mean M_0 and standard deviation σ_μ^{int}

$$M_i \sim \mathcal{N}(M_0, (\sigma_\mu^{\text{int}})^2). \tag{10.15}$$

This σ_μ^{int} is the intrinsic dispersion of the absolute magnitudes which remains in the SNe Ia absolute magnitudes even after correction for stretch and colour, σ_μ^{int} characterizes the scatter in absolute magnitudes remaining in the stretch corrected light curves.

The cosmological parameters $\mathcal{C} = \{\Omega_m, \Omega_\Lambda \text{ or } w, h\}$ are unknown as are the SNe Ia global fit parameters α, β, these are the parameters we would like to infer. If $\mathcal{C}$ and z_i were known, then we could deterministically specify μ_i using Eq. (10.7). Each SNe Ia also has its own stretch parameter x_{1i} and colour parameter c_i, drawn from their parent distributions. In this work, we model the distributions of the true stretch and colour parameters as Gaussians parametrised each by a mean $(c_\star, x_\star)$ and a variance (R_c^2, R_x^2) as

$$c_i \sim \mathcal{N}(c_\star, R_c^2), \quad x_{1i} \sim \mathcal{N}(x_\star, R_x^2). \tag{10.16}$$

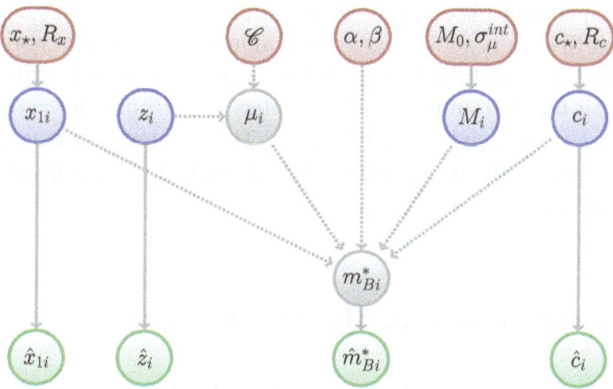

Fig. 10.1 Graphical representation of the Bayesian hierarchical network showing the deterministic (dashed) and probabilistic (solid) connections between variables in our Bayesian hierarchical model (BHM). Variables of interest are in red, latent (unobserved) variables are in blue and observed data (denoted by hats) are in green.

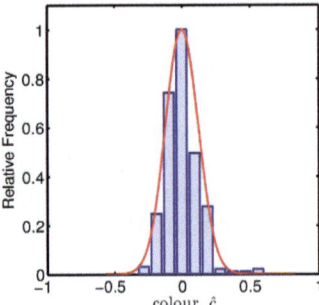 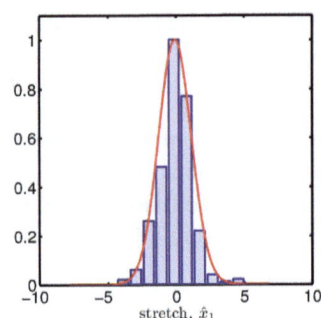

Fig. 10.2 Histogram of observed stretch parameters $\hat{x}_{1i}$ and observed colour parameters $\hat{c}_i$ from the 288 SNIa from [8], compared with a Gaussian fit (red curve).

The choice of a Gaussian distribution for the latent variables $\underline{c}$ and $\underline{x}_1$ is justified by the fact that the observed distribution of $\underline{\hat{c}}$ and $\underline{\hat{x}}_1$, shown in Fig. 10.2 for the actual SNIa sample described in section 10.6 below, is fairly well described by a Gaussian. As shown in Fig. 10.2, there might be a hint for a heavier tail for positive values of $\underline{\hat{c}}$, but this does not fundamentally invalidate our Gaussian approximation. It would be easy to expand our method to consider other distributions, for example mixture models of Gaussians to describe a more complex population or a distribution with heavier tails, if non-Gaussianities in the observed distribution should make such modelling necessary. In our work, we consider the simple uni-modal Gaussians given by Eq. (10.16).

If the true values discussed so far were known, then we could deterministically specify m^*_{Bi}. But we do not have the latent or true values (blue circles), we only have the corresponding observed or fitted values $\hat{z}_i$, $\hat{c}_i$, $\hat{x}_{1i}$, $\hat{m}^*_{Bi}$. The true redshift z_i of each SNe Ia is subject to a small amount of Gaussian noise giving us a slightly different observed redshift $\hat{z}_i$. Likewise the true stretch, colour and maximum magnitudes are also subject to uncertainties in the fitting process, giving us the measured values $\hat{c}_i$, $\hat{x}_{1i}$, $\hat{m}^*_{Bi}$ (green circles). The problem of parameter inference we now have is how to obtain the parameters of interest $\mathcal{C}$, α, β given the observed or measured data $\hat{z}_i$, $\hat{c}_i$, $\hat{x}_{1i}$, $\hat{m}^*_{Bi}$.

10.5 Bayesian Hierarchical Model: Calculation

Having described the Bayesian Hierarchical Model (BHM) up in terms of the graphical representation of the Bayesian hierarchical network, we will now present the details of the calculation. The purpose of this calculation is to determine the joint posterior probability of the parameters of interest $\Theta = \{\mathcal{C}, \alpha, \beta, \sigma^{int}_\mu\}$ given that we have the measured values D, that is, we wish to determine $p(\Theta|D)$.

10.5.1 A Note on Notation

Throughout this work we use the notation:

$$x \sim \mathcal{N}(m, \sigma^2) \tag{10.17}$$

to denote a random variable x being drawn from an underlying Gaussian distribution with mean m and variance σ^2. In vector notation, m is replaced by a vector $\underline{m}$, while σ^2 is replaced by the covariance matrix Σ

$$\underline{x} \sim \mathcal{N}(\underline{m}, \Sigma), \tag{10.18}$$

where $\underline{x}$ has the probability density function

$$p(\underline{x}) \equiv |2\pi\Sigma|^{-\frac{1}{2}} \exp\left[-\frac{1}{2}(\underline{x}-\underline{m})^T \Sigma^{-1}(\underline{x}-\underline{m})\right]. \tag{10.19}$$

We use the compressed notation to write the probability density function as

$$\mathcal{N}_{\underline{x}}(\underline{m}, \Sigma) \equiv |2\pi\Sigma|^{-\frac{1}{2}} \exp\left[-\frac{1}{2}(\underline{x}-\underline{m})^T \Sigma^{-1}(\underline{x}-\underline{m})\right]. \tag{10.20}$$

10.5.2 Calculation Expressed in Matrix Notation

We re-write Eq. (10.15) and (10.16) in matrix notation as:

$$\underline{M} \sim \mathcal{N}(\underline{M}_0, \Sigma_\Delta), \tag{10.21}$$
$$\underline{c} \sim \mathcal{N}(c_\star \cdot \mathbf{1}_n, \text{diag}(R_c^2 \cdot \mathbf{1}_n)) \tag{10.22}$$
$$\underline{x}_1 \sim \mathcal{N}(x_\star \cdot \mathbf{1}_n, \text{diag}(R_x^2 \cdot \mathbf{1}_n)) \tag{10.23}$$

where

$$\underline{M} = (M_1, \ldots, M_n) \in \mathbb{R}^n, \tag{10.24}$$
$$\underline{M}_0 = M_0 \cdot \mathbf{1}_n \in \mathbb{R}^n, \tag{10.25}$$
$$\Sigma_\Delta = \text{diag}((\sigma_\mu^{\text{int}})^2 \cdot \mathbf{1}_n) \in \mathbb{R}^{n \times n}. \tag{10.26}$$

Having introduced $3n$ latent (unobserved) variables $(\underline{c}, \underline{x}_1, \underline{M})$, where n is the number of SNe in the sample, the fundamental strategy of our method is to link them to underlying population parameters via Eqs. (10.15) and (10.16), then to use the observed noisy estimates to infer constraints on the population parameters of interest (alongside the cosmological parameters), while marginalizing out the unobserved latent variables.

In doing so we are following the methodology of [12] and essentially placing a prior on the range of the latent (true) quantities $\underline{c}, \underline{x}_1, \underline{M}$.

$$p(\underline{c}|c_\star, R_c) = \mathcal{N}_{\underline{c}}(c_\star \cdot \mathbf{1}_n, \text{diag}(R_c^2 \cdot \mathbf{1}_n)) \tag{10.27}$$

$$p(\underline{x}_1|x_\star, R_x) = \mathcal{N}_{\underline{x}_1}(x_\star \cdot \mathbf{1}_n, \text{diag}(R_x^2 \cdot \mathbf{1}_n)). \tag{10.28}$$

It is necessary to apply this prior because two conditions are fulfilled (1) there are errors on both the measured values $\hat{\underline{c}}, \hat{\underline{x}}_1$ and (2) the errors on $\hat{\underline{c}}, \hat{\underline{x}}_1$ are large compared with the range of $\hat{\underline{c}}, \hat{\underline{x}}_1$. Failure to apply this prior on the range of the latent quantities when these two conditions are met results in a biassed recovery of the SNe Ia population parameters α, β. For further discussion of this crucial step, and an example with a linear toy model, see [3].

We chose to model the probability of the true absolute magnitudes $\underline{M}$ also as a Gaussian

$$p(\underline{M}|M_0, \sigma_\mu^{\text{int}}) = \mathcal{N}_{\underline{M}}\left(M_0 \cdot \mathbf{1}_n, \text{diag}((\sigma_\mu^{\text{int}})^2 \cdot \mathbf{1}_n)\right). \tag{10.29}$$

Notice that there are two levels of specification or choice here: (1) The choice of the underlying distributions, described by Eqs. (10.21), (10.23), (10.22) and (2) The choice of the priors on those latent parameters. Throughout this work we assume that (1) are Gaussian, and when trialling the method with simulated data, we choose Gaussians for (1). We also chose to use Gaussian priors for (2). Of course the natural choice is to match the shape of prior with the shape of the underlying distribution, this is possible when using simulated data, but with the real data the precise shape of the unknown distribution is unknown but assumed Gaussian. An interesting question is what happens when a different shaped prior is chosen from the underlying distribution, e.g. what happens if a uniformly distributed population of $\underline{c}$ are used with a Gaussian prior on $\underline{c}$ – a question which we will investigate in future work.

The absolute magnitude M_i is related to the observed B-band magnitude $\hat{m}_B^*$ and the distance modulus μ by Eq. (10.14), which can be rewritten in vector notation as

$$\underline{m}_B^* = \underline{\mu} + \underline{M} - \alpha \underline{x}_1 + \beta \underline{c}. \tag{10.30}$$

The above relation is *exact*, i.e. $\underline{M}, \underline{x}_1, \underline{c}$ are here the latent variables (not the observed quantities), while $\underline{m}_B^*$ is the true value of the B-band magnitude (also unobserved). This is represented by the dotted (deterministic) arrows connecting the variables in Fig. 10.1.

We seek to determine the posterior pdf for the parameters of interest $\Theta = \{\mathcal{C}, \alpha, \beta, \sigma_\mu^{\text{int}}\}$, while marginalizing over the unknown population mean absolute magnitude, M_0. From Bayes theorem, the marginal posterior for Θ is given by

$$p(\Theta|D) = \int dM_0\, p(\Theta, M_0|D) = \int dM_0 \frac{p(D|\Theta, M_0) p(\Theta, M_0)}{p(D)}, \tag{10.31}$$

where $p(D)$ is the Bayesian evidence (a normalizing constant) and the prior $p(\Theta, M_0)$ can be written as

$$p(\Theta, M_0) = p(\mathcal{C}, \alpha, \beta) p(M_0, \sigma_\mu^{\text{int}}) = p(\mathcal{C}, \alpha, \beta) p(M_0|\sigma_\mu^{\text{int}}) p(\sigma_\mu^{\text{int}}). \quad (10.32)$$

We take a uniform prior on the variables $\mathcal{C}, \alpha, \beta$ (on a sufficiently large range so as not to truncate likelihood, except in the case of $\Omega_m > 0$ which rules out an unphysical choice of parameter), as well as a Gaussian prior for $p(M_0|\sigma_\mu^{\text{int}})$, since M_0 is a location parameter of a Gaussian (conditional on σ_μ^{int}). Additionally, we apply a prior which excludes that part of parameter space which is the 'no Big Bang' region in the Ω_m, Ω_Λ plane.

The prior on M_0 is

$$p(M_0|\sigma_\mu^{\text{int}}) = \mathcal{N}_{M_0}(M_m, \sigma_{M_0}^2), \quad (10.33)$$

where the mean of the prior ($M_m = -19.3$ mag) is taken to be a reasonable value based on observations of nearby SNe Ia, and the variance ($\sigma_{M_0} = 2.0$ mag) is sufficiently large so that the prior is very diffuse and non-informative (the precise choice of mean and variance for this prior does not impact on our numerical results). Finally, the appropriate prior for σ_μ^{int} is a Jeffreys' prior, i.e. uniform in $\log \sigma_\mu^{\text{int}}$, as σ_μ^{int} is a scale parameter.

The likelihood $p(D|\Theta, M_0) = p(\hat{\underline{c}}, \hat{\underline{x}}_1, \hat{\underline{m}}_B^*|\Theta, M_0)$ can be expanded out an rewritten as:

$$p(\hat{\underline{c}}, \hat{\underline{x}}_1, \hat{\underline{m}}_B^*|\Theta, M_0)$$
$$= \int d\underline{c} \, d\underline{x}_1 \, d\underline{M} \; p(\hat{\underline{c}}, \hat{\underline{x}}_1, \hat{\underline{m}}_B^*|\underline{c}, \underline{x}_1, \underline{M}, \Theta, M_0) p(\underline{c}, \underline{x}_1, \underline{M}|\Theta, M_0) \quad (10.34)$$
$$= \int d\underline{c} \, d\underline{x}_1 \, d\underline{M} \; p(\hat{\underline{c}}, \hat{\underline{x}}_1, \hat{\underline{m}}_B^*|\underline{c}, \underline{x}_1, \underline{M}, \Theta)$$
$$\times \int dR_c \, dR_x \, dc_\star \, dx_\star \;\; p(\underline{c}|c_\star, R_c) p(\underline{x}_1|x_\star, R_x) p(\underline{M}|M_0, \sigma_\mu^{\text{int}})$$
$$\times p(R_c) p(R_x) p(c_\star) p(x_\star) \quad (10.35)$$

In the first line, we have introduced a set of $3n$ latent variables, $\{\underline{c}, \underline{x}_1, \underline{M}\}$, which describe the *true* value of the colour, stretch and absolute magnitude for each SNIa. Since these variables are unobserved, we need to marginalize over them. In the second line, we have replaced $p(\underline{c}, \underline{x}_1, \underline{M}|\Theta, M_0)$ by the priors on the latent $\{\underline{c}, \underline{x}_1, \underline{M}\}$ given by Eq. (10.29) and Eqs. (10.27–10.28), (assumed separable) and marginalized out the population parameters $\{R_c, R_x, c_\star, x_\star\}$

$$p(\underline{c}, \underline{x}_1, \underline{M}|\Theta, M_0) = \int dR_c \, dR_x \, dc_\star \, dx_\star \; p(\underline{c}|c_\star, R_c) p(\underline{x}_1|x_\star, R_x)$$
$$\times p(\underline{M}|M_0, \sigma_\mu^{\text{int}}) p(R_c) p(R_x) p(c_\star) p(x_\star) \quad (10.36)$$

(we have also dropped M_0 from the likelihood, as conditioning on M_0 is irrelevant if the latent $\underline{M}$ are given). If we further marginalize over M_0 (as in Eq. (10.31)), including the prior on M_0), the expression for the effective likelihood, Eq. (10.35), then becomes

$$p(\hat{\underline{c}}, \hat{\underline{x}}_1, \hat{\underline{m}}_B^* | \Theta) = \int d\underline{c} \, d\underline{x}_1 \, d\underline{M} \, p(\hat{\underline{c}}, \hat{\underline{x}}_1, \hat{\underline{m}}_B^* | \underline{c}, \underline{x}_1, \underline{M}, \Theta)$$
$$\times \int dR_c \, dR_x \, dc_\star \, dx_\star \, dM_0 \, p(\underline{c}|c_\star, R_c) p(\underline{x}_1|x_\star, R_x)$$
$$\times p(\underline{M}|M_0, \sigma_\mu^{\text{int}}) p(R_c) p(R_x) p(c_\star) p(x_\star) p(M_0|\sigma_\mu^{\text{int}}). \quad (10.37)$$

The term $p(\hat{\underline{c}}, \hat{\underline{x}}_1, \hat{\underline{m}}_B^* | \underline{c}, \underline{x}_1, \underline{M}, \Theta)$ is the conditional probability of observing values $\{\hat{\underline{c}}, \hat{\underline{x}}_1, \hat{\underline{m}}_B^*\}$ if the latent (true) value of $\underline{c}, \underline{x}_1, \underline{M}$ and of the other cosmological parameters were known. From Fig. 10.1, $\underline{m}_B^*$ is connected only deterministically to all other variables and parameters, via Eq. (10.30). Thus we can replace $\underline{m}_B^* = \mu + \underline{M} - \alpha \cdot \underline{x}_1 + \beta \cdot \underline{c}$ and write

$$p(|\underline{c}, \underline{x}_1, \underline{M}, \Theta) = \prod_{i=1}^n \mathcal{N}_{[\hat{c}, \hat{x}_1, \hat{m}_B^*]}(\mu_i + M_i - \alpha \cdot x_{1i} + \beta \cdot c_i, \hat{C}_i) \quad (10.38)$$

$$= |2\pi \Sigma_C|^{-\frac{1}{2}} \exp\left(-\frac{1}{2}[(X - X_0)^T \Sigma_C^{-1} (X - X_0)]\right) \quad (10.39)$$

where $\mu_i \equiv \mu_i(z_i, \Theta)$ and we have defined

$$X = \{X_1, \ldots, X_n\} \in \mathbb{R}^{3n}, \quad X_0 = \{X_{0,1}, \ldots, X_{0,n}\} \in \mathbb{R}^{3n}, \quad (10.40)$$
$$X_i = \{c_i, x_{1,i}, (M_i - \alpha x_{1,i} + \beta c_i)\} \in \mathbb{R}^3, \quad X_{0,i} = \{c_i, x_{1,i}, \hat{m}_{Bi}^* - \mu_i\} \in \mathbb{R}^3, \quad (10.41)$$

as well as the $3n \times 3n$ block covariance matrix[1]

$$\Sigma_C = \begin{pmatrix} \hat{C}_1 & 0 & 0 & 0 \\ 0 & \hat{C}_2 & 0 & 0 \\ 0 & 0 & \ddots & 0 \\ 0 & 0 & 0 & \hat{C}_n \end{pmatrix}. \quad (10.42)$$

Finally we explicitly include redshift uncertainties in our formalism. The observed apparent magnitude, $\hat{m}_B^*$, on the left-hand-side of Eq. (10.38), is the value at the observed redshift, $\hat{z}$. However, μ in Eq. (10.38) should be evaluated at the true (unknown) redshift, $\underline{z}$. As above, the redshift uncertainty is included by introducing

---

[1] Notice that we neglect correlations between different SNIa, which is reflected in the fact that Σ_C takes a block-diagonal form. It would be however very easy to add arbitrary cross-correlations to our formalism (e.g. coming from correlated systematic within survey, for example zero point calibration) by adding such non-block diagonal correlations to Eq. (10.42).

the latent variables $\underline{z}$ and integrating over them:

$$p(\underline{\hat{c}}, \underline{\hat{x}}_1, \underline{\hat{M}} | \underline{c}, \underline{x}_1, \underline{M}, \Theta) = \int d\underline{z}\ p(\underline{\hat{c}}, \underline{\hat{x}}_1, \underline{\hat{M}} | \underline{c}, \underline{x}_1, \underline{M}, \underline{z}, \Theta) p(\underline{z}|\underline{\hat{z}}) \quad (10.43)$$

where we model the redshift errors $p(\underline{z}|\underline{\hat{z}})$ as Gaussians:

$$\underline{\hat{z}} \sim \mathcal{N}(\underline{z}, \Sigma_z) \quad (10.44)$$
$$p(\underline{z}|\underline{\hat{z}}) = \mathcal{N}_{\underline{z}}(\underline{\hat{z}}, \Sigma_z) \quad (10.45)$$

with a $n \times n$ covariance matrix:

$$\Sigma_z = \text{diag}(\sigma_{z_1}^2, \ldots, \sigma_{z_n}^2). \quad (10.46)$$

It is now necessary to integrate out all latent variables and nuisance parameters from the expression for the likelihood, Eq. (10.37). This can be done analytically, as all necessary integral are Gaussian.

10.5.3 Integration over Intrinsic Redshifts

In order to perform the multi-dimensional integral over $\underline{z}$, we Taylor expand $\underline{\mu}$ around $\underline{\hat{z}}$ (as justified by the fact that redshift errors are typically small: the error from 300 km/s peculiar velocity is $\sigma_{z_i} = 0.0012$, while the error from spectroscopic redshifts from SNe themselves is $\sigma_{z_i} = 0.005$, see [8]):

$$\mu_j = \mu(z_j) \quad (10.47)$$
$$= 5\log_{10}\left(\frac{D_L(z_j)}{\text{Mpc}}\right) + 25 \quad (10.48)$$
$$\approx \mu(\hat{z}_j) + 5(\log_{10} e) \frac{\partial_{z_j} D_L(z_j)}{D_L(z_j)}\bigg|_{\hat{z}_j} (z_j - \hat{z}_j). \quad (10.49)$$

With this approximation we can now carry out the multi-dimensional integral of Eq. (10.43), obtaining

$$p(\underline{\hat{m}}_B^* | \underline{c}, \underline{x}_1, \underline{M}, \Theta)$$
$$= |2\pi \Sigma_m|^{-\frac{1}{2}} \exp\left[-\frac{1}{2}(\underline{\hat{m}}_B^* - (\underline{\mu} + \underline{M} - \alpha \cdot \underline{x}_1 + \beta \cdot \underline{c}))^T \Sigma_m^{-1}\right.$$
$$\left. \times (\underline{\hat{m}}_B^* - (\underline{\mu} + \underline{M} - \alpha \cdot \underline{x}_1 + \beta \cdot \underline{c}))\right] \quad (10.50)$$

where from now on, $\underline{\mu} = \underline{\mu}(\hat{\underline{z}})$ and

$$\sigma_{m_B^* i} \to \sigma_{m_B^* i}^{\text{raw data}} + f_i \sigma_{zi} f_i \tag{10.51}$$

$$f = \text{diag}(f_1, \ldots, f_n) \tag{10.52}$$

$$f_i = 5 \log_{10}(e) \left. \frac{D_L'(z_i)}{D_L(z_i)} \right|_{\hat{z}_i} \tag{10.53}$$

$$= \frac{5 \log_{10}(e)}{D_L(\hat{z}_i)} \left[\frac{D_L(\hat{z}_i)}{1+z_i} + \frac{c}{H_0}(1+\hat{z}_i) \right.$$

$$\times \cos n \left\{ \sqrt{|\Omega_\kappa|} \int_0^{\hat{z}} dz' \left[(1+z')^3 \Omega_m + \Omega_{\text{de}}(z) + (1+z)^2 \Omega_\kappa \right]^{-1/2} \right\}$$

$$\left. \times ((1+z')^3 \Omega_m + \Omega_{\text{de}}(z) + (1+z)^2 \Omega_\kappa)^{-1/2} \right]. \tag{10.54}$$

Strictly speaking, one should integrate over redshift in the range $0 \leq z_i < \infty$, not $-\infty < z_i < \infty$, which would result in the appearance of Gamma functions in the final result. However, as long as $\frac{\sigma_{z_i}}{z_i} \ll 1$ (as is the case here), this approximation is expected to be excellent.

10.5.4 Integration over Latent $\{\underline{c}, \underline{x}, \underline{M}\}$

From Eq. (10.35) and using the expression in Eq. (10.38), we wish to integrate out the latent variables

$$Y = \{Y_1, \ldots, Y_n\} \in \mathbb{R}^{3n}, \tag{10.55}$$

$$Y_i = \{c_i, x_{1,i}, M_i\} \in \mathbb{R}^3, \tag{10.56}$$

We therefore recast expression (10.38) as

$$p(\hat{\underline{c}}, \hat{\underline{x}}_1, \hat{\underline{m}}_B^* | \underline{c}, \underline{x}_1, \underline{M}, \Theta) = |\Sigma_C|^{-\frac{1}{2}} \exp\left(-\frac{1}{2}[(AY - X_0)^T \Sigma_C^{-1}(AY - X_0)]\right) \tag{10.57}$$

where we have defined the block-diagonal matrix

$$A = \text{diag}(T, T, \ldots, T) \in \mathbb{R}^{3n \times 3n} \tag{10.58}$$

with

$$T = \begin{bmatrix} 1 & 0 & 0 \\ 0 & 1 & 0 \\ \beta & -\alpha & 1 \end{bmatrix} \begin{bmatrix} c_i \\ x_i \\ M_i \end{bmatrix} \tag{10.59}$$

10 Bayesian Hierarchical Model of SNe Ia Data

The prior terms appearing in Eq. (10.37), namely $p(\underline{c}|c_\star, R_c)p(\underline{x}_1|x_\star, R_x)$ $p(\underline{M}|M_0, \sigma_\mu^{int})$, may be written as

$$p(\underline{c}|c_\star, R_c)p(\underline{x}_1|x_\star, R_x)p(\underline{M}|M_0, \sigma_\mu^{int})$$
$$= |2\pi\Sigma_P|^{-\frac{1}{2}} \exp\left(-\frac{1}{2}[(Y - Y_*)^T \Sigma_P^{-1}(Y - Y_*)]\right) \quad (10.60)$$

where

$$S^{-1} = \text{diag}(R_c^{-2}, R_x^{-2}, (\sigma_\mu^{int})^{-2}) \in \mathbb{R}^{3\times 3} \quad (10.61)$$
$$\Sigma_P^{-1} = \text{diag}(S^{-1}, S^{-1}, \ldots, S^{-1}) \in \mathbb{R}^{3n\times 3n} \quad (10.62)$$

$$Y_* = \underline{J} \cdot \underline{b} \in \mathbb{R}^{3n\times 1} \quad (10.63)$$

$$\underline{J} = \begin{bmatrix} 1 & 0 & 0 \\ 0 & 1 & 0 \\ 0 & 0 & 1 \\ \vdots & \vdots & \vdots \\ 1 & 0 & 0 \\ 0 & 1 & 0 \\ 0 & 0 & 1 \end{bmatrix} \in \mathbb{R}^{3n\times 3}, \quad (10.64)$$

$$\underline{b} = \begin{bmatrix} c_* \\ x_* \\ M_0 \end{bmatrix} \in \mathbb{R}^{3\times 1}. \quad (10.65)$$

Now the integral over $dY = d\underline{c}\, d\underline{x}_1\, d\underline{M}$ in Eq. (10.37) can be performed, giving:

$$\int dY\, p(\hat{\underline{c}}, \hat{\underline{x}}_1, \hat{\underline{m}}_B^*|\underline{c}, \underline{x}_1, \underline{M}, \Theta)p(\underline{c}|c_\star, R_c)p(\underline{x}_1|x_\star, R_x)p(\underline{M}|M_0, \sigma_\mu^{int})$$
$$= |2\pi\Sigma_C|^{-\frac{1}{2}}|2\pi\Sigma_P|^{-\frac{1}{2}}|2\pi\Sigma_A|^{\frac{1}{2}}$$
$$\times \exp\left(-\frac{1}{2}[X_0^T \Sigma_C^{-1} X_0 - Y_0^T \Sigma_A^{-1} Y_0 + Y_*^T \Sigma_P^{-1} Y_*]\right) \quad (10.66)$$

where

$$\Sigma_A^{-1} = A^T \Sigma_C^{-1} A + \Sigma_P^{-1} \in \mathbb{R}^{3n\times 3n}, \quad (10.67)$$
$$\Sigma_A^{-1} Y_0 = A^T \Sigma_C^{-1} X_0 + \Sigma_P^{-1} Y_*, \quad (10.68)$$
$$Y_0 = \Sigma_A(A^T \Sigma_C^{-1} X_0 + \Sigma_P^{-1} Y_*)\Sigma_A(\Delta + \Sigma_P^{-1} Y_*), \quad (10.69)$$
$$\Delta = A^T \Sigma_C^{-1} X_0 \in \mathbb{R}^{3n\times 1}. \quad (10.70)$$

Substituting Eq. (10.66) back into Eq. (10.37) gives:

$$p(\hat{\underline{c}}, \hat{\underline{x}}_1, \hat{\underline{m}}_B^*|\Theta) = \int dR_c \, dR_x \, dc_\star \, dx_\star \, |2\pi \Sigma_C|^{-\frac{1}{2}} |2\pi \Sigma_P|^{-\frac{1}{2}} |2\pi \Sigma_A|^{\frac{1}{2}}$$
$$\times \exp\left(-\frac{1}{2}[X_0^T \Sigma_C^{-1} X_0 - Y_0^T \Sigma_A^{-1} Y_0 + Y_*^T \Sigma_P^{-1} Y_*]\right)$$
$$\times p(R_c) p(R_x) p(c_\star) p(x_\star) p(M_0|\sigma_\mu^{\text{int}}). \tag{10.71}$$

10.5.5 Integration over Population Variables $\{c_\star, x_\star, M_0\}$

The priors on the population variables $\underline{b} = \{c_\star, x_\star, M_0\}$ in Eq. (10.71) can be written as:

$$p(\underline{b}) = p(c_\star) p(x_\star) p(M_0|\sigma_\mu^{\text{int}})$$
$$= |2\pi \Sigma_0|^{-\frac{1}{2}} \exp\left(-\frac{1}{2}(\underline{b} - \underline{b}_m)^T \Sigma_0^{-1} (\underline{b} - \underline{b}_m)\right) \tag{10.72}$$

where

$$\Sigma_0^{-1} = \begin{bmatrix} 1/\sigma_{c_*}^2 & 0 & 0 \\ 0 & 1/\sigma_{x_*}^2 & 0 \\ 0 & 0 & 1/\sigma_{M_0}^2 \end{bmatrix} \tag{10.73}$$

and

$$\underline{b}_m = \begin{bmatrix} 0 \\ 0 \\ M_m \end{bmatrix} \in \mathbb{R}^{3 \times 1} \tag{10.74}$$

Thus Eq. (10.71) can be written as:

$$p(\hat{\underline{c}}, \hat{\underline{x}}_1, \hat{\underline{m}}_B^*|\Theta)$$
$$= \int dR_c \, dR_x \, d\underline{b} \, |2\pi \Sigma_C|^{-\frac{1}{2}} |2\pi \Sigma_P|^{-\frac{1}{2}} |2\pi \Sigma_A|^{\frac{1}{2}} |2\pi \Sigma_0|^{-\frac{1}{2}} p(R_c) p(R_x)$$
$$\times \exp\left(-\frac{1}{2}[X_0^T \Sigma_C^{-1} X_0 - (\Sigma_A(\Delta + \Sigma_P^{-1} \underline{J} \cdot \underline{b}))^T \Sigma_A^{-1} (\Sigma_A(\Delta + \Sigma_P^{-1} \underline{J} \cdot \underline{b}))\right.$$
$$\left. + \underline{b}^T \underline{J}^T \Sigma_P^{-1} \underline{J} \underline{b} + (\underline{b} - \underline{b}_m)^T \Sigma_0^{-1} (\underline{b} - \underline{b}_m)]\right)$$
$$= \int dR_c \, dR_x \, |2\pi \Sigma_C|^{-\frac{1}{2}} |2\pi \Sigma_P|^{-\frac{1}{2}} |2\pi \Sigma_A|^{-\frac{1}{2}} |2\pi \Sigma_0|^{-\frac{1}{2}} p(R_c) p(R_x)$$
$$\times \exp\left(-\frac{1}{2}[X_0^T \Sigma_C^{-1} X_0 - \Delta^T \Sigma_A \Delta - k_0^T K^{-1} k_0 + \underline{b}_m^T \Sigma_0^{-1} \underline{b}_m]\right)$$
$$\times \int d\underline{b} \, \exp\left(-\frac{1}{2}[(\underline{b} - k_0)^T K^{-1} (\underline{b} - k_0)]\right)$$
$$\tag{10.75}$$

where

$$K^{-1} = -\underline{J}^T \Sigma_P^{-1} \Sigma_A \Sigma_P^{-1} \underline{J} + \underline{J}^T \Sigma_P^{-1} \underline{J} + \Sigma_0^{-1} \in \mathbb{R}^{3 \times 3}, \quad (10.76)$$

$$K^{-1} k_0 = \underline{J}^T \Sigma_P^{-1} \Sigma_A \Delta + \Sigma_0^{-1} \underline{b}_m \in \mathbb{R}^{3 \times 1}, \quad (10.77)$$

$$k_0 = K(\underline{J}^T \Sigma_P^{-1} \Sigma_A \Delta + \Sigma_0^{-1} \underline{b}_m). \quad (10.78)$$

We can now carry out the Gaussian integral over $\underline{b}$ in Eq. (10.71), obtaining our final expression for the effective likelihood,

$$p(\hat{\underline{c}}, \hat{\underline{x}}_1, \hat{\underline{m}}_B^* | \Theta)$$
$$= \int d\log R_c \, d\log R_x \, |2\pi \Sigma_C|^{-\frac{1}{2}} |2\pi \Sigma_P|^{-\frac{1}{2}} |2\pi \Sigma_A|^{\frac{1}{2}} |2\pi \Sigma_0|^{-\frac{1}{2}} |2\pi K|^{\frac{1}{2}}$$
$$\times \exp\left(-\frac{1}{2}[X_0^T \Sigma_C^{-1} X_0 - \Delta^T \Sigma_A \Delta - k_0^T K^{-1} k_0 + \underline{b}_m^T \Sigma_0^{-1} \underline{b}_m]\right), \quad (10.79)$$

where we have chosen an improper Jeffreys' prior for the scale variables R_c, R_x:

$$p(R_c) \propto R_c^{-1} \Rightarrow p(R_c) dR_c \propto d\log R_c, \quad (10.80)$$

and analogously for R_x. These two remaining nuisance parameters cannot be integrated out analytically, so they need to be marginalized numerically. Hence, R_c, R_x are added to our parameters of interest and are sampled over numerically, and then marginalized out from the joint posterior.

The expression for the effective likelihood given by Eq. (10.79) is the major result presented in this chapter. Having shown how this effective likelihood is motivated and arrived at, we will now present some numerical trials in which the effective likelihood is tested using simulated data.

10.6 Numerical Trials with Simulated Data

10.6.1 Description of the Real SNe Ia Data Sets

The simulated data sets used in the numerical trials are modeled on the (then) recent compilation of 288 SNIa from [8], which presents analysis of new data from SDSS-II along with publicly available data from four existing surveys. The Kessler et al. [8] compilation comprises of:

- SDSS: 103 SNe [8]
- ESSENCE: 56 SNe [13, 14]
- SNLS: 62 SNe [9]
- Nearby Sample: 33 SNe [15]
- HST: 34 SNe [16–19]

The compiled set of 288 SNe Ia were analysed by [8] using both the SALT-II method and the MLCS method. In the following, we are exclusively employing the results of their SALT-II fits and use those as the observed data set for the purposes of our current work, as described in the previous section. More refined procedures could be adopted, for example by simulating lightcurves from scratch, using e.g. the publicly available package SNANA [20]. In this work we chose a simpler approach, which is to simulate the SALT-II fit results in such a way to broadly match the distributions and characteristics of the real data set used in [8].

10.6.2 Description of the Simulated SNe Ia Data Sets

The numerical values of the parameters used for he simulated data sets are shown in Table 10.1. We adopt a flat ΛCDM cosmological model as fiducial cosmology. The α and β global fit parameters are chosen to match the best-fit values reported in [8], while the distributional properties of the colour and stretch correction match the observed distribution of their total SN sample. For each survey, we generate a number of SNe matching the observed sample, and we model their redshift distribution as a Gaussian, with mean and variance estimated from the observed distribution within each survey. The observational errors of $m_B^*, \underline{c}, \underline{x}_1$ are again drawn from Gaussian distributions whose means and variances have been matched to the observed ones for each survey. Finally, the simulated data (i.e. the simulated SALT-II fits results $\hat{\underline{m}}_B^*, \hat{\underline{c}}, \hat{\underline{x}}_1$) are generated by drawing from the appropriate distributions

Table 10.1 Input parameter values used for the fiducial model in the generation of the simulated SNe SALT-II data sets.

Parameter	Symbol	True Value
Matter energy density parameter	Ω_m	0.3
Dark energy density parameter	Ω_Λ	0.7
Dark energy equation of state	w	-1
Spatial curvature	Ω_κ	0.0
Hubble expansion rate	H_0 [km/s/Mpc]	72.0
Mean absolute magnitude of SNe	M_0 [mag]	-19.3
Intrinsic dispersion of SNe magnitude	σ_μ^{int} [mag]	0.1
Stretch correction	α	0.13
Colour correction	β	2.56
Mean of distribution of $\underline{x}_1$	$x_\star$	0.0
Mean of distribution of $\underline{c}$	$c_\star$	0.0
s.d. of distribution of $\underline{x}_1$	R_x	1.0
s.d. of distribution of $\underline{c}$	R_c	0.1
Observational noise on m_B^*	$\sigma_{m_{Bi}^*}$	Depending on survey
Observational noise on $\underline{x}_1$	$\sigma_{x_1 i}$	Depending on survey
Observational noise on $\underline{c}$	σ_{ci}	Depending on survey
Correlation between $\underline{x}_1$ and $\underline{c}$	$\sigma_{x_1 i, ci}$	0.0

centered around the latent variables. For simplicity, we have set to 0 the off-diagonal elements in the correlation matrix (10.42) in our simulated data, and neglected redshift errors. None of these assumptions have a significant impact on our results. In summary, our procedure for simulating data for each survey is as follows:

1. Draw a value for the latent redshift z_i from a normal distribution with mean and variance matching the observed ones. As we neglect redshift errors in the simulated data for simplicity (since the uncertainty in z is subdominant in the overall error budget), we set $\hat{z}_i = z_i$.
2. Compute μ_i using the fiducial values for the cosmological parameters $\mathcal{C}$ and the above z_i from Eq. (10.7).
3. Draw the latent parameters x_{1i}, c_i, M_i from their respective distributions (in particular, including an intrinsic scatter $\sigma_\mu^{\text{int}} = 0.1$ mag in the generation of M_i).
4. Compute m_{Bi}^* using x_{1i}, c_i, M_i and the SALT-II relation Eq. (10.14).
5. Draw the value of the standard deviations $\sigma_{x_1 i}, \sigma_{c_i}, \sigma_{m_i}$, from the appropriate normal distributions for each survey type. A small, z_i-dependent stochastic linear addition is also made to $\sigma_{x_1 i}, \sigma_{c_i}, \sigma_{m_i}$, to mimic the observed correlation between redshift and error.
6. Draw the SALT-II fit results from $\hat{x}_{1i} \sim \mathcal{N}(x_{1i}, \sigma_{x_1 i})$, $\hat{c}_i \sim \mathcal{N}(c_i, \sigma_{c_i})$ and $\hat{m}_{Bi}^* \sim \mathcal{N}(m_{Bi}^*, \sigma_{m_i})$.

As shown in Fig. 10.3, the simulated data from our procedure have broadly similar distributions to the to real ones. The two notable exceptions are the overall vertical shift observed in the distance modulus plot, and the fact that our simulated data cannot reproduce the few outliers with large values of the variances (bottom panels). The former is a consequence of the different absolute magnitude used in

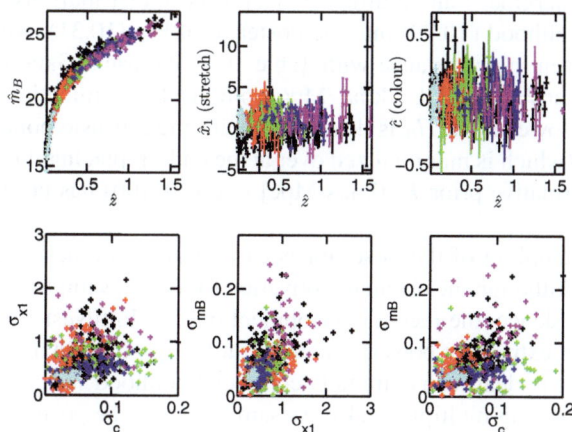

Fig. 10.3 An example realization of our simulated data sets (coloured according to survey), superimposed on real data (black). Colour code for simulated data survey: nearby sample (cyan), ESSENCE (green), SDSS (red), SNLS (blue) and HST (magenta).

our simulated data (as the true one is unknown). However, the absolute magnitude is marginalized over at the end, so this difference has no impact on our inferences. The absence of outliers is a consequence of the fact that our simulation is a pure phenomenological description of the data, hence it cannot encapsulate such fine details. While in principle we could perform outlier detection with dedicated Bayesian procedures, we do not pursue this issue further in this paper. We stress once more that the purpose of our simulations is not to obtain realistic SNIa data. Instead, they should only provide us with useful mock data sets coming from a known model so that we can test our procedure. More sophisticated tests based on more realistically generated data (e.g. from SNANA) are left for future work.

10.6.3 Numerical Sampling

After analytical marginalization of the latent variables, we are left with the following eight parameters entering the effective likelihood of Eq. (10.79):

$$\{\Omega_m, \Omega_\kappa \text{ or } w, H_0, \sigma_\mu^{\text{int}}, \alpha, \beta, R_c, R_x\}. \tag{10.81}$$

As mentioned above, in keeping with the literature we only consider either flat Universes with a possible $w \neq -1$ (the ΛCDM model), or curved Universes with a cosmological constant ($w = -1$, the wCDM model). Of course it is possible to relax those assumptions and consider more complicated cosmologies with a larger number of free parameters if one so wishes (notably including evolution in the dark energy equation of state).

Of the parameters listed in Eq. (10.81), the quantities R_c, R_x are of no interest and will be marginalized over. As for the remaining parameters, we are interested in obtaining their marginal 1 and 2-dimensional posterior distributions. This is done by inserting the likelihood (10.79) into the posterior of Eq. (10.31), with priors on the parameters chosen in accordance with Table 10.2. We use a Gaussian prior on the Hubble parameter $H_0 = 72 \pm 8$ km/s/Mpc from local determinations of the Hubble constant [21]. However, as H_0 is degenerate with the intrinsic population absolute magnitude M_0 (which is marginalized over at the end), replacing this Gaussian prior with a less informative prior $H_0[\text{km/s/Mpc}] \sim \mathcal{U}(20, 100)$ has no influence on our results.

Numerical sampling of the posterior is carried out via a nested sampling algorithm [22–25]. Although the original motivation for nested sampling was to compute the Bayesian evidence, the recent development of the MultiNest algorithm [24, 25] has delivered an extremely powerful and versatile algorithm that has been demonstrated to be able to deal with extremely complex likelihood surfaces in hundreds of dimensions exhibiting multiple peaks. As samples from the posterior are generated as a by-product of the evidence computation, nested sampling can also be used to obtain parameter constraints in the same run as computing the Bayesian evidence. In this paper we adopt the publicly available MultiNest algorithm [24] to obtain

Table 10.2 Priors on our model's parameters used when evaluating the posterior distribution. Ranges for the uniform priors have been chosen so as to generously bracket plausible values of the corresponding quantities.

Parameter	ΛCDM	wCDM
Ω_m	Uniform: $\mathcal{U}(0.0, 1.0)$	Uniform: $\mathcal{U}(0.0, 1.0)$
Ω_κ	Uniform: $\mathcal{U}(-1.0, 1.0)$	Fixed: 0
w	Fixed: -1	Uniform: $\mathcal{U}(-4, 0)$
H_0 [km/s/Mpc]	$\mathcal{N}(72, 8^2)$	$\mathcal{N}(72, 8^2)$
	Common priors	
σ_μ^{int} [mag]	Uniform on $\log \sigma_\mu^{\text{int}}$: $\mathcal{U}(-3.0, 0.0)$	
M_0 [mag]	Uniform: $\mathcal{U}(-20.3, -18.3)$	
α	Uniform: $\mathcal{U}(0.0, 1.0)$	
β	Uniform: $\mathcal{U}(0.0, 4.0)$	
R_c	Uniform on $\log R_c$: $\mathcal{U}(-5.0, 2.0)$	
R_x	Uniform on $\log R_c$: $\mathcal{U}(-5.0, 2.0)$	

samples from the posterior distribution of Eq. (10.31). We use 4000 live points and a tolerance parameter 0.1, resulting in about 8×10^5 likelihood evaluations.

We also wish to compare the performance of our BHM with the usually adopted χ^2 minimization procedure. To this end, we fit the simulated data using the χ^2 expression of Eq. (10.10). In order to mimic what is done in the literature as closely as possible, we chose a value of σ_μ^{int} then minimize the χ^2 w.r.t. the fit parameters $\vartheta = \{\Omega_m, \Omega_\kappa \text{ or } w, H_0, M_0, \alpha, \beta\}$. We update the value of σ_μ^{int} then repeat the minimization as described below, until a value of $\chi^2/\text{dof} = 1$ is obtained. The steps in the process can be enumerated as follows:

1. Select a trial value for σ_μ^{int}.
2. Minimise the χ^2 given in Eq. (10.10) by simultaneously fitting for the cosmology and SNe Ia parameters $\vartheta = \{\Omega_m, \Omega_\kappa \text{ or } w, H_0, M_0, \alpha, \beta\}$.
3. Evaluate χ^2/dof at minimum (i.e. the best fit point). If $\chi^2/\text{dof} > 1$ select a higher trial value for σ_μ^{int}, if $\chi^2/\text{dof} < 1$ select a lower trial value for σ_μ^{int}, repeat from the minimization step (ii) onwards.
4. Stop the process of minimization and iterative updating of σ_μ^{int} when a value of $\chi^2/\text{dof} = 1$ is obtained.

Once we have obtained the global best fit point, we derive 1- and 2-dimensional confidence intervals on the parameters by profiling (i.e. maximising over the other parameters) over the likelihood

$$L(\vartheta) = \exp\left(-\frac{1}{2}\chi(\vartheta)^2\right), \tag{10.82}$$

with χ^2 given by Eq. (10.10). According to Wilks' theorem, approximate confidence intervals are obtained from the profile likelihood as the regions where the χ^2

Table 10.3 Change in $\Delta\chi^2$ required in 1D profile likelihood for 1σ and 2σ confidence intervals.

Likelihood content	68.3% (1σ)	95.4% (2σ)
$\Delta\chi^2$	1.00	4.00

increases by $\Delta\chi^2$ from its minimum value, where $\Delta\chi^2$ can be computed from the chi-square distribution with the number of degree of freedoms corresponding to the number of parameters of interest and is given in standard look-up tables. The appropriate $\Delta\chi^2$ values for the 1D likelihoods are shown in Table 10.3. Obtaining reliable estimates of the profile likelihood using Bayesian algorithms (such as MultiNest) is a considerably harder numerical task than mapping out the Bayesian posterior. However, it has been shown that MultiNest can be successfully used for this task even in highly challenging situations [26], provided the number of live points and tolerance value used are adjusted appropriately. For our χ^2 scan, we adopt 10^4 live points and a tolerance of 0.1. We have found that those values give accurate estimates of the profile likelihood more than 2σ into the tails of the distribution for an 8 dimensional Gaussian toy model (whose dimensionality matches the case of interest here). With these MultiNest settings, we gather 1.5×10^5 samples, from which the profile likelihood is derived.

Our implementation of the χ^2 method is designed to match the main features of the fitting procedure usually adopted in the literature (namely, maximisation of the likelihood rather than marginalization of the posterior, and iterative determination of the intrinsic dispersion), although we do not expect that it exactly reproduces the results obtained by any specific implementation. Its main purpose is to offer a useful benchmark against which to compare the performance of our new Bayesian methodology.

10.6.4 Parameter Reconstruction

We compare the cosmological parameters reconstructed from the standard χ^2 method and our Bayesian approach in Fig. 10.4 for a typical data realization. The left-hand-side panel shows constraints in the $\Omega_m - \Omega_\Lambda$ plane for the ΛCDM model, both from our Bayesian method (filled regions, marginalized 68.3% and 95.4% posterior) and from the standard χ^2 method (red contours, 68.3% and 95.4% confidence regions from the profile likelihood). In the right-hand-side panel, constraints are shown in the $w - \Omega_m$ plane for a flat wCDM model Universe. In a typical reconstruction, our Bayesian method produced considerably tighter constraints on the cosmological parameters of interest than the usual χ^2 approach. Our constraints are also less biased w.r.t. the true value of the parameters, an important advantage that we further characterize below.

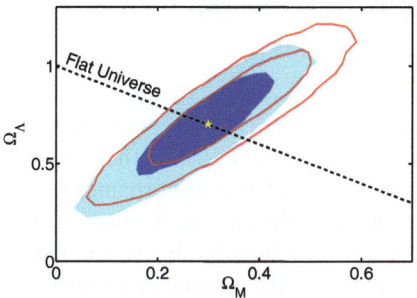
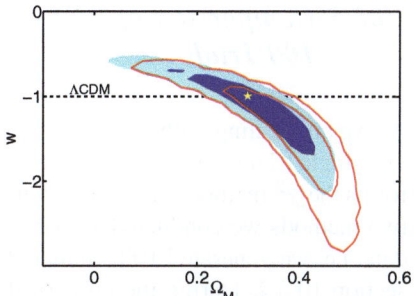

Fig. 10.4 Reconstruction of cosmological parameters from a simulated data set encompassing 288 SNIa, with characteristics matching presently available surveys (including realization noise). Blue regions contain 95.4% and 68.3% of the posterior probability (other parameters marginalized over) from our BHM method, the red contours delimit 95.4% and 68.3% confidence intervals from the standard χ^2 method (other parameters maximised). The yellow star indicates the true value of the parameters. The left panel assumes $w = -1$ while the right panel assumes $\Omega_\kappa = 0$. Notice how out method produced considerably less biassed constraints on the parameters.

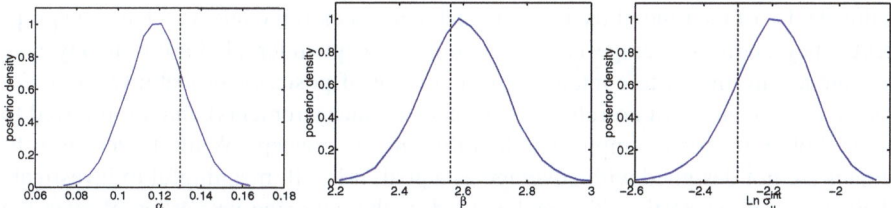

Fig. 10.5 Marginalised posterior for the stretch correction α, colour correction β parameter and logarithm of the intrinsic dispersion of SNe, $\log \sigma_\mu^{int}$, from a simulated data set from our Bayesian method. The vertical, dashed line gives the true value for each quantity.

Our BHM further produces marginalized posterior distributions for all the other parameters of the fit, including the global SNe Ia parameters α and β and the intrinsic dispersion of the SNe. The 1D marginal posteriors for those quantities are shown in Fig. 10.5. The recovered posterior means lie within 1σ of the true values. Notice that we do not expect the posterior mean to match exactly the true value, because of realization noise in the simulated data. However, as shown below, our method delivers less biased estimates of the parameters, and a reduced mean squared error compared with the standard χ^2 approach. The stretch correction α is determined with 8% accuracy, while the colour correction parameter β is constrained with an accuracy better than 3%. A new feature of our method is that it produces a posterior distribution for the SN population intrinsic dispersion, σ_μ^{int} (right-hand-side panel of Fig 10.5). This allows to determined the intrinsic dispersion of the SNIa population to typically about 10% accuracy.

10.6.5 Comparison of Performance of the Two Methods Over 100 Trials

As we are dealing with a system subject to Gaussian statistical noise, the results of one single trial are not sufficient to validate the claim that the BHM method outperforms the χ^2 method. In order to demonstrate further the relative performance of the two methods we conducted a series of trials using 100 different realizations of the data, i.e. we generated 100 simulated data sets each with 288 SNe Ia, as detailed in section 10.6.2. During the numerical sampling phase of the parameter reconstruction, several trials failed due to computational problems, in these cases the results for the relevant trials for both the BHM and χ^2 methods were omitted from the final analysis. We are interested in comparing the average ability of both methods to recover parameter values that are precise, accurate and as much as possible unbiassed with respect to their true values, as well as to establish the coverage properties of the credible and confidence intervals.

Coverage is defined as the probability that an interval contains (covers) the true value of a parameter, in a long series of repeated measurements. The defining property of a e.g. 95.4% Frequentist confidence interval is that it should cover the true value 95.4% of the time; thus, it is reasonable to check if the intervals have the properties they claim. Coverage is a Frequentist concept: intervals based on Bayesian techniques are meant to contain a given amount of posterior probability for a *single measurement* (with no reference to repeated measurements) and are referred to as credible intervals to emphasize the difference in concept. While Bayesian techniques are not designed with coverage as a goal, it is still meaningful to investigate their coverage properties. To our knowledge, the coverage properties of even the standard χ^2 method (which, being a frequentist method would ideally be expected to exhibit exact coverage) have never been investigated in the SN literature.

We generate 100 realizations of the simulated data from the fiducial model of Table 10.1 as described in section 10.6.2, and we analyse them using our BHM method and the standard χ^2 approach, using the same priors as above, given in Table 10.2. We quantify the performance of the two methods in two ways: in terms of the precision (i.e. error bar size) and in terms of accuracy (i.e. distance of reconstructed parameter value from true parameter value). For each parameter of interest θ, we compare the precision by evaluating the relative size of the posterior 68.3% range from our BHM method, $\sigma_\theta^{\text{BHM}}$, compared with the 68.3% confidence interval from the χ^2 method, $\sigma_\theta^{\chi^2}$, which is summarized by the quantity $\mathcal{S}_\theta$ which shows the percentage change in error bar size with respect to the error bar derived using the χ^2 method

$$\mathcal{S}_\theta \equiv \left(\frac{\sigma_\theta^{\text{BHM}}}{\sigma_\theta^{\chi^2}} - 1 \right) \times 100. \tag{10.83}$$

A value $\mathcal{S}_\theta < 1$ means that our BHM method delivers tighter error bars on the parameter θ, so is more precise. A histogram of this quantity for the variables of interest is shown in Fig. 10.6, from which we conclude that our BHM method gives

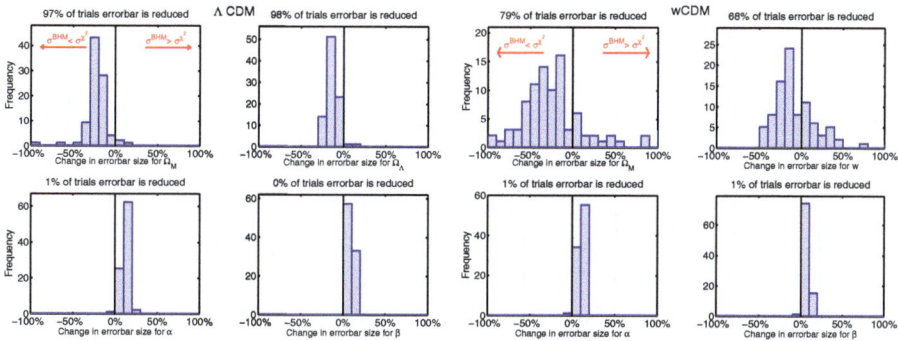

Fig. 10.6 Histograms of the quantity defined in Eq. (10.83), comparing of the error bars on each parameter from our method and from the standard χ^2 approach for 100 realization, for the ΛCDM model (left) and the wCDM model (right). A change in error bar size of -10% indicates BHM error bars are 10% smaller than χ^2 error bars. A change in error bar size of $+10\%$ indicates BHM error bars are 10% larger than χ^2 error bars. Our BHM method generally delivers smaller errors on the cosmological parameters (top row) so is more precise, but larger errors on the SNe Ia global fit parameters α, β (bottom row) so is less precise.

smaller error bars on Ω_m, Ω_Λ and w in almost all cases. However the uncertainty on α, β is larger from our method than from the χ^2 approach in most data realizations, as expected from [12].

Precision and tight error bars are good, but not if they come at the expense of a less accurate reconstruction. To evaluate the accuracy of each method, we build the following test statistic for each reconstruction:

$$\mathcal{T}_\theta \equiv |\overline{\theta}_{\text{BHM}}/\theta_{\text{true}} - 1| - |\theta_{\chi^2}^{\text{bf}}/\theta_{\text{true}} - 1|, \qquad (10.84)$$

where $\overline{\theta}_{\text{BHM}}$ is the posterior mean recovered using our BHM method, $\theta_{\chi^2}^{\text{bf}}$ is the best-fit value for the parameter recovered using the standard χ^2 approach and θ_{true} is the true value for that parameter. $\mathcal{T}_\theta$ can be interpreted as follows: for a given data realization, if the reconstructed posterior mean from our BHM is closer to the true parameter value than the best-fit χ^2, then $\mathcal{T}_\theta < 0$, which means that our method is more accurate than χ^2. A histogram of the distribution of $\mathcal{T}_\theta$ across the 100 realizations, shown in Fig. 10.7, can be used to compare the two methods: a negative average in the histogram means that the BHM outperforms the usual χ^2. For all of the parameters considered, our BHM method is more accurate than the χ^2 method, outperforming χ^2 about 2/3 of the time. Furthermore, the reconstruction of the intrinsic dispersion σ_μ^{int} is better with our BHM method almost 3 times out of 4. We emphasize once more that our methodology also provides an estimate of the uncertainty in the intrinsic dispersion, not just a best-fit value as the χ^2 approach.

We can further quantify the improvement in the statistical reconstruction by looking at the bias, defined as

$$\text{bias} = \langle \hat{\theta} - \theta_{\text{true}} \rangle \qquad (10.85)$$

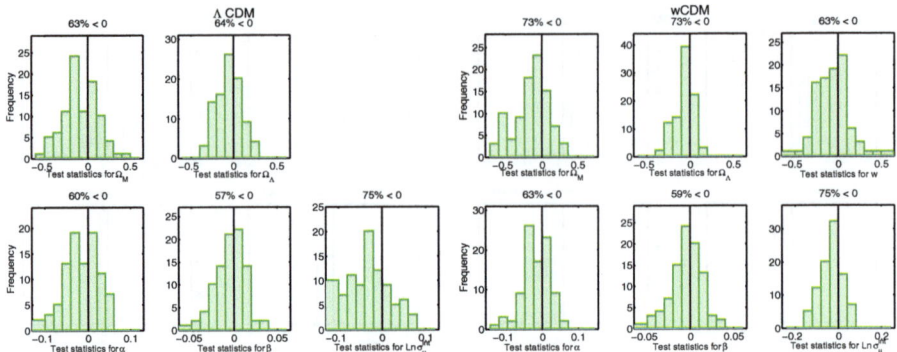

Fig. 10.7 Histograms of the test statistics defined in Eq. (10.84), comparing the long-term performance of the two methods for the parameters of interest in the ΛCDM model (left) and the wCDM model (right). A predominantly negative value of the test statistics means that our method gives a parameter reconstruction that is closer to the true value than the usual χ^2, i.e. less biased. For the cosmological parameters (top row), our method outperforms χ^2 about 2 times out of 3.

and mean squared error (MSE) for each parameter, defined as

$$\text{MSE} = \text{bias}^2 + \text{Var}, \tag{10.86}$$

respectively, where the expectation is taken by averaging over the observed values in our 100 simulated trials, $\hat{\theta} = \overline{\theta}_{\text{BHM}}$ ($\hat{\theta} = \theta^{\text{bf}}_{\chi^2}$) for the BHM (for the χ^2 approach) and Var is the observed parameter variance. The bias is the expectation value of the difference between estimator and true value, and tells us by how much the estimator systematically over or under estimates the parameter of interest. The MSE measures the average of the squares of the errors, i.e. the amount by which the estimator differs from the true value for each parameter. A smaller bias and a smaller MSE imply a better performance of the method. The results for the two methods are summarized in Table 10.4, which shows how our method reduces the bias by a factor $\sim$2–3 for most parameters, while reducing the MSE by a factor of $\sim$2. The only notable exception is the bias of the EOS parameter w, which is larger in our method than in the χ^2 approach.

Finally, in Fig. 10.8 we plot the coverage of each method for 68.3% and 95.4% intervals. Error bars give an estimate of the uncertainty of the coverage result, by giving the binomial sampling error from the finite number of realizations considered, evaluated from the binomial variance as $Np(1 - p)$, where $N = 100$ is the number of trials and p is the observed fractional coverage. Both method slightly undercover, i.e. the credible region and confidence intervals are too short, although the lack of coverage is not dramatic: e.g., the typical coverage of the 1σ (2σ) intervals from our method is $\sim$60% (90%). Our method shows slightly better coverage properties than the χ^2 method, while producing considerably tighter and less biassed constraints (as demonstrated above). This further proves that the tighter intervals recovered by our method do not suffer from bias w.r.t the true values.

Table 10.4 Comparison of the bias and mean squared error for our Bayesian method and the usual χ^2 approach. The columns labelled "Improvement" give the factor by which our Bayesian method reduces the bias and the MSE w.r.t. the χ^2 approach.

Parameter		Bias			Mean squared error		
		Bayesian	χ^2	Improvement	Bayesian	χ^2	Improvement
ΛCDM	Ω_m	−0.0188	−0.0183	1.0	0.0082	0.0147	1.8
	Ω_Λ	−0.0328	−0.0223	0.7	0.0307	0.0458	1.5
	α	0.0012	0.0032	2.6	0.0001	0.0002	1.4
	β	0.0202	0.0482	2.4	0.0118	0.0163	1.4
	σ_μ^{int}	−0.0515	−0.1636	3.1	0.0261	0.0678	2.6
wCDM	Ω_m	−0.0177	−0.0494	2.8	0.0072	0.0207	2.9
	Ω_Λ	0.0177	0.0494	2.8	0.0072	0.0207	2.9
	w	−0.0852	−0.0111	0.1	0.0884	0.1420	1.6
	α	0.0013	0.0032	2.5	0.0001	0.0002	1.5
	β	0.0198	0.0464	2.3	0.0118	0.0161	1.4
	σ_μ^{int}	−0.0514	−0.1632	3.2	0.0262	0.0676	2.6

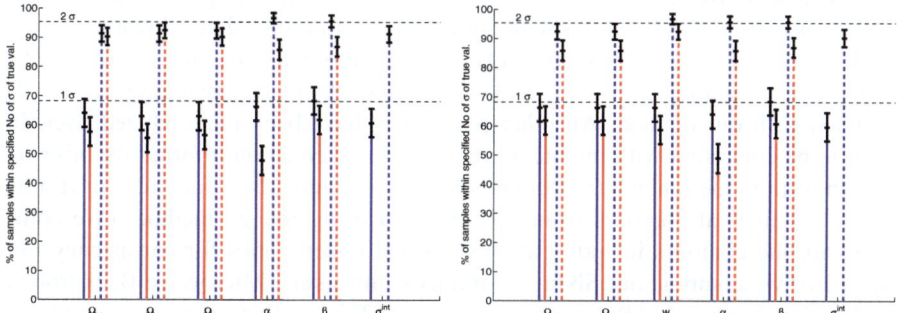

Fig. 10.8 Coverage of our method (blue) and standard χ^2 (red) for 68.3% (solid) and 95.4% (dashed) intervals, from 100 realizations of simulated data for the ΛCDM model (left) and the wCDM model (right). While both methods show significant undercoverage for all parameters, our method has a comparable coverage to the standard χ^2, except for w. Coverage values for the intrinsic dispersion σ_μ^{int} are not available from the χ^2 method, as it does not produce an error estimate for this quantity.

To summarise, the results from our numerical trials with simulated data show that:

1. In general, our BHM method gives more precise constraints on the cosmological parameters, but gives less precise constraints on the SNe Ia global fit parameters α, β
2. In 60–70% of trials, our BHM method recovers a more accurate value of the parameter of interest.
3. Our BHM method is less biassed than the χ^2 method, except in the reconstruction of the w parameter

4. Both methods undercover, with our BHM method giving slightly better coverage.

10.7 Cosmological Constraints from Current SNIa Data Using BHM

We now apply our BHM to fitting real SN data. We use the SALT-II fits result for 288 SNIa from [8], which have been derived from five different surveys described briefly in section 10.6.1. Our method only includes statistical errors according to the procedure described in section 10.5, coming from redshift uncertainties (arising from spectroscopic errors and peculiar velocities), intrinsic dispersion (which is determined from the data) and full error propagation of the SALT-II fit results. Systematic uncertainties play an important role in SNIa cosmology fitting, and (although not included in this study) can also be treated in our formalism in a fully consistent way. We comment on this aspect further below, though we leave a complete exploration of systematics with our BHM to a future, dedicated work.

We show in Fig. 10.9 the constraints on the cosmological parameters $\Omega_m - \Omega_\Lambda$ (left panel, assuming $w = -1$) and $w - \Omega_m$ (right panel, assuming $\Omega_\kappa = 0$) obtained with our method. All other parameters have been marginalized over. In order to be consistent with the literature, we have taken a non-informative prior on H_0, uniform in the range [20, 100] km/s/Mpc. The figure also compares our results with the statistical contours from [8], obtained using the χ^2 method. (We compare with the contours including only statistical uncertainties for consistency.) In Fig. 10.10 we combine our SNIa constraints with Cosmic Microwave Background (CMB) data from WMAP 5-yrs measurements [27] and Baryonic Acoustic Oscillations (BAO) constraints from the Sloan Digital Sky Survey LRG sample [28],

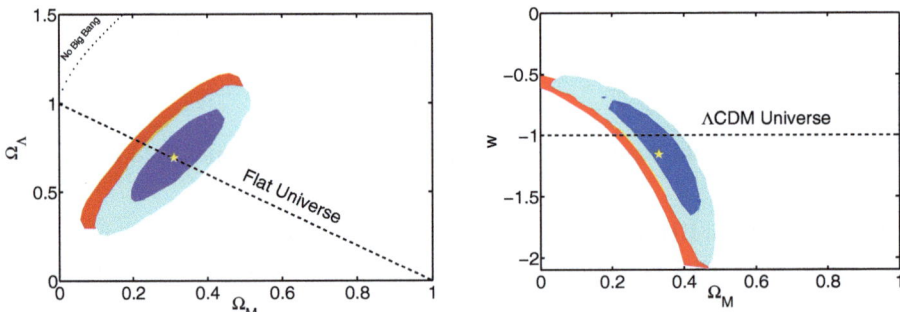

Fig. 10.9 Constraints on the cosmological parameters Ω_m, Ω_Λ (left panel, assuming $w = -1$) and w, Ω_m (right panel, assuming $\Omega_\kappa = 0$) from our Bayesian method (light/dark blue regions, 68.3% and 95.4% marginalized posterior), compared with the statistical errors from the usual χ^2 approach (yellow/red regions, same significance level; from [8]). The yellow star gives the posterior mean from our analysis.

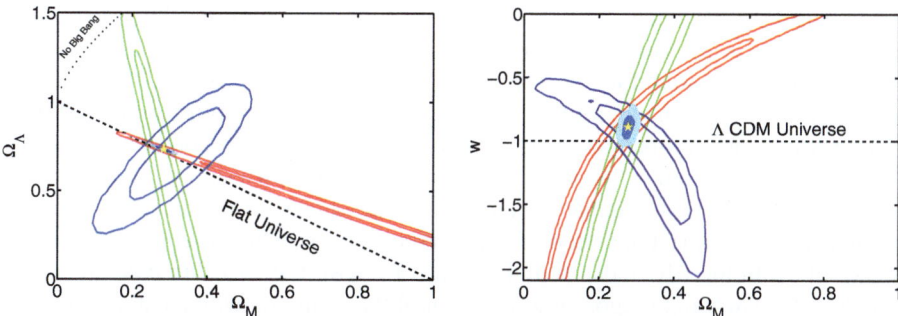

Fig. 10.10 Combined constraints on the cosmological parameters Ω_m, Ω_Λ (left panel, assuming $w = -1$) and w, Ω_m (right panel, assuming $\Omega_\kappa = 0$) from SNIa, CMB and BAO data. Red contours give 68.3% and 95.4% regions from CMB alone, green contours from BAO alone, blue contours from SNIa alone from our Bayesian method. The filled regions delimit 68.3% and 95.4% combined constraints, with the yellow star indicating the posterior mean.

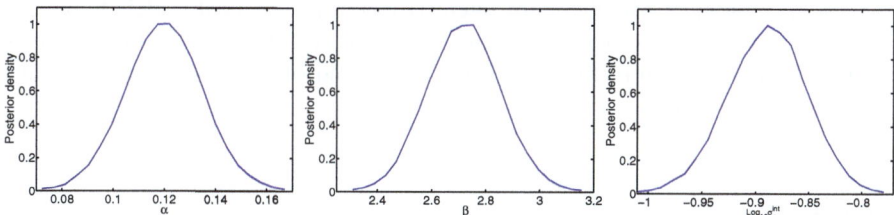

Fig. 10.11 Marginalised posterior for the stretch correction α, colour correction β parameter and logarithm of the intrinsic dispersion of SNe, $\log \sigma_\mu^{\text{int}}$ from current SNIa data.

using the same method as [8]. The combined SNIa, CMB and BAO statistical constraints result in $\Omega_m = 0.28 \pm 0.02, \Omega_\Lambda = 0.73 \pm 0.01$ (for the ΛCDM model) and $\Omega_m = 0.28 \pm 0.01, w = -0.90 \pm 0.05$ (68.3% credible intervals) for the wCDM model. Although the statistical uncertainties are comparable to the results by [8] from the same sample, our posterior mean values present shifts of up to $\sim 2\sigma$ compared to the results obtained using the standard χ^2 approach. This is a fairly significant shift, which can be attributed to our improved statistical method, which exhibits a reduced bias w.r.t. the χ^2 approach.

Fig. 10.11 shows the 1D marginalized posterior distributions for the SNe Ia global fit parameters α, β and for the intrinsic dispersion σ_μ^{int}. All parameters are well constrained by the posterior, and we find $\alpha = 0.12 \pm 0.02, \beta = 2.7 \pm 0.1$ and a value of the intrinsic dispersion (for the whole sample) $\sigma_\mu^{\text{int}} = 0.13 \pm 0.01$ mag. Kessler et al. [8] find values for the intrinsic dispersion ranging from 0.08 (for SDSS-II) to 0.23 (for the HST sample), but their χ^2 method does not allow them to derive an error on those determinations. With our method, it would be easy to derive constraints on the intrinsic dispersion of each survey – all one needs to do is to replace Eq. (10.15) with a corresponding expression for each survey. This introduces one pair of population parameters $(M_0, \sigma_\mu^{\text{int}})$ for each survey. In the same way, one could study whether the intrinsic dispersion evolves with redshift. A detailed study of these issues is left for future work.

The value of α found in [8] is in the range 0.10–0.12, depending of the details of the assumptions made, with a typical statistical uncertainty of order $\sim$0.015. These results are comparable with our own. As for the colour correction parameter β, constraints from [8] vary in the range 2.46–2.66, with a statistical uncertainty of order 0.1–0.2. This stronger dependence on the details of the analysis seems to point to a larger impact of systematic uncertainties for β, which is confirmed by evidence of evolution with redshift of the value of β ([8], Fig. 39). Our method can be employed to carry out a rigorous assessment of the evolution with redshift of colour corrections. A possible strategy would be to replace β with a vector of parameters $\beta_1, \beta_2, \ldots$, with each element describing the colour correction in a different redshift bin. The analysis proceeds then as above, and it produces posterior distributions for the components of β, which allows to check the hypothesis of evolution. Finally, in such an analysis the marginalized constraints on all other parameters (including the cosmological parameters of interest) would automatically include the full uncertainty propagation from the colour correction evolution, without the need for further *ad hoc* inflation of the error bars. These kind of tests will be pursued in a forthcoming publication.

10.8 Conclusions

The primary aim of the work presented in this chapter was to address certain deficiencies in the existing methods for extracting the cosmological parameters from SNe Ia data in conjunction with the SALT-II lightcurve fitter. The two main motivations were the lack of an appropriate framework for assessing the unknown intrinsic dispersion σ_μ^{int} and its uncertainty, and the incompatibility of existing parameter reconstruction methods with methods of Bayesian model selection. In order to address these dual problems, we have derived a new and fully Bayesian method for parameter inference based on a Bayesian Hierarchical Model, BHM.

The main novelty of our method is that it produces an effective likelihood that propagates uncertainties in a fully consistent way. We have introduced an explicit statistical modelling of the absolute magnitude distribution of the SNIa population, which for the first time allows one to derive a full posterior distribution of the SNIa intrinsic dispersion.

We have tested our method using simulated data sets and found that it compares favourably with the standard χ^2 approach, both on individual data realizations and in the long term performance. Statistical constraints on cosmological parameters are significantly improved, while in a series of 100 simulated data sets our method outperforms the χ^2 approach at least 2 times out of 3 for the parameters of interest. We have also demonstrated that our method is less biassed and has better coverage properties than the usual approach.

We applied our methodology to a sample of 288 SNIa from multiple surveys. We find that the flat ΛCDM model is still in good agreement with the data, even under our improved analysis. However, the posterior mean for the cosmological parameters exhibit up to 2σ shifts w.r.t. results obtained with the conventional χ^2 approach. This is a consequence of our improved statistical analysis, which benefits from a reduced bias in estimating the parameters.

While in this chapter we have only discussed statistical constraints, our method offers a new, fully consistent way of including systematic uncertainties in the fit. As our method is fully Bayesian, it can be used in conjunction with fast and efficient Bayesian sampling algorithms, such as MCMC and nested sampling. This will allow to enlarge the number of parameters controlling systematic effects that can be included in the analysis, thus taking SNIa cosmological parameter fitting to a new level of statistical sophistication. The power of our method as applied to systematic errors analysis will be presented in a forthcoming, dedicated work.

At a time when SNIa constraints are entering a new level of precision, and with a tenfold increase in the sample size expected over the next few years, we believe it is timely to upgrade the cosmological data analysis pipeline in order to extract the most information from present and upcoming SNIa data. This work represents a first step in this direction.

Acknowledgements This work was partially supported by travel grants by the Royal Astronomical Society and by the Royal Society. MCM was partially supported by a Royal Astronomical Society grant. GDS and PV were supported by a grant from the US-DOE to the CWRU theory group, and by NASA grant NNX07AG89G to GDS. PV was supported by CWRU's College of Arts and Sciences.

References

1. Riess, A.G., Filippenko, A.V., Challis, P., Clocchiatti, A., Diercks, A., Garnavich, P.M., Gilliland, R.L., Hogan, C.J., Jha, S., Kirshner, R.P., Leibundgut, B., Phillips, M.M., Reiss, D., Schmidt, B.P., Schommer, R.A., Smith, R.C., Spyromilio, J., Stubbs, C., Suntzeff, N.B., Tonry, J.: Observational Evidence from Supernovae for an Accelerating Universe and a Cosmological Constant. Astron. J. **116**, 1009–1038 (1998). DOI 10.1086/300499
2. Perlmutter, S., Aldering, G., Goldhaber, G., Knop, R.A., Nugent, P., Castro, P.G., Deustua, S., Fabbro, S., Goobar, A., Groom, D.E., Hook, I.M., Kim, A.G., Kim, M.Y., Lee, J.C., Nunes, N.J., Pain, R., Pennypacker, C.R., Quimby, R., Lidman, C., Ellis, R.S., Irwin, M., McMahon, R.G., Ruiz-Lapuente, P., Walton, N., Schaefer, B., Boyle, B.J., Filippenko, A.V., Matheson, T., Fruchter, A.S., Panagia, N., Newberg, H.J.M., Couch, W.J., The Supernova Cosmology Project: Measurements of Omega and Lambda from 42 High-Redshift Supernovae. Astrophys. J. **517**, 565–586 (1999). DOI 10.1086/307221
3. March, M.C., Trotta, R., Berkes, P., Starkman, G.D., Vaudrevange, P.M.: Improved constraints on cosmological parameters from Type Ia supernova data. Mon. Not. R. Astron. Soc. **418**, 2308–2329 (2011). DOI 10.1111/j.1365-2966.2011.19584.x
4. Wang, L., Goldhaber, G., Aldering, G., Perlmutter, S.: Multicolor Light Curves of Type Ia Supernovae on the Color-Magnitude Diagram: A Novel Step toward More Precise Distance and Extinction Estimates. Astrophys. J. **590**, 944–970 (2003). DOI 10.1086/375020

5. Conley, A., Goldhaber, G., Wang, L., Aldering, G., Amanullah, R., Commins, E.D., Fadeyev, V., Folatelli, G., Garavini, G., Gibbons, R., Goobar, A., Groom, D.E., Hook, I., Howell, D.A., Kim, A.G., Knop, R.A., Kowalski, M., Kuznetsova, N., Lidman, C., Nobili, S., Nugent, P.E., Pain, R., Perlmutter, S., Smith, E., Spadafora, A.L., Stanishev, V., Strovink, M., Thomas, R.C., Wood-Vasey, W.M., Supernova Cosmology Project: Measurement of Ω_m, Ω_Λ from a Blind Analysis of Type Ia Supernovae with CMAGIC: Using Color Information to Verify the Acceleration of the Universe. Astrophys. J. **644**, 1–20 (2006). DOI 10.1086/503533
6. Mandel, K.S., Narayan, G., Kirshner, R.P.: Type Ia Supernova Light Curve Inference: Hierarchical Models in the Optical and Near-infrared. Astrophys. J. **731**, 120–+ (2011). DOI 10.1088/0004-637X/731/2/120
7. Mandel, K.S., Narayan, G., Kirshner, R.P.: Type Ia Supernova Light Curve Inference: Hierarchical Models in the Optical and Near Infrared. arXiv:1011.5910 (2010)
8. Kessler, R., Becker, A.C., Cinabro: First-Year Sloan Digital Sky Survey-II Supernova Results: Hubble Diagram and Cosmological Parameters. Astrophys. J. Suppl. Ser. **185**, 32–84 (2009). DOI 10.1088/0067-0049/185/1/32
9. Astier, P., Guy, J.: The Supernova Legacy Survey: measurement of OmegaM, OmegaL and w from the first year data set. Astron. Astrophys. **447**, 31–48 (2006). DOI 10.1051/0004-6361:20054185
10. Kowalski, M., Rubin, D., Aldering, G., Agostinho, R.J., Amadon, A.: Improved Cosmological Constraints from New, Old, and Combined Supernova Data Sets. Astrophys. J. **686**, 749–778 (2008). DOI 10.1086/589937
11. Tripp, R.: A two-parameter luminosity correction for Type IA supernovae. Astron. Astrophys. **331**, 815–820 (1998)
12. Gull, S.: Bayesain Data Analysis: Straight-line fitting. Maximum Entropy and Bayesian Methods pp. 511–518 (1989)
13. Miknaitis, G., Pignata, G.: The ESSENCE Supernova Survey: Survey Optimization, Observations, and Supernova Photometry. Astrophys. J. **666**, 674–693 (2007). DOI 10.1086/519986
14. Wood-Vasey, W.M., Miknaitis, G., Stubbs, C.W.: Observational Constraints on the Nature of Dark Energy: First Cosmological Results from the ESSENCE Supernova Survey. Astrophys. J. **666**, 694–715 (2007). DOI 10.1086/518642
15. Jha, S., Riess, A.G., Kirshner, R.P.: Improved Distances to Type Ia Supernovae with Multicolor Light-Curve Shapes: MLCS2k2. Astrophys. J. **659**, 122–148 (2007). DOI 10.1086/512054
16. Garnavich, P.M., Kirshner, R.P., Challis, P.a.: Constraints on Cosmological Models from Hubble Space Telescope Observations of High-z Supernovae. Astrophys. J., Lett. **493**, L53+ (1998). DOI 10.1086/311140
17. Knop, R.A., Aldering, G., Amanullah, R., Astier, P.: New Constraints on Ω_M, Ω_Λ, and w from an Independent Set of 11 High-Redshift Supernovae Observed with the Hubble Space Telescope. Astrophys. J. **598**, 102–137 (2003). DOI 10.1086/378560
18. Riess, A.G., Strolger, L.: Type Ia Supernova Discoveries at $z > 1$ from the Hubble Space Telescope: Evidence for Past Deceleration and Constraints on Dark Energy Evolution. Astrophys. J. **607**, 665–687 (2004). DOI 10.1086/383612
19. Riess, A.G., Strolger, L.: New Hubble Space Telescope Discoveries of Type Ia Supernovae at $z \geq 1$: Narrowing Constraints on the Early Behavior of Dark Energy. Astrophys. J. **659**, 98–121 (2007). DOI 10.1086/510378
20. Kessler, R., Bernstein, J.P., Cinabro, D., Dilday, B., Frieman, J.A., Jha, S., Kuhlmann, S., Miknaitis, G., Sako, M., Taylor, M., Vanderplas, J.: SNANA: A Public Software Package for Supernova Analysis. Publ. Astron. Soc. Pac. **121**, 1028–1035 (2009). DOI 10.1086/605984
21. Freedman, W.L., Madore, B.F., Gibson, B.K., Ferrarese, L., Kelson, D.D., Sakai, S., Mould, J.R., Kennicutt Jr., R.C., Ford, H.C., Graham, J.A., Huchra, J.P., Hughes, S.M.G., Illingworth, G.D., Macri, L.M., Stetson, P.B.: Final Results from the Hubble Space Telescope Key Project to Measure the Hubble Constant. Astrophys. J. **553**, 47–72 (2001). DOI 10.1086/320638

22. Skilling, J.: Nested Sampling. In: Fischer, R. Preuss, R., Toussaint, U. V. (eds.) American Institute of Physics Conference Series, *American Institute of Physics Conference Series*, vol. 735, pp. 395–405 (2004). DOI 10.1063/1.1835238
23. Skilling, J.: Nested sampling for general Bayesian computation. Bayesian Analysis **1**, 833–861 (2006)
24. Feroz, F., Hobson, M.P.: Multimodal nested sampling: an efficient and robust alternative to Markov Chain Monte Carlo methods for astronomical data analyses. Mon. Not. R. Astron. Soc. **384**, 449–463 (2008). DOI 10.1111/j.1365-2966.2007.12353.x
25. Feroz, F., Hobson, M.P., Bridges, M.: MULTINEST: an efficient and robust Bayesian inference tool for cosmology and particle physics. Mon. Not. R. Astron. Soc. **398**, 1601–1614 (2009). DOI 10.1111/j.1365-2966.2009.14548.x
26. Feroz, F., Cranmer, K., Hobson, M., de Austri, R.R., Trotta, R.: Challenges of Profile Likelihood Evaluation in Multi- Dimensional SUSY Scans. arXiv:1101.3296 (2011)
27. Komatsu, E., Dunkley, J., Nolta, M.R., Bennett, C.L., Gold, B., Hinshaw, G., Jarosik, N., Larson, D., Limon, M., Page, L., Spergel, D.N., Halpern, M., Hill, R.S., Kogut, A., Meyer, S.S., Tucker, G.S., Weiland, J.L., Wollack, E., Wright, E.L.: Five-Year Wilkinson Microwave Anisotropy Probe Observations: Cosmological Interpretation. Astrophys. J. Suppl. Ser. **180**, 330–376 (2009). DOI 10.1088/0067-0049/180/2/330
28. Eisenstein, D.J., Zehavi, I., Hogg, D.W., Scoccimarro, R., Blanton, M.R., Nichol, R.C., Scranton, R., Seo, H.J., Tegmark, M., Zheng, Z., Anderson, S.F., Annis, J., Bahcall, N., Brinkmann, J., Burles, S., Castander, F.J., Connolly, A., Csabai, I., Doi, M., Fukugita, M., Frieman, J.A., Glazebrook, K., Gunn, J.E., Hendry, J.S., Hennessy, G., Ivezić, Z., Kent, S., Knapp, G.R., Lin, H., Loh, Y.S., Lupton, R.H., Margon, B., McKay, T.A., Meiksin, A., Munn, J.A., Pope, A., Richmond, M.W., Schlegel, D., Schneider, D.P., Shimasaku, K., Stoughton, C., Strauss, M.A., SubbaRao, M., Szalay, A.S., Szapudi, I., Tucker, D.L., Yanny, B., York, D.G.: Detection of the Baryon Acoustic Peak in the Large-Scale Correlation Function of SDSS Luminous Red Galaxies. Astrophys. J. **633**, 560–574 (2005). DOI 10.1086/466512

MIX
Papier aus verantwortungsvollen Quellen
Paper from responsible sources
FSC® C105338

If you have any concerns about our products,
you can contact us on
ProductSafety@springernature.com

In case Publisher is established outside the EU,
the EU authorized representative is:
**Springer Nature Customer Service Center GmbH
Europaplatz 3, 69115 Heidelberg, Germany**

Printed by Libri Plureos GmbH
in Hamburg, Germany